B V C D M G A Z H J P Y J K

S L J G V U D B W H K P

S F H Y K L P I O D0970231

S F W H T U J N I

G E Q U J K M I L G W

F J Q A T R V U O K L H F D

R U N I E D A T A B A S E U

T O F S C U N I E B M E D O

V M H A D R E A M S N C U M

Y N I E B M H A V O G V O N

H G S C U N I E B M H B O G

B H A O C G S C U N I U M H

J I E D U A X D A G S O N I

U S C U G V H J I W N K I A

V K E P A Z C N P K M F R T

O B E F T H E T G E N T I E

S V T C U R L O S T V H S B

A V O F Q U E S T E B I A U

E H S P L N E Q U J B X I R

C U H M A N H A M V O A C L

G O V H V N M L Q E P R V E

V O I B M X T B Q P N E B M

U N H O T O Z N V B E B U O

O G S I C A T A L O G E O V

M H A H U M A N I T Y C M S

I B E N X F P K Q P K C N H

G C S E I N J P L W O A G I

DATABASE OF DREAMS

DATABASE OF DREAMS

The Lost Quest to Catalog Humanity

REBECCA LEMOV

Yale

UNIVERSITY PRESS

New Haven & London

Published with assistance from the foundation established in
memory of Amasa Stone Mather of the Class of 1907,
Yale College.

Yale University Press books may be purchased in quantity for
educational, business, or promotional use. For information, please
e-mail sales.press@yale.edu (U.S. office) or
sales@yaleup.co.uk (U.K. office).

Endpaper art by Tom Starr.

Set in Janson type by IDS Infotech, Ltd.
Printed in the United States of America.

Library of Congress Control Number: 2015940162
ISBN 978-0-300-20952-5 (cloth : alk. paper)

A catalogue record for this book is available from the British Library.

This paper meets the requirements of ANSI/NISO Z39.48–1992
(Permanence of Paper).

10 9 8 7 6 5 4 3 2 1

To Penelope, my mother, and Ivy, my daughter

The study of dreams is particularly difficult, for we cannot examine dreams directly, we can only speak of the memory of dreams. And it is possible that the memory of dreams does not correspond exactly to the dreams themselves. . . . If we think of the dream as a work of fiction—and I think it is—it may be that we continue to spin tales when we wake and later when we recount them.

—JORGE LUIS BORGES, lecture at the Teatro Coliseo, Buenos Aires, 1977

We often take the shadow of things for substance.

—ROBERT HOOKE, *Micrographia*, 1665

Contents

Acknowledgments

More than a decade ago, when I was living in Oakland, California, I was under the impression that if one found a book lying on the street or stacked in a free pile near the sidewalk, one should probably read it because it might carry a message. Indeed, I did so when I found the book *The Captive Mind*, by Czeslaw Milosz (1953), around the end of the last century, in a red jacket, abandoned on the sidewalk. In the book, Milosz describes the fate of "human materials," and this idea stuck with me through the end of graduate school, all sorts of life events, the early years of teaching, and the eight years it's taken me to write this book. Under particular conditions and certain systems, Milosz seems to say, humans function as both materials and living beings or as subjects and objects. My first book was about the constraint of human materials. This is a book about human materials also. The archive I'm calling the "database of dreams" was at its core a collection of just such materials—sometimes also called by their collectors "human documents"—and was part of a larger movement to collect the same. In the process of telling the story of the archive housing these documents and the lives they represent, I have incurred many human debts and drawn on many friendships.

First are the people from around the world who contributed to the archive I've written about. These are the subjects whose stories and dreams I have referred to, drawn from, or retold in fragmented fashion. Most are deceased; many may not have known that their dreams or other materials were being stored in a social-scientific project. Others worked willingly with those who came and engaged in the project or subsidiary projects. I owe a great debt to them.

I also owe a large debt to the anthropologists and psychologists who collected these materials and later collected the collections. First of all, the late Bert Kaplan, whose children, Josh and Emily, have been generous with their time and memories; whose granddaughter Hannah provided me with a preliminary bibliography of her grandfather's papers; and most particularly, his wife Hermia, who has been a wonderful friend, a gracious host, and delightful to talk to. As in the case of the subjects, most of the scientists who contributed to the data collection are now deceased (George and Louise Spindler, A. I. Hallowell, and Dorothy Eggan, to name some of the most important to this story), but I have been immeasurably aided by the recollections of colleagues who knew them or who were affiliated with these projects and whom I was privileged to interview: G. William Domhoff, Richard Randolph, Brewster Smith, and Howard Becker. I would like to acknowledge the contributors to the *Microcard Publications of Primary Records in Culture and Personality* for their generosity in adding their data sets to a collective experimental inquiry, from which I was able to draw. I would also like to thank H. David Brumble III for allowing me to quote from his communications. Robert LeVine has been especially generous with his time over the past seven years, in both Berlin and Massachusetts. In addition, archivists and microfilm experts at the University of Chicago, the Library of Congress, Harvard University, the American Philosophical Society, UCLA, UC Santa Barbara, and the National Academy of Sciences, as well as the National Archives, have been generous in helping me.

Friends who agreed to read this book occupy a lofty, nearly angelic circle of gratitude, especially my three first full-manuscript readers, Sophia Roosth, Jeremy Greene, and Cathy Gere, who offered encouragement and ideas at key moments. At a later stage, Allan Brandt, Anne Harrington, and Charles Rosenberg offered immensely helpful commentary and ideas. I discussed this project with many interlocutors at great length or briefly and was greatly helped by the interlocution: Christine von Oertzen, John Tresch, James Delbourgo, Heidi Voskuhl, Jimena Canales, Natasha Schüll (in Berkeley, Santa Fe, and Cambridge), David Sepkoski, Teri Chettiar,

Joanna Radin, Alisha Rankin, Fernando Vidal, Elena Aronova, Etienne Benson, Lisa Gitelman, Bregje van Eekelen, Aude Fauvel, Jelena Martinovic, Chris Kelty, Hannah Landecker, Eric Hounshell, Josh Berson, Orit Halpern, Sarah Richardson, Alex Czisar, Joel Isaac, Paul Erickson, Judy Klein, Thomas Sturm, Latif Nasser, Michael Gordin, and David Kaiser. It's been inspiring to run into Dan Rosenberg after some time. Good friends Oriana Walker, Julie Livingston, Margo Boenig-Lipstin, Nasser Zakariya, Liz Murphy, Junko Kitanaka, Mark Ostow, Emily Shea, Josephine Fenger, Laura Hadded, Gretchen Rubin, John Carson, Andrew Jewett, Healan Gaston, Ahmed Ragab and Soha Bayoumi, and Ian and Crate Miller have given ongoing inspiration. Scott Berryman generously offered to read this in manuscript, and he is missed. Mentors, including Paul and Susan Fleischman, have helped my family and me orient ourselves in New England, and Gair and Rick Crutcher are lifelong friends even though we left the Northwest. Colleagues from my home department have been more than generous over the years and have supported me for too many lunchtime discussions to count: Peter Galison, Charles Rosenberg, Steven Shapin, Janet Browne, Anne Harrington, Katy Park, Everett Mendelsohn, Allan Brandt, Shigehisa Kuriyama, Naomi Oreskes, and Sheila Jasanoff. I'm grateful for the chance to work with David Jones, Evelynn Hammonds, and Jean-François Gauvin.

The late Alan Dundes remains a source of inspiration for me because of the boundlessly generous way he treated his students. Likewise David Hollinger—a nearly ideal professor. And the experience of working with Paul Rabinow as my adviser in graduate school continues to resonate in all of my projects. Other teachers, including Leslie Brisman, George Fayen, John Hollander, Nir Tiomkin, David Garrigues, Ruth Balis, Walter Bartman, Stefania Pandolfo, Mariane Ferme, S. N. Goenka, and Lawrence Cohen, have earned my longtime gratitude.

Audiences heard precursors of this book's chapters at the University of Pennsylvania, Harvard's STS Circle, the University of Toronto, Johns Hopkins, UCLA, the Max Planck Institute for the History of Science (MPIWG) in Berlin, Erasmus University (Netherlands), and the University of Lausanne, and their feedback has helped me improve the book a great deal.

To Raine Daston I owe endless thanks for support over the past years, including two years of scholarly residence in Berlin, and for general bearing up and encouragement; my participation in the life at the MPIWG, Department II, has been a scholarly and personal gift.

Sarah Burnes, of the Gernert Company, is the ideal reader, and I'm very grateful to be working with her. Steve Wasserman, my editor at Yale, is startlingly brilliant and fun to talk to and fortunately for me an adherent of "bespoke publishing." I also appreciate Eva Skewes's assistance, as well as that of others at the press, and that of my copyeditor, Bojana Ristich. And my faculty assistant, Linda Schneider, has been a constant prod to action.

My parents have generously read this manuscript in various versions and, in addition to kindly bringing me into the world, have been amazingly encouraging even in the face of my propensity to study odd topics. My father, Michael, has offered many "big picture" reminders, as well as unstinting help, and my brother, Doug, has been ever supportive and willing to make quantitative calculations when called upon. Palo Coleman, my partner and a professional "solution-based thinker," is the ideal person to force to listen to yet another reiteration of a chapter, and I feel lucky to share my life with him. To the bond of love that joins my mother, Penelope, and my daughter, Ivy, this book is dedicated.

DATABASE OF DREAMS

Introduction

THIS book is about a lost history of data and a secret collection of dreams. It tells the story of a group of industrious people who piled up a "mountain of data" in the middle of the twentieth century. Not only was the size of the mountain notable—although today, of course, it is dwarfed by countless amalgamations of data—but also it was made of peculiar "stuff."

When a Hopi grandmother dreamed of white chickens in a snow-filled evergreen forest one night in 1949; when a young man from the northeast Pakistan frontier saw visions of water snakes; when four German exchange students and several patients in a Lebanese mental hospital answered psychological test questions, looking at inkblots, drawing pictures, filling out sentences—all this information entered researchers' records and merged in a single archive. Soon, materials that would almost certainly not have been kept, issuing mainly from people who did not generally write in journals (if they wrote at all), rested, preserved, in a vast bank of data. It was the most intimate of data mines, made of thousands and thousands of Rorschach test protocols, as well as sets of fleeting thoughts, random asides, irreverent inquiries and sad memories, life stories and dreams. Even today, the Library of Congress holds copies of these fragments of seemingly unremarkable human lives in condensed form. Built in 1956, the archive's materials dated back to earlier decades of the twentieth century, and its subjects' lives often dated back even further, to the end of the nineteenth century. By the time of its last installment in 1963, it formed a unique enterprise, neither simply a large lump sum of data, nor a purpose-driven agglomeration gathered to test a particular hypothesis, but a collection of collections, a massive clearinghouse holding a global array of data sets, a sort of memory machine. Its strange combination of old and new technologies, its great size, and its (in time) greater obscurity made it a data device that no historian has examined before in published writing.

The "database of dreams" is one of the most promising and yet strangely forgotten undertakings in American social science—a dizzyingly ambitious 1950s-era project to capture people's dreams in large amounts and store them in an experimental data bank. Over sixty researchers pooled their data in one place and chose one format in which to do the pooling: the Microcard, in 1955 the latest in micropublishing technologies, capable of reducing a normal page to one-twenty-fourth of its original size and storing it on opaque cards at the attractive price of just over half a cent per page.[1] Perhaps their "nominal cost" relative to traditional publishing suggested cheapness; yet Microcards offered security against the threats of dust, dampness, war, and other forms of neglect to which the book was susceptible: "Should Microcards be damaged or destroyed by fire or flood, exact duplicates can be quickly prepared from negatives which the Microcard Corporation, upon request, keeps on file," assured their brochure.[2] Backed by such guarantees, the mass of collected materials constituted "a vast scientific resource."[3] It was believed that very small bits—data sets culled from hundreds of workers—when put together en masse and miniaturized by advanced machines made a grand vision possible. But the "database of dreams" as I argue, was not simply this innovation (of miniaturizing the massive stocks of collected materials from field stations, ethnographic sites, behavioral labs, and potentially almost any social or cultural situation). It was a combination of techniques and tools that came together over a concentrated period of time to make an odd yet strangely successful (if later obscure) device.

Its makers—especially its primary mover, psychologist Bert Kaplan, and his group, the Committee on Primary Records in Culture and Personality of the National Research Council—

envisioned the project as a "master plan" for how to save and harness data across the social and behavioral sciences (especially anthropology, sociology, and psychology), and even though only one part of the plan was carried out—namely, *Microcard Publications of Primary Records in Culture and Personality*, conceived originally as a pilot project of twenty to thirty thousand pages—their design and indeed their dream was far bigger. Participants were in fact pioneers of data, to apply a retrospective label. Nearly alone, they worried about what would become of social researchers' data sets, so carefully harvested via "Herculean efforts," so carelessly provided for.[4] In 1955, a California-based child-study expert remarked exhaustedly on the expense and effort of researchers in gathering their data on children's doll-playing and block-stacking games, as well as other behavior. Clearly it would be desirable for others to share in the fruits; yet, she declared to her interlocutor, Kaplan, "Our data has been very expensive to collect and we should make it available. . . . Our problem is that we have so darned much data."[5] Was there a standard method that would manage, save, store, hold, and circulate this excessive amount?

The device described here was the first iteration of its originators' dream of total data management in the social sciences. Data sets from all sorts of unlikely contexts, Kaplan's group agreed, were in danger of inadequate preservation, and the group hoped to make a home for such research that would allow scientists of the present and future freely to access it. To mention the fact that only today are such repositories coming into being and that researchers' up-to-the-minute debates concern these very problems is to suggest just how prescient these pioneers were sixty years ago and how poignant their project appears in retrospect. It is tempting to place the Kaplan group's effort in the tradition of literary works such as William Wordsworth's 1850 *Prelude*, intended to be mere prefatory material for something much longer and greater but which ended up instead serving as the work itself. Likewise, despite its enormous prescience, the pilot project, *Microcard Publications of Primary Records in Culture and Personality*, became its own endpoint.

The project began with little fanfare in the summer of 1947 as foot-wearying research for a doctoral dissertation. Bert Kaplan, then a Harvard graduate student, set out from Cambridge to the New Mexico desert to study four American Indian tribes living near the small farming town of Ramah, where a sign welcomed visitors to "The Pinto Bean Capital of the World." Kaplan was young, tall, shy, and gangly, giving some people the impression that he was physically unrelated to his own body. When he was in high school, due to his height, he was invited out of the blue to try out for the football team, for which he played affably during the year. When, at the start of the next year, a new coach "somehow did not realize that I was on the team" (as he recalled decades later), Kaplan proved equally unflappable and "just let the matter drop."[6] Even at an early age, Kaplan had a gift for getting along with people and a team-spirit generosity.

When he arrived at Ramah, he had just returned from World War II service as an army psychologist. With only a college degree in psychology from Brooklyn College—though with rave reviews from his mentor, the humanist psychologist Abraham Maslow—his army orders at first had him on a crew maintaining B-26's "despite [my] being pretty uncertain what a wrench was." His stint in the mechanics squadron was cut short with a dispatch to officer candidate school. After a brief time as supply officer, he was "somehow picked out by some equipment of the computer to be a psychologist" and shipped off to Okinawa and other places in the far reaches of the Pacific Front to treat soldiers suffering from traumatic neurosis, a malady now known as Post-Traumatic Stress Disorder.[7] Kaplan loved being an army psychologist. Given his own ward, he tried out psychological testing; hypnotic trance induction; the occasional dose of sodium amytal, a bitter, barbiturate-like powder once believed to function as a truth serum (administered by Kaplan's colleague, who was a psychiatrist); and psychoanalytic talk therapy to get emotionally distraught soldiers to unburden themselves. Kaplan noticed that having the chance to tell their stories helped the men he was treating more than almost any other approach. He grew more interested in storytelling and psychological testing and in the vagaries of personality differences, different responses to the same situation. He also seemed to enjoy the multiplying of methods rather than fastening on a favorite one, an eclecticism that would prove useful later.

With some finagling on his part, Kaplan received an offer, while he was still at the Pacific front, to join Harvard's PhD program in social relations. As his wife Hermia recalled it, Harvard admitted him somewhat half-heartedly, claiming (in her words), "We don't ordinarily take people with your academic background, but because there's a war on, we'll accept you."[8] This reluctant invitation perhaps added to Kaplan's initial lack of confidence in graduate school. The son of a Tammany Hall "captain" of East European Jewish immigrant stock, he was not at all "to the manner born," much less to the manor, as were many at Harvard. Nonetheless, when there, Kaplan embarked on ambitious research using the most up-to-date methods. He and Hermia set off for New Mexico in the summer of his second year under the wing of Clyde Kluckhohn, a powerful professor at Harvard.

Their destination in New Mexico was a rich area for social scientific inquiry due to the fact that at least four different subgroups—Zuni, Navaho, Hopi, and Spanish-American—lived side by side within a twenty-five-mile contained area. Instead of the usual anthropology-style fieldwork, Kaplan went to Ramah to do just one thing: give tests. To be precise, he planned to give several kinds of highly exacting projective psychological tests and then gather a mini-encyclopedia of results in an unprecedentedly short timespan. Although he focused on veterans of the recent war—one of his colleagues was writing about American Indian vets, and Kaplan intended, by sharing his data, to contribute to this study— he also wanted to gather full sets of test results encompassing whole villages. What kind of tests? Kaplan had four in hand: the Rorschach inkblot test; the Thematic Apperception Test (TAT), a picture-based test he administered in two standard varieties; and a sentence-completion test of thirty-six questions.

Kaplan went on to write his dissertation on the four cultures he had tested but increasingly found his attention drawn to the matter of data collection and storage—a concern that today might be referred to as the problem of the "perishable format."[9] It worried him that many researchers generated significant amounts of data—new, never-before-seen kinds of data detailing the inner lives of subjects (peasants, primitives, poor people) who were otherwise reluctant to tell their stories, at least to strangers and scientists—but few took pains over the fate of such data. Kaplan himself had spent arduous hours giving tests and collecting results; his friends gathered dream materials and life histories with zeal. In the manner of a reverse milkman, one anthropologist in Tanzania was known to go around the village each morning before breakfast to collect villagers' dreams before they went on with their day and forgot them. Other colleagues stacked up long hours observing nursery school children play with blocks or primates groom each other and coding the component "behavioral items" in regularized field notes. Some came across incidental "found" data—the 500-page memoir of a female heroin addict the pioneering Chicago School sociologist Howard Becker recorded in the late 1940s, for example. Others used their classrooms to generate data—not least 453 dreams of Harvard undergraduates. These last were just two of the data sets Kaplan's committee targeted (although never realized) for its most ambitious collection. Kaplan's insight was to see that the future of these reams of information was not provided for. Unless researchers planned, their data, no matter how valuable and how meticulously rounded up, might disappear. Over the next few years, he thought up a plan to secure potentially all psychological, anthropological, and sociological data through the highest-tech methods he could find.

Cobbling together old and new technologies, a select group of American social scientists at mid-twentieth century, led by Kaplan and a committee of data-driven innovators, built what was in effect a machine to capture the contents and feel of other people's experiences. Unlike European and American psychoanalysts before them, they did not seek the deep meaning of a dream or track it through the thickets of the unconscious. They treated dreams and memories as "stuff," as material reality. They wanted to gather records of exactly what fades away and in effect prevent it from doing so.

Quantities of dreams typed out as records and miniaturized in micropublished format to serve the needs of future scientists was a new concept, something even Freud had not thought to do. It was also almost entirely new in history. Certainly, dream catalogs existed (for example, the sixteenth-century *Chinese Encyclopedia of Dreams*), and others had attempted to capture dreams in significant amounts (Jesuits collected Iroquois dreams in the seventeenth century, Jungians collected all kinds of dreams in the twentieth, and in 1963 the British writer J. B. Priestley collected evidence of precognitions, much of it stored in the dreams he urged television viewers of his popular broadcasts to mail in to him), but these were almost always gathered with an eye to supporting a particular theory or cosmology, from the "wish theory" of the Iroquois to the nature of the unconscious to speculations about the non-chronological nature of time. In each case, the dreams gathered up were said to be particularly *meaningful*, auguries, spectacular. At the turn of the twentieth century, psychiatric hospitals such as Massachusetts General and Bellevue collected dreams as part of an attempt to record more fully than ever before the psychic life of patients.[10] What was different in this later collection, rivaled perhaps only by the Mass Observation project in Britain, was the goal of collecting ordinary, everyday, unremarkable dreams from ordinary, everyday, unremarkable people—and these from cultures around the world. Neither medical patients nor seers, these subjects offered access to the often mundane and sometimes strange variety their offhand remarks and life stories described. Here was a database that held no age-old techniques or exotic rites but was made of passing thoughts, unremarkable dreams, and people's unknown lives. Some told dreams of the ordinary, such as grocery lists, and others told dreams of nighttime assignations with deities and of special powers gained. The dreams were not only "dreams themselves" but symbolic of the types of materials researchers would target—the dream-like, the penetrating, the possessed and the dispossessed. In the way that dreams happen every night and some days, and yet are often fantastic, this was about collecting the worldly and the otherworldly where they met.

How did investigators in 1947 begin to collect the uncollectable? How did they isolate the "intimate laws" of the "irresponsible incense of the imagination" (to borrow from Jorge Luís Borges)? How, at the same time, did they expand the scope of totality, even as the project of total knowledge-gathering bore, as we will see, a built-in tendency toward alienation? As we will see, in the gathering up of subjective materials, one can glimpse a "haunted" quality of the pursuit of objectivity: in marching on, collecting more and more, the specter of inevitable loss also asserts itself. This process casts a spell under which we continue to operate, in our database-aided search for an ever more total accounting. One can ask: What is lost and gained in the process of trying to recover something akin to Borges's "irrecoverable colors of the sky"?[11]

This archive was by 2005 or 2010 unexpectedly hard to locate in physical form, even in libraries where online listings claimed to hold a copy of it. Footnotes indicated its existence. Card catalogs assured its presence in scattered university microfilm rooms and in the Library of Congress. Yet very often I arrived only to find it either not available that day—temporarily lost, hard to find, mis-filed, head scratchingly missing, or simply not readable due to the scarcity of capable machines. "Not much demand for these re-cords!" a librarian at the Library of Congress remarked once when I requested it. Initially the archival assistant was unable to find it after considerable rummaging in the stacks. Such elusiveness emphasized a key finding of my research that would emerge over time: these data, and the technology that sustained and trans-formed them, had entered a curious state of limbo, one into which unfavored or once-but-no-longer-favored data often fall. This state has a dynamics of its own. It is not stable at all. "Latent life," a phrase the Catholic priest and scientist Basile Luyet used in the 1940s to describe a biological being that is neither fully alive nor absolutely dead, can apply to paperwork as well. Preserved in hard-to-access paper file folders or, later, in a Microcard archive—or, even later, on 3.5″ floppy disks—latent files can enter an in-between state, neither lost nor found. Such states of suspended animation become more and more common, as seen in such phe-nomena as "atomic trash," "zombie satellites," and scholarly anxiet-ies of the sort Pacific anthropologist Roderick Ewens recently expressed: "I'm sure I'm not the only person that still has informa-tion locked away on the odd disk in a bottom drawer, probably never to be accessed again."[12]

Pushing at my thoughts while pursuing this archive was an ever more persistent fact: the nature of memory itself was chang-ing. Joan Didion and John Gregory Dunne used to save in a small box the quixotic sentences their daughter spoke when she was learning to talk, odd scraps oddly piled in, the box sitting on Dunne's desk. Didion told the story from the point of view of her grief at the death of her husband and the soon-to-be-fatal illness of her daughter. Everything passes. And where was the box? Those who had filled it were no longer there, and the details of what they had said were also fragile. A change was taking place in terms of what was done with the scraps of paper of the world. Who today does not experience such change? "Small details that were once captured in dim memories or fading scraps of paper are now preserved forever in the digital minds of computers, in vast databases with fertile fields of personal data."[13] The question became: What about that forever?

The "database of dreams" was a pioneering exercise in the forever storage of intimate details. Today there are many more such databases, powered by machines that are far more powerful than the Rube Goldberg–like invention of these years. Yet the fate of what this earlier effort stored is a parable for our time. Its very existence, the parts that turned out to be fragile and the parts that ended up enduring, equally raise important questions about what it means to be human in an increasingly data-based world. That the experimental hard-won data it held got lost in many ways—though never entirely—is a symptom of our own future buried in the present. And yet this database holds thousands of pages of irreplaceable "raw" data from societies spread over the globe, data that had been gathered with care over previous decades of fieldwork, testing, and subject interviews. Somehow, this great project—great in ambition and great in scale—has disappeared almost entirely from the historical record and also from the memory of many of its disciplinary inheritors.[14] This near expunging is itself a historical phenomenon worth considering.

On one level this book is about the lists and charts; tests and standardized sheets; opaque cards, reports, and dissertations; public policy statements and census-style paperwork—what all together have been called "little tools of knowledge"—that make up and made possible this one-of-a-kind archive. Such "little tools" might seem to be trifles. Does it really matter what kinds of notebooks Dorothy Eggan's Hopi subjects used to write down their dreams?

Does it make a difference that Kaplan chose the Microcard rather than a normal filing cabinet or the preexisting Human Relations Area Files (HRAF) taxonomic system—or the dreaded microfiche, for that matter—as his vehicle? How are life histories taken down? By means of which tools and which rules? What happens when people stutter, burp, ask for more tea or more money, or refuse to go on? And can the whole technique count, itself, as a new technology? Can it do what scientists such as Robert Park asked of them—to reveal "what goes on behind the faces of men"? In total, the book forms a prosopography, or a portrait of a group of actors whose collective history can begin to "construct an intelligible picture of society and politics."[15] It is also, in effect, a prosopography of technologies, including tools both little and big.

What made such a collection possible in the middle of the twentieth century? How did researchers suppose they could render the elusiveness of dreams and the richness of inner psychological states as a string of micropublished numbers, images, lists, records, narratives, and charts—in short, as data? Was there nothing so remote or odd that it could not be transformed in this way? What is and what is not, what should and should not be, database-able? This book answers these questions by following the scientific logic and tool-based action that led some researchers to target ever more elusive and difficult-to-work-with materials for their files. Through the efforts of a cast of behavioral researchers working in the 1940s, 1950s, and early 1960s, a strange hybrid device developed that could take old-fashioned materials documenting old and (purportedly) disappearing ways of life, render them as data via the latest psychological and sociological techniques, and cache them in futuristic micro-reduction cards so that they could be read on a network of advanced gadgets around the world. The idea, and eventually the reality, was that any researcher could stand in New York or Basel and "hack into" assorted dreams of people far away— a Paraguayan tribesman or a Pathan villager.

The effort involved "suites of technologies"—combined techniques, new and old.[16] In order to understand how Kaplan's data bank worked and later didn't work, then, the chapters below follow the "careers" of a succession of techniques, technologies, and tests as they emerged and eventually, through the assiduous work of Bert

Kaplan, merged. Ultimately they formed a creole and compound new thing in the world of data storage. What lies ahead, then, are accounts including the birth of and fashion for projection tests (for example, the Rorschach), the emergence of miniaturizing photographic techniques, the life of the life-history method, the proceedings of dream-extraction forays, the varieties of fieldwork approaches, and the way interpersonal encounters could be leveraged as data production. All these would merge to make the Kaplan collection possible.

For the purposes of this book, I call this invention—really a set of interlocking inventions and borrowings—the "database of dreams," bearing in mind that its official and best-known name was the significantly less poetic *Microcard Publications of Primary Records in Culture and Personality.* Its primary home was the seemingly unromantic Primary Records Committee, whose members inhabited a network of departments, hotel meeting spots, field sites, and encampments all over the world. Also keep in mind that the topic here is not simply the *built* archive but the unbuilt one. My claim is that it was a system of interlocking printing technologies, reading devices, psychological techniques, fieldwork methods, "little tools" of knowledge, and human subjects and objects, all of which worked together to produce an untoward capability, a machine for making the invisible visible and the intangible tangible—at least for a short time in the middle of the last century. This book thus shows how technology can embed itself in subjectivity and how subjectivity shapes and is shaped by technology.

What would only become recognized as a full-fledged database in the 1980s has of course retrospectively shaped conceptions of what even a primitive "bank of data" looks like.[17] Here I want to expand the possibilities of the past and of construing the history of media and information. To be considered in the lineage of what makes a database, then, a data-storage device need not toe the line of the standard histories in terms of either espousing digital over analog or symbolizing success over failure. Thus I use the phrase "database of dreams" figuratively to evoke the larger project as well its individual components, including especially the Microcard archive that is the subject of intensive scrutiny in the pages that follow. I do not mean to claim that it was an electronic dual-processing database of the sort that became available in "primitive form" in 1964, fully eight years after the Primary Records Committee began its work, nor that it was something like a relational database of the 1970s or an object-oriented, object-relational database of the 1990s.[18] This is a story of succession. The "database of dreams," in contrast, is a way of tracing a different genealogy to pursue the question of how data and data storage came to be central to social life and personal accounting. It asks about the place of dreams within data. How has this apparently dream-free substrate called "data" with its clouds and servers, its banks and bases, come to be the site of memory—decisions about what is kept and how—as well as worries over what is lost over the course of a life? As we will see, the banking of data in a haphazard and improvisational form preceded and sheds light on such apparently uniform and penetrating technologies.

Most of this sum total of now forgotten personal data was drawn from people in the Philippines, the South Seas of the Pacific, the Great Basin of the United States (as well as many additional American Indian reservations), and a range of other places once seen as "far away." The data came frequently from colonial possessions that were in the process of being handed off or reconstituted or gaining independence, places where anthropologists had worked in tandem with or in opposition to, or sometimes merely alongside, colonial administrators and settler areas. One could find out, for example, what changes Menominee Indians living in the Wisconsin woods in 1947 thought the next ten years would bring to their lives on and off the reservation, as in anthropologist Louise Spindler's data set, "61 Rorschachs and 15 Expressive Autobiographic Interviews of Menomini Indian Women." What did one woman daydream about during the long afternoon hours at home when her husband was logging? How did she score on an array of psychological tests? Another subject from this same data set, known as Case 1 in the records, reflected on a baby she had lost six months before: "Old men used to understand babies. Now, nobody understands their language. That baby used to talk all the time." As in this example, the data contained a record, at times, of people experiencing a deep loss and profound change—of culture, or life-ways, or within their own families.[19] Fieldworkers in these places focused on groups seen as non-literate (though some members had been taught to write in colonial schools while others had never even seen a photograph or printed page before).

Despite the fact that the data collection was global in scope, a large portion—approximately one-half to two-thirds, depending on how it is counted—came from American Indian tribal groups, who appealed to researchers as both distant (because they symbolized age-old traditions) and nearby (because a researcher could visit them in a two- or three-day drive from Chicago or New York). Likely, too, their inner data were necessary in the second birth of an American national consciousness in the postwar years. American Indians' experiences of loss, fragmentation, persistence, endurance, strength, assertiveness, and suffering seemed to promise advanced insights into the conditions of modern life at a moment of acute cultural crisis due to lack of infrastructure, environmental depredation, shifts in political policies, and the crosswinds of great global battles such as World War II. Within anthropology, there was a long tradition of treating the American southwest as a laboratory. Here the decades-old laboratory became a site to focus on a new kind of intensively collected data, called subjective materials, and new cutting-edge ways of storing them.

One of the themes of the book is the imperfections and ambiguities that arise in the search to find and extract secret information, in ever more various kinds, from various kinds of people. My wish is to explore the intersecting effects of technology and subjectivity in history. Often, too, in the history of the human sciences, the role of technological experimentation and a tool-centered methodological spirit has received scant attention, and although this is changing in some recent work, I have tried to bring these two concerns together in a new way.[20] In locating and to some extent resurrecting data that have long been neglected and out of sight (though never hidden), I have been careful to keep the data donors' names anonymized or coded, to bypass overly intimate revelations—because after all the book is not about these but rather the technologies of access—and to respect both of the parties involved in this once celebrated and later forgotten research.

Here one can find the most fleeting thoughts and exorbitant night- and daytime visions. "Human documents," they were called. Their story is also the story of what "humanness" came to mean in an age of rapid change in technological and social conditions. Call it a database of consciousness, a repository of humankind's most elusive ways of being human, or an anthropological archive; through it a veritable sluice of social knowledge flowed from seemingly unlikely encounters. This book is about those encounters—between scientists and subjects, between knowledge and machines—as well as the data that flowed out of them and the ways these were preserved and not preserved.

CHAPTER ONE

Paperwork of the Inner Self

A LITTLE-KNOWN turning point in the prosecution of World War II war crimes occurred in 1945 at Nuremberg. Sitting on his prison cot was Hermann Göring, recently captured Reichsfeldmarschall, founder of the Sturmabteilung (SA), creator of the first concentration camps, and a man who, not many weeks before, had been poised next in command to Adolf Hitler. He was taking the Rorschach test. Examining the inkblot on Card III, he displayed emotions ranging from delight to anger to amusement in a few seconds. At first he was sure that he saw the black blots as two charming cartoon figures in high collars, "but the red spots ... I can't figure that out." He got impatient and snapped his forefinger at the three red spots on the card, as though to brush them off, the test giver noted. Then Göring continued, "What these things are ... damned if I know.... They are debating over something ... maybe two doctors arguing over the inner organs of a man." He laughed. When another American psychiatrist retested him some weeks later, Göring exclaimed, "Oh those crazy cards again."[1]

The great crime of National Socialism had just played out, and the victors wanted to know in a deeper way what it was they had been fighting. Was there something one could call a "Nazi mind"? Paper records of the test protocols were equivalent to lab results. As Göring's first Rorschach test administrator, the American psychiatrist Douglas M. Kelley, remarked, "I had at Nuremberg the purest known Nazi-virus cultures—twenty-two clay flasks as it were—to study, and with but a short time in which to work."[2] The Rorschach test and other such tests, experts generally agreed, were the most direct way for them to go to work quickly and efficiently at the most profound level. The fact that such tests could not be used in criminal prosecution (at least not yet) makes the point even more strongly. During this period they were for research and therapy only.

It is not clear what, if anything, Göring's response to Card III or the other cards revealed. Some argued the test was invalidated by the unorthodox conditions under which it was given. Some claimed a distinct pathology revealed itself, not so much in the content—the argumentative dissection Göring saw, for example— but in his own irritation at finding red blots among the black ones, a response to color that could indicate a dangerous irascibility and an uncontrolled emotional life. In the case of Göring, a certain emotional lability might have already been clear from his on-again off-again morphine addiction and his wont to wear outlandish fur-and-feathered costumes. Others held there was a lack of pathology at all in the evidence of the test—at least no pathology sufficient to explain the actions of a man capable of such heinous crimes.[3]

What is beyond dispute is that Göring's Rorschach records remained in a repository for future study, along with those of twenty-one other Nuremberg detainees, soon joined by the records of 209 largely Danish Nazi collaborators from the Copenhagen War Crimes Trial. Most of the Nuremberg accused also submitted to the Thematic Apperception Test (TAT) and the Weschler-Bellevue Intelligence Test before they, or at least a goodly proportion of them, underwent hanging. These "Nazi protocols" became memoria for future generations, and, the question of their scientific usefulness aside for the moment, they functioned a bit like relics (or the opposite), fetishes from a soon-to-be execution.

In keeping with this status of prospective reliquary, the paper copies of the Nazi test results ended up, within a year of their subjects' sentences, as rare objects themselves. Initially, through 1946 and 1947, they circulated freely among the United States' most skilled Rorschach interpreters, ten of whom received invitations to attempt an ultimate diagnostic judgment, but the project was soon shelved, copies were misfiled and mislaid, and this collaborative evaluation never came to fruition. After an initial rush to publish by the two on-site psychologists, both of whom admitted to having reached only provisional conclusions and expressed the hope that future studies would be more objective, silence held sway.[4] For about three decades, the Nazi Rorschachs all but disappeared. One copy sat in the Chicago files of a prominent Rorschach expert (Samuel Beck, who received them direct from Kelley) while at least one other copy remained with Molly Harrower (originally in charge of the expert evaluation program). It was as if the paperwork were lost because the materials ceased for so long to circulate, and the tests became the stuff of legend and rumor. This temporary scarcity or difficulty of locating them in physical form was a kind of latency.[5] On the first page of the first publication to come out of Nuremberg, Kelley had spoken repeatedly of "securing the material," "securing the data," and (yet again) "material secured," and his words conveyed reassuring finality.[6] However, the paperwork's subsequent travails indicated that simply securing the tests was not a final act, as many had presumed, but that the protocols had an unfolding fate, moving in and out of availability.

After almost thirty years of latency, the protocols emerged once more, and several studies took up the beckoning hope that once and for all a final say could be had. Two camps emerged. One set of researchers found significant similarities among the Nazi results, enough to generalize that their psyches indeed indicated pathological development. Among other things, the unusually high number of botanical images (for example, dahlias, daffodils, roses) found in the Nazi records were said to link them to an undeveloped sense of human relatedness.[7] Another set found no regularities at all and concluded—as was their preexisting view, uncoincidentally—that it was the social, cultural, and political context that had forged and shaped the behavior of these otherwise quite typical men. No toxic environment, no Nazis.[8] Perhaps most confidently, in the 1980s a third wave of interpretation arose with the advent of two computer programs capable of automatically assessing protocols. Not only these programs, but also the accretion, in the intervening years, of a large protocol pool from schizophrenics, outpatients, and "normals," now allowed a more thoroughly comparative evaluation of the Nazi records to be made. Running the eight most "serious" Nazi records (that is, serious Nazis, not serious responses) through the John Exner and Eric Zillmer computer assessment programs shed new light on old hypotheses. The eight emerged with a mean score of 3.37 on the Rorschach Schizophrenia Index (SCZI)—schizophrenia likelihood—but this had little meaning because the scores were so widely dispersed. Only one qualified as depressed: Hans Frank, who achieved a score of 3 on the Rorschach Depression Index (DEPI) and who was one of the few to express a sense of responsibility for his actions, stating at the trial that "a thousand years will pass and this guilt of Germany will not be erased." Although a few seemed possibly suicidal, this interestingly did not include Göring, who was shortly to commit suicide but whose suicide constellation (S-CON) was only 4 of a possible 11 factors. Taking into further account the Reichsfeldmarschall's high affectivity score (AFR) and ego strength (EGO), the computer program described him as a "person who is very attracted to being around emotional stimuli," in addition to engaging in a narcissistic overvaluation of the self, observations seemingly if retrospectively buttressed by his declaring before the trial started that he would have the courage to confine his defense to three simple words: "Kiss my ass."[9] The basic finding of these computer-assisted analyses was that there was no common finding: no common denominator emerged, but rather a lot of variety.

So it was that in addition to the eleven published volumes of documents that filled six freight cars when originally shipped to Nuremberg and the twenty volumes of trial proceedings, there issued this unstable repository of the war criminals' inner lives, rendered as paper protocols. By following these documents, discovering their fates, and describing the subsequent conditions under which these tests flourished even as others failed, one can also begin to discern from what their unique if elusive power derived. This retrospectively odd power must be emphasized in telling the story of what I am calling the "database of dreams," for that archive could not have come into existence without these tests—or rather, not so much the tests themselves but the claims that were made about the tests' penetrating powers and the need to preserve their resulting data against loss, misplacement, and unfashionability.

In the end, the fifty-two-year-old Göring ate cyanide tablets less than an hour before his scheduled execution. He had developed rapport with psychiatrist Kelley, then in his mid thirties, and asked him to look after his family following his death. (Some have suspected Kelley passed Göring the pills, though the source has never been found, but Kelley by this time had returned to California.) In a sad afterlife of this relationship, Kelley, subject to mood swings and increasing instability after Nuremberg, committed suicide in front of his three children and wife on New Year's Day 1958 by swallowing cyanide powder, which he apparently carried around with him habitually in his pocket. A note from Göring's coprisoner, propagandist Alfred Rosenberg, was in his papers: "I regret your departure from Nuremberg, as do the comrades confined with me. I thank you for your human behavior and also for your attempt to understand our reasons." Ever since, heavy-handed journalists have invoked a heart-of-darkness thesis for what happened. Kelley had looked pure evil in the face, the thesis goes, and was forever changed. ("Had he carried back from those twenty-two cells something darker than psychiatric reports?" asked one reporter. "Was there a part of him that was hollowed out by living inside the minds of men who killed so easily?")[10] Yet twice-baked rhetoric aside, it is tempting to leave open for questioning the elements contributing to Kelley's death: How much does change in the process of trying to measure those things that exceed, elude, or defy measurement, things at the very edge of unmeasurability? How might one be changed?

The 1940s, '50s, and '60s were the heyday of the projective tests' powers in the United States, and it is no coincidence that their administration to the world's most notorious accused war criminals bookended the period. In 1962, an Israeli psychologist administered the Rorschach to Adolf Eichmann before his trial in Jerusalem, these efforts qualifying him as one among the parade of "soul experts" whom Hannah Arendt derided in a *New Yorker* account and later in *Eichmann in Jerusalem* for finding, supposedly, nothing amiss with Eichmann's "perfectly normal" psyche.[11] The point here is not to reevaluate these results, nor to comment on the evaluations of others, nor even to mention the latter-day carping at Arendt's "very ill-conceived" claims about what they said, but to mark simply the

fact of their existence as a resource on file.[12] That the highest ene-
mies of state underwent such psychological tests indicates, if nothing else, the apex of influence to which the tests had risen, and their
collected results were akin to a national resource on the nature of
human nature. It is hard to imagine the equivalent today. If Osama
Bin Laden had been captured, would a prime order of business have
been to give him the Rorschach or TAT tests? It is easier to envi-
sion him submitting to an fMRI of his brain function. Between the
testing of Hermann Göring in 1945 and Adolf Eichmann in 1962,
however, psychoanalysis flourished, and the projective test was king
of technologies for seeing into the inner self.

With the addition of Eichmann's protocol, the Nazi Rorschachs,
although intended to be a secure data set, continued to slip in and
out of circulation and high-security status, just as the diagnoses they
occasioned ranged up and down scales of pathology and banality.
Anything but stable, never easily located, and even today not defini-
tively interpreted—despite dispiritingly regular claims of their having arrived at last at a "final undisputed diagnosis"[13]—the tests, as
paper, raise questions relevant to the problem of data, and specifically the data collection that is my subject here—the "database of
dreams"—lending insight into how its techniques grew and spread
and why it was needed.

The Rorschach psychodiagnostic test (by Hermann Rorschach,
originally 1921) and the TAT (by Henry A. Murray and Christiana
Morgan, originally 1935) were the number one and two tests
respectively, and their rise to these positions in the projective test
movement illuminates the unlikely yet, in retrospect, nearly inevitable ways researchers recruited them to do what could not yet be
done: look directly into the mind and heart of a human being. It
was vital for future endeavor to plumb the depths of the human
psyche to know exactly what a certain inkblot or picture held for a
certain person (a bear? a clown? an eviscerated body on a dissection table?). How that person saw a card was a roadmap to how the
person saw him- or herself and the world.

In an unusual twist that characterized few other diagnostic
techniques, Hermann Rorschach's test, published as *Psychodiagnostik*,
was rumored to have resulted in its author's death from heartbreak.

With an inaugural publication run of only a little over a thousand copies, it nonetheless sold so disappointingly that Rorschach, who had hoped for a breakthrough to revolutionize the practice of psychotherapy, fell into despondency. He had been preparing the test for years. At Münsterlingen in Switzerland during his early training as a resident he had been known for adopting strange but oddly productive methods. A monkey, deliberately set loose, climbed around the clinic, and he recorded his patients' revealing responses to its apery. At times he experimented with turning the clinic into a theater, projecting shadow-puppet shows and allowing his patients to dress up and perform, their antics also revealing for their files and subsequent diagnoses. He also had patients draw and posted their results on the walls. Eventually he settled on the inkblot as a desirably ambiguous stimulus that could reveal the preoccupations, and indeed the whole human truth, of the person at hand. The history of the inkblot form in Europe hardly began with Rorschach, however, as it extended many decades back to its employment as a parlor game, an artistic impetus (to Leonardo, Botticelli, and Victor Hugo), an accompaniment to poetry (in Justinus Kerner's 1857 dark volume *Kleksographien*), and even an intelligence test by Alfred Binet, all of which preceded and inspired Rorschach's adoption of the form.[14] Notably, these predecessors found that the qualities of an inkblot, whose forms combined the accidental with the hyperdefined, were stimulating. By the time Rorschach embraced the inkblot, he had passed through the Krjukovo clinic, a private sanitarium in Moscow, and another in Berne before he finally took up a practice near the Swiss border with Austria at Herisau, where he was more systematic than before. There he tested 405 subjects, patients, and colleagues on a set of different inkblots made from folding a page in half. Form, color, and movement were his three broad parameters for interpretation: it was not only what one found in the blots, but also *how* one found them—processes of perception—that these revealed.

These chosen parameters were why Göring's response years later to the red spots on Card III could be considered important: a dominant "color" response, labeled "C" or "CF," indicated a childlike lack of emotional control and tendency toward extraversion, whereas a dominant "form" response, labeled "F" or "FC," suggested introversion and a more integrated emotional life. In answering Rorschach's question, "What might this be?," the patient proceeded through the inkblots from the purely black-and-white Card I; to the black, white, and red Cards II and III; to the riotously tropical palette of Cards VIII through X. Some evinced a hypersensitivity to color that Rorschach called "color shock." Such a response suggested possible psychotic tendencies. On the other hand, the descrying of movement, or "M," in an image—as in Göring's high-collared dancers—indicated high intelligence. A final parameter was whether one saw a whole image (W), parts of a whole (D, or common detail), or incredibly minute detail (Dd), each of which slid down a scale of increasing likelihood of schizophrenia. An answer's content (seeing a bat or a bird, say) also counted, though not quite as much as the other parameters, and a response might earn either an "A" for Animal or an "H" for Human or then again "Orig" for an unusual answer. At the heart of the test was deriving the subject's underlying "experiential type," or *Erlebnistyp*, by calculating the ratio of movement to color (M:C). Seeing more movement in the cards meant that one was in possession of a mature emotional life, whereas seeing color indicated, to put it mildly if over simply, that one was still stuck in childish ways.

Even after Rorschach streamlined his title to the one-word *Psychodiagnostik*, six publishers turned the work down, and the one who at last agreed to put it out was half-hearted, had trouble funding even the limited run, and insisted Rorschach cut five of the fifteen inkblots he had originally intended.[15] Despite the arbitrariness of this decision, based on impecuniousness rather than a grand design, the ten cards became totemic. Rorschach's final inkblots were, in fact, the outcome of years of experimentation with haphazard blots he made from folding a piece of paper in two over a small ink puddle; however, he ultimately decided the blots worked best, and he made them with carefully brushed-on ink and (most probably) watercolors to create a range of "chance forms." Despite the fact that Rorschach stipulated the cards should in future be printed on special paper made under highly specific humidity and temperature conditions so as to duplicate the inkblots of the 1921 first edition, those original blots themselves were deemed by one eminent scholar as a "total failure," marked by the printer's "slovenly work," which was full of errors. What error, and, more important, how can an inkblot meant to embody an accidental happenstance be wrong? In the following way: it seems that Rorschach's inkblots, as he originally painted them, contained little "shading" within the black blots, but when the first edition came out, quite a bit of "intriguing unanticipated variation in the shading" revealed itself (as another scholar has described it, putting a more positive spin on the incident). In his last article, Rorschach himself suggested this newly evident shading was significant—it would surely influence subjects' responses—but since no explanatory alterations appeared in the *Psychodiagnostik* itself, his remarks went unheeded at first, and the question of "shading," which was in coming decades to spark relentless and fierce controversy among Rorschach aficionados, remained as yet uncontroversial.[16]

Not long after his work's largely unheralded debut, Rorschach succumbed to sudden-onset peritonitis and died a disappointed man at thirty-seven. Although his technique gained some posthumous support in Europe (adopted by Jungians, crucially) and began to spread in Japan due to an "accidental" discovery a browsing psychologist named Yuzaburo Uchida made in 1925 when he stumbled across a copy of *Psychodiagnostik* sitting on the shelves of a Tokyo bookstore, it was only in the United States that it could be said, initially, to flourish.[17]

Most directly, the reason for its overseas flourishing was the arrival in July 1934 of impresario émigré Bruno Klopfer in Brooklyn, accompanied by his son Walter and carrying the Rorschach in his luggage. Until his decision to flee Nazi rule, Klopfer had been a well-placed Jewish-Bavarian psychologist with a successful practice in Berlin. There, in addition to his practice, he hosted a five-year run of a popular radio program that advised parents all over Germany on matters of child rearing, answering their letters and experimenting in new forms of dialogue. The show "was quite an impudence," Klopfer recalled toward the end of his life in an interview, "because people were still accustomed to . . . give *lectures* over the radio, and it wasn't at all customary to do it that way [sitting and talking]." Early success in broadcasting offered a hint to the effect Klopfer would later have on students and adherents: he was able to draw people in to his projects through a give-and-take approach, or,

as one of his students recalled, "He made you feel you were collabo-
rating with him in the unraveling of a human puzzle."[18] As Berlin
life became untenable for Klopfer and his wife and son, they fled
Germany and were forced to separate, the son staying with his
father. Stopping for a year in Switzerland on the invitation of Carl
Jung (who issued it in response to Klopfer's visa problems), Klopfer
gained experience in the use of the Rorschach for personnel selection at the famed Burghölzi clinic. Once relocated, with the family
reunited in a small Brooklyn apartment and the Rorschach cards
safely tucked away, he took a job as research assistant in Columbia
University's department of anthropology under the legendary Franz
Boas, a hiring decision that, though the job paid remarkably poorly,
would have fateful consequences.

As it happened, by the mid-1930s American social scientists
were eager to know more about the rumored capacities of the
Rorschach test, for few had proper training in the use of this exotic
psychometric device and its ritualized procedures. Predating
Klopfer's rather dramatic U.S. landing was a small in situ circle of
dedicated Rorschachers, all of whom learned from the psychologist
David Levy, an American who had trained in Switzerland in the
mid-1920s with Emil Oberholzer, Rorschach's executor and colleague. Thus, Klopfer was not exactly the first to carry the
Rorschach across the Atlantic, but he did so with aplomb, under
circumstances that drew attention, at least in the telling, to the
physicality of the cards and the adventure of their crossing. He
would become the most insistent institution builder and, not incidentally, would have by far the most influence with those who
eventually built the data clearinghouse described here, as well as
like-minded experimentalists, including far-voyaging anthropologists, sociologically inclined fieldworkers, and outside-of-the-box
psychologists.[19]

Klopfer's arrival at Columbia was a godsend, and when graduate students discovered his intimate expertise with the test as well
as his possession of an actual set of authentic cards, the news "went
through the department like wildfire."[20] Intensive training sessions
sprang up—not initially at Columbia itself, due to resistance from
stodgy traditionalists, but rather in Klopfer's apartment, in empty
churches, and in the kitchens of other adherents. During this fervid

time certain obvious flaws in Rorschach's research design were remedied, discussion (it is said) extended late into the night, and a cadre of stalwart devotees emerged. Klopfer, as founder (or what Marguerite Hertz calls "leading spirit") in 1936 of the mimeographed *Rorschach Research Exchange* and the fledgling Society for Projective Techniques and co-author of an influential how-to manual for the Rorschach published in the United States, befriended anthropologists and was instrumental in the spread of the technique to this discipline with its in-built access to "other kinds of minds." (Klopfer's co-author in the soon-to-be-classic Rorschach manual was Douglas M. Kelley, who would go on to conduct those fateful Nazi Rorschach sessions with Göring.)[21]

Meanwhile, internecine battles ensued in which rivals pitted themselves against Klopfer for control of the Rorschach so that eventually, by the 1940s, the test's American-based enthusiasts split into five mutually antagonistic communities, each of which systematized the Rorschach in its own way, and each of which had different ideas about the relationship of science to interpretation. Some saw themselves above all as atheoretical mathematizers (Zygmunt Piotrowski), some as statisticians (Beck), and some as sympathetic standardizers (Hertz), and each camp tended to cast aspersions at the others, exaggerating their respective inclinations, such that Klopfer was pegged as having "an emphasis on extreme subjectivity."[22] A telling detail about Klopfer's particular style is that he developed out of necessity—poor eyesight—a method of memorizing Rorschach responses rather than writing them down, and he could accurately recall entire tests from memory as if he had "internalized" them. When asked to consult another subject's written Rorschach protocol in order to interpret it, he would hold the record extremely close to his eyes. To the onlooker he appeared to be smelling the paper "and in some mysterious fashion combining visual and olfactory clues in his subsequent interpretation"—or so legend had it.[23]

Drawn to subtlety, Klopfer also championed sensitivity to "shading," that onetime printer's error that now came replicated in each edition of the cards. Shading, he felt, revealed how a subject organized his need for affection, and he pioneered a whole vocabulary to describe people's responses. "Shading evasion," for example, indicated emotional avoidance. If a test taker responded to Card IV by saying, "About the only thing I could see would be something under water," this constituted avoidance of shading and likely indicated "reluctance to accept one's need for affection . . . stemming from early experiences of rejection and deprivation." Many shading-related behaviors could be found, Klopfer stipulated, including shading denial and shading insensitivity, and eventually "Klopfer introduced the largest number of shading response categories in the literature, i.e., 12," as one rival Rorschach researcher acknowledged.[24]

Klopfer's charismatic take on the Rorschach, one that relied on a personal line of transmission, a gift for organizing, a visceral connection to the cards, and an atmosphere of suasion, contrasted with the more rigorously quantitative and depersonalized approaches others embraced. So bitter did the resulting enmity among Rorschach rivals become, especially in the case of Klopfer versus the number-crunching but reputedly less personally winning Samuel Beck, that they refused to appear in the same room together or read each other's work.

Professional dust-ups aside, in the early postwar period, the tendency of the test was almost always in the direction of more traction, more authority, more credence. It spread especially within American juridical, clinical, and "pure research" circles. Even as experts administered the test to subjects exhibiting an extreme range of human behavioral possibilities—on the one hand, notorious negative achievers such as (eventually) psychosexual killer Jeffrey Dahmer; on the other, exemplars of excellence such as Franklin D. Roosevelt, Linus Pauling, and Albert Einstein—it achieved wide acceptance for use in day-to-day child custody cases and human resources departments. Did you get the job? Will you have rights to see your child two, three, four (or zero) days a week? It could depend on, as one disgruntled divorced father put it, whether you saw a butterfly or a bat. Each year, hundreds of thousands "or perhaps millions" of people continue even in the twenty-first century to take the test.[25] Throughout the test's years of growth, a technical claim of epistemological cogency and object-related transparency of vision prevailed and offered a self-justifying rationale for the test's further use; at the very least, researchers were assembling a systematic database of millions of responses. As a nurse who worked early on with Rorschach at Herisau noted, the staff could "penetrate by way of the test into the world of the mentally ill to an amazing extent."[26] Worded differently but in essence the same, this claim would surface again and again as the projective test movement gained ground around the world, and its expression even at the outset, within the test's scene of origin, should be marked, for it spoke of the need, and the felt achievement of, *penetration*—not just of the mad, but also of all that was far off or difficult to talk to.

The Rorschach test truly came of age in the mid- to late-1930s, and the TAT, its only real rival in the realm of projective instruments, observed similar timing, born in the middle of that decade and gaining much ground by the end. The timing, as we will see, was not merely a coincidence.

This test, too, had its adherents. Whereas the Rorschach was the product of a single father, the TAT emerged from a non-fertile but romantic relationship between two people, one an artist, the other a psychologist. Henry A. Murray was a New England Puritan-stock, Harvard-educated biochemist who turned psychologist after reading *Moby Dick* and meeting Carl Jung during a European tour. On the ship's voyage over, Murray discovered his initials were the same as Melville's, H. A. M., a coincidence that he found mystically significant, and he went on, as a scholarly sidelight, to help found the revival of Melville studies in the United States. The then neglected Melville moved him because Murray felt personally connected to his blue-blood predecessor's unflagging efforts to understand and depict the human quest for meaning.

Murray and his lover, Christiana Morgan, another Boston Brahmin, met separately with Jung in the late 1920s—her visionary drawings from this time were immortalized in a special set of seminars in the Jungian oeuvre[27]—and decided to pursue their interest in the "dyad" as the root of all human relationships by inventing a psychological test based on an emotionally resonant series of pictures. By 1935, when they first published an article describing their test, titled "A Method for Investigating Fantasies," Murray had gained the directorship of the Harvard Psychological Clinic, and

Morgan was an assistant there. They solidified their invention in a 1938 volume from Harvard University Press called *The Thematic Apperception Test* and followed up with the reissuing in 1943 of the manual, again with Harvard, by which time the test was a star, a new light in the field of personality psychology.[28] The two also were the first to use the term "projective technique," a formulation that became the launching board for the projective movement. Just as inkblot tests aplenty preceded the Rorschach, so too did several see-a-picture, tell-a-story tests (such as Horace Brittain's in 1907, Walter Libby's in 1908, L. P. Clark's in 1926, and Louis Schwartz's in 1932) preexist the TAT, which, however, easily cast these forerunners into the shadows. Arguably, it was the step of joining the tests together under the rubric of "projection" that catapulted them to a level of success others had failed to reach.[29]

Morgan and Murray offered a way of exploring the least accessible unconscious contents of the personality in themselves. The Rorschach in their view elicited relatively simple responses. To begin with "Looks like …," as the Rorschach did, augured only "quasi-projections" or pseudo-projections based on the surface perceptions of the subject. In contrast, they felt, the TAT could access *apperception*—that is, the secret machineries by which the fantasy life and its imaginative fancies guided people's lives. The test was to be a way of making the invisible visible and retrieving the irretrievable in some manifest form: "My idea," Murray said in a later interview, "was to illuminate the unconscious processes— that were repressed—of which the subject was not aware. That was the whole point of it."[30]

Murray was in fact making a critique, by means of the operations of the test itself, of a significant portion of professional psychologists in his time. He was disgusted with what he saw others doing: colleagues racing to qualify as experimentalists by the endless running of rats through mazes "had trained in incapacity. They were trained to have tunnel vision." Obsessed with quantifying and being precise about carefully delimited areas of human functioning, his cohort shied from the unruly, the "darker, blinder areas of the psyche."[31] Yet it was not that Murray rejected exactitude and statistics. Paradoxically, his and Morgan's test, with its claims of new penetrating powers, itself was to become the object of a full-scale rush to quantify and standardize during World War II and the Cold War.

In order to enter this unmapped terrain of fantasy, the two invented their test during intensive months (1933–1934) working together in the Harvard "Psycho Clinic," the institution of which Murray had suddenly assumed directorship due to the untimely death of its founder, the wealthy Harvard benefactor Dr. Morton Prince. This promotion from mere research assistant to head occurred despite Murray's lack of training in relevant areas, for aside from his personal connection with Carl Jung, Murray had spent his professional years as a physiological chemist doing embryological research on chicks at the Rockefeller Institute. Not only Murray's precipitous rise but also the very existence of the well-heeled Psycho Clinic rubbed some people at Harvard the wrong way— usually less well-connected people with fewer independent sources of income and a greater timidity around the unconventional. A high-level employee at the clinic, Morgan was at the start of engaging with Murray on a forty-year-long quasi-sadomasochistic, quasi-scientific investigation of male-female "dyadic" relations via their joint sexual and intellectual life. Capes were worn, a stone tower was built, and both felt the result of their efforts would be a conjoint masterpiece, a "love [that] was going to be a turning point in world history and culture."[32] Beyond their respective families, spouses, children; their books written and paintings painted; their colleagues mentored and friendships cultivated lay the one-on-one connection they developed first in the spirit of dizzy trysters and later with distinctly Ahab-like grit. Their original investigatory site, the clinic, was by all accounts a redoubt for exiles from a straitened Harvard psychological tradition dominated in those years by the chair of the department, a man perhaps infelicitously but accurately named Boring (Edwin Boring, to be precise).

Each picture in Morgan and Murray's series of thirty-one cards came from a current popular magazine photograph or an illustrated pulp novel. Morgan, a skilled draftsman, stripped away many indexical details indicating story or context so that each black-and-white drawing, adapted, became something new: the portent of an ominous but unknown future event. A mood of angst, hard to describe in words but easy to locate in the series, descended.

Although the pictures were said to be full of "ambiguous" stimuli, the dominant tone was decidedly ominous, bringing to mind Jung's analysis of Morgan's own dilemmas: "She is constantly fighting against something overpowering that comes from below," Jung remarked in the course of his four-year-long seminar analyzing Morgan's personal drawings. The TAT images shared this quality. The test's two creators described Picture 6, for example, this way: "The silhouette of a man's figure against a bright window. The rest of the picture is totally dark." Likewise Picture 12 portrayed a struggle with an unnamed dark force: "A young man helplessly clutched from behind by two hands, one on each of his shoulders. The figure of his antagonist is invisible." Elsewhere in the series a man clung to a rope in midair, a boy huddled next to a revolver, and a girl stood alone, her expression "obviously" one of terror and anxiety.[33]

The test was simple at first: show a picture to a patient or subject and ask him or her to tell one story per card. (The specific instructions were to describe "what's happening, what led up to it, the outcome, and the thoughts and feelings of the characters.") The analyst subsequently analyzed the accumulated stories, and this constituted the entirety of the test. "As a rule, the subject leaves the test happily unaware that he has presented the psychologist with what amounts to an X-Ray picture of his inner self," observed Murray. By getting the subject to focus on an indeterminate yet emotionally saturated phenomenon, the perceptive interpreter—"one with 'double hearing,'" as the researchers put it—will see that the subject "is exposing certain inner forces and arrangements, wishes, fears, and traces of past experiences."[34] There was no fail-proof method given for interpreting the test, much less tabulating its results. Success depended on the hermeneutic gifts of the test giver. The TAT, thus, was a powerful if not sure-fire way of looking inside someone's skull. Soon, followers would heighten and generalize further the claims for its prowess at doing so. And soon, adherents would claim for the data it produced the status of dreams.

Card I in what came to be called the standard Murray TAT series showed a boy looking at a violin that lay on a table. The original was a Samuel Lumière photograph of Yehudi Menuhin as a boy, published as "A Violin Genius of 10" in an article in the January 1928 *Musical Observer*. In its second life as a Morgan-rendered drawing for the TAT, its contrasts amplified, its identifying features expunged, the image was, as with all projective stimuli, undetermined. What was the boy thinking or doing? was the question. A "typical" response was to say that the boy was looking at the violin hoping to avoid practicing it and wishing he was outside playing baseball. As it turned out, a typical Japanese response was quite different: the boy was wishing he could afford lessons and yearning to hold the violin in many cases. When the card was shown to a group of young Navajo Indian men during Bert Kaplan's visit to the reservation in 1947, an army veteran named Eddie, recently returned from the war, described the boy as a "country boy," around thirteen years old, who is drawn to music but later loses interest and "he just quits this music and later on he gets interested in the agriculture."[35] Quite a few young Navajo men saw this card as depicting a boy who was trying to "fix up" a broken violin or guitar. (Cross-cultural testing is the subject of the next chapter, and the tricky undertaking of giving tests to non-mainstream groups will be addressed there.)

But typical responses were less interesting than . . . interesting responses. One manual gave the following as an example of a productive answer: a "depressive psychotic" described the picture as showing a boy giving up forever on the violin because a string had broken. Another guide provided the example of a hard-drinking fifty-two-year-old Hollywood film editor who constructed a story thus:

S[ubject]: Now from this I'm supposed to tell you what?
E[xaminer]: [Repeats instructions.]
S: He has just finished practicing and . . . and he is sitting there reflecting . . . over his violin . . . on a score which he's just tackled. Is that enough?
E: Make up more of a story. . . . How does he feel? . . .
S: I should say he feels a little . . . hmmm, disturbed, no, not disturbed; well we'll [mumbles something], we'll say a little disturbed by the fact that he hadn't brought off, what will we say, the Scarlatti exercise to his satisfaction. He is a

sensitive, thoughtful child who, like myself, needs a haircut. You can leave that out if you wish. Okay, that takes care of Buster. Oh, you put everything down [noticing verbatim recording].[36]

The TAT analyst interpreted this as evidence of someone self-conscious about making up a story ("well . . . we'll say . . ."), proud of his elite cultural knowledge (for it was rare to mention, as in the "Scarlatti" specification, a particular piece of music), afraid of punishment for his oedipal strivings (the boy's failure and the storyteller's attempt to avoid even so moderate a description of affect as "disturbed"), passively identifying with the feminine but eager to avoid sentimentality (the "sensitive, thoughtful child" rechristened "Buster"), and desperate, perhaps, to conceal from others the toll years of drinking had taken on his cognitive abilities (his opening question concerning instructions). From the whole TAT result, including the deliberate sequencing of cards, the analyst concluded that the patient was attempting to deal with his fears as follows:

(1) by a detached, cynical, worldly manner, through which he attempts to exercise great control and to be above any intense feeling—a maneuver by which he appears to be then neither masculine nor feminine but only uncommitted; the type of stories he introduces, the labels he gives them, and his strenuous avoidance of sentimentality combine to indicate that he is probably contending with intense sentimentality with feminine and passive aspects. Fear of affect (anxiety, guilt, depression, sentimentality) is generally pronounced.

(2) by multiple, shifting, insubstantial identifications in which he is a man of culture, one of the cognoscenti of the entertainment industry, politically aware, socially and economically elite, and a first-rate low brow as well as high brow. In addition, his choices of metaphor and imagery tend to underscore the markedly oral orientation evident in the content.[37]

A single test, even a test such as this one issuing from a reluctant subject with a sassy attitude, could produce a great deal of information.

Years later, as it happened, Menuhin recalled what he was thinking while the photo was taken: "Actually, I was gazing in my usual state of being half absent in my own world and half in the present. I have usually been able to 'retire' in this way. I was also thinking that my life was tied up with the instrument and would I do it justice?"[38] Menuhin's reminiscence makes the patent point that there could be no "right" answers to a projective test image and that, indeed, all the ways in which a respondent might be "wrong" about what was really going on in the picture added to the test giver's power and multiplied possibilities for interpreting his response. The following sample exchange, from a prominent Rorschach expert, gave examiners a sense of how to instruct subjects—or rather not instruct them—in their responses:

S: (After giving a response.) Is that the kind of thing you want?
E: Yes, whatever it looks like to you.
S: Is that the right answer?
E: There are all sorts of answers.[39]

Such a preternaturally calm demeanor, which examiners were to cultivate, was only one element of a veritable choreography of gestures, affects, attitudes, and instructions necessary to the scientifically proper giving and scoring of such tests. (And as we will see, the calm demeanor turned out not to work as well in cross-cultural settings.) As Peter Galison recently observed, the elaborate protocol of the tests (the Rorschach here being one example among many) involved not only a set of pictures and certain answers given, but also "a system of charts, tables, and graphs, of scripted questions, calculated indexes, and [by the 1980s] downloadable computer programs." Projective tests were more than simply a set of inkblots or drawings on cards and some ideas about scoring. They involved procedural rituals, standardized steps, and elaborate instructions about everything from how far away the subject must be from the image (arm's length) to how not to lead or coerce the subject so that in the end the whole apparatus would work like an impersonal but very accurate machine for extracting samples of the "self."[40]

In meeting the most extreme human behavior or the greatest mysteries of social and cultural existence during these years, the logical step to take, it seemed, was to administer a scientific test or a battery of tests. On the one hand, mid-twentieth-century Americans' love of hi-tech gadgetry in arenas as diverse as spy craft and the culinary arts had something to do with this. By 1966 even Frigidaires were sold on the basis of the "Space Age Advance" they embodied, even as U.S. espionage experts favored a spectacular futurism and fell into "technophilic hubris" with their procedural arsenal of invisible ink, drug-laced artifacts, and exploding cats. On the other, perhaps even more compelling hand, there was the scientific-imprimatur factor: specific "X-ray"-like capacities attributed to tests such as the Rorschach lent them, for a critical period of around twenty years, the sense that they could do two contradictory things that simple observation could not. First, like a powerful "invisible hand" of science, they could exert experimental or experiment-like control over a situation—and thus deliver good, reliable results. Such results could be compared across studies: anyone who ever took the Rorschach in a standardized manner could be compared to anyone else who took it. In its mix of usefulness and outlandishness, this set of values, particularly concentrated in the projective test movement, could be called the "practical spectacular."

Second, many Rorschachers and TAT enthusiasts sought techniques that were more than mere mechanical gauges providing simple measures or trait lists or aspiring to apply nineteenth-century-style mechanical objectivity to the depths of subjectivity. Promoters self-consciously wanted to go beyond the psychometrics of Alfred Binet, Lewis Terman, Edward Thorndike, and Louis Thurstone, which permitted standardized measurement of intelligence, social attitudes, aptitudes, and other dimensions of behavior but remained limited, derided by some critics as mere "pencil-and-paper" measures.[41] What was wanted was a way to gain access to what was not immediately visible to the naked eye nor easily ticked off on a checklist, such as the hidden impulses, complexes, drives, and emotional mazes postulated to make up the inner life.

Not just any test would do, of course. To possess these capacities, a test had to be "projective"—that is, it had to target the way
individuals "project" their own preoccupations and ways of seeing
the world into neutral or ambiguous scenarios (such as an inkblot
or an odd cartoon). The less distinct the scenario, the more projection occurred. This sense of how projection works derived most directly from Freud, who used the term "projection" *(Projektion)* in
1895, 1896, 1911, and 1913, initially to describe a paranoid way of
reacting ("internal perceptions replaced by external perceptions")
and later to describe how all people, sane and insane, can be said to
"make" the worlds they see and in which they live. Projection is
how this happens. "We should feel tempted to regard this remarkable process [of projection] as . . . being absolutely pathognomic . . .
if we were not opportunely reminded [that] . . . in fact it has a regular share assigned to it in our attitude to the external world."
Projection, as Freud characterized it, entailed a distinct and often
forceful movement out, an "ejection," a shifting, or even a shooting
forth, of internal concerns onto external targets. It was, he argued,
a "remarkable process"—and yet it was entirely, literally banal, for
it constituted the warp and woof of daily existence in all its ongoing distortions and small or large self-deceptions. Indeed it was
exactly the way we "[built] up the external world," said Freud,
through projections that "should by rights remain part of the *internal* world."[42] (Freud saw parallels among dream work, creative
writing, and the "projected creations" of primitive men.)

Borrowing from Freud but expanding on (and ultimately eras-
ing) his influence, a movement arose to promote projective tests as
the answer to an impasse in the human sciences: how to "get at"
the very essence of human meaning, subjectivity itself.

In attempting to import the quality of objectivity to their own
tools, behavioral scientists were, if anything, more thorough and
more bold than natural scientists, with whose work objectivity was
more easily linked.[43] It was not a simple mechanical objectivity
they sought but a malleable, subtle one that incorporated not only
judgment but also intuition and counted its object not only as a
thing but also an ever-changing complex alignment and realign-
ment of qualities. Working holistically, projective testers revealed
key patterns that showed how, within the ongoing flow of everyday

experience for each person, access to a hidden yet not hidden truth lay available to the proper technique. Here again the practical-spectacular held sway. There is a longer history of such hybridizing through technologies, although it is sometimes ignored. Recently, John Tresch's *The Romantic Machine* revived a series of strange nineteenth-century technologies that put relationships above essences and rendered nature as part of the social fabric. From Auguste Comte's Cosmogram to Charles Fourier's Phalanstery to Pierre Leroux's Pianotype, romantic machines as Tresch describes them make them sound like eerie precursors to projective techniques. Just as these nineteenth-century machines (built by technophile mystics) incorporated emotion, aesthetics, and individual needs into their workings, so did twentieth-century projective tests (built by modernist mystics) decenter "the human" in order to include interactive forces of projection and subjective making. "Any attempt to depict the world, and especially to conceive of the cosmos as a whole . . . must include recognition of the human activity involved in shaping the world picture."[44] Along similar lines, projective test workers met the demand to scientifically capture human nature by swirling in the shifting mechanics of projection. (As we will see, projection was both a human activity and a technological effect.) In this way, they created a more "human" version of objectivity, one saturated in human activity. This also forms part of an against-the-grain story of technology.

Objectivity itself has been a topic of reevaluation recently in the history of science, yielding the main insight that "objectivity has a history," one that renders it a forever morphing set of values and practices rather than an unchanging ideal. Lorraine Daston asked in a 1999 talk at the University of Chicago, "Can Objectivity Have a History?" and she and Galison have answered in the affirmative in a series of articles and books, from a number of angles. Once, objectivity appeared as a kind of North Pole of science—that heroic point on the map toward which the enterprise was forever striving, forever attempting to arrive, and therefore a part of science that could safely escape historical attention. In recent decades, historians of science have come to see objectivity as changeable and surprisingly plastic. To put it simply, it meant something different to be "objective" in 1610 than it did in 1910, and this difference also depended on whether one was a crystallographer, tulip aficionado, or budding dream collector. Yet there was a developmental pattern discernible in objectivity. As a way of looking at the world, objectivity over time developed "its own coherence and rhythm, as well as its own distinctive patterns of explanation."[45]

The realm of subjectivity, too, according to recent scholarship, is ripe for reevaluation. If objectivity has a history, so too does subjectivity. Steven Shapin argued recently in an essay titled "The Sciences of Subjectivity" that it is about time for scholars who study science to see subjectivity not just as a wayward element that acts to disrupt objective processes—a view according to which subjectivity is the part of scientific inquiry that is "inchoate, arbitrary, unstable, and endlessly varying," the part that is constantly "contaminating" the workings of objectivity, and in the end "what we're sadly stuck with if we don't watch out." Instead, one can focus on subjectivity as a domain of active knowledge making and an "explicitly framed topic of inquiry."[46] Insight from these inquiries is doubly applicable within the projective test movement, for the nexus of objectivity and subjectivity is precisely where such tests emerged. When considered within the history of the human and behavioral sciences, they cast particularly good light.

Projective test workers (as they called themselves) frequently spoke of objectivity as their goal and even at times of "objectifying the subjective."[47] Through their tests they strived for a "view from nowhere"—also known as aperspectival objectivity—while at the same time actively embracing the dynamics of projection. This approach might seem like a contradiction (a total escape from perspective versus a total embrace of perspective), but it was one they welcomed. Projection was a counterweight to the problems inherent in what has been called "perspectivity." At its heart is a quest to escape perspective by finding a position of having no position, achieving a stance of hovering above phenomena, or—as in the 2003 film *Kitchen Stories*, depicting social scientific objectivity practiced by Swedish social researchers stationed among elderly Norwegian farmers—sitting in a tall chair in the kitchen as if you weren't there at all. In apparent contrast to perspectivity, projection is the mechanism by which the "self" operates in a patterned way on reality and experience. The process of projection posits a mobile, mutable, ever-moving self that is more a series of trajectories and encounters than it is a solid, unchanging "thing." In a projective test, a stimulus card such as an inkblot becomes a spot where the self moves and becomes visible as a pattern of interactions. It is glimpsed—there!—at the place where the card or paper tool meets the user or test taker. Projection is thus perhaps a response to a crisis within the pursuit of social scientific objectivity. It entails a shifting relationship between subjectivity (as both domain and point-of-view) and objectivity (as both domain and goal). The "projective" part of projective tests allowed them to alter their relationship to objectivity. It tended to allow more complex possibilities for how objectivity manifested itself in the middle of the twentieth century.[48]

And here is where the story picks up momentum. For it seems that although Rorschach never himself used the term "projection" or "projective test" to describe his technique, which (as mentioned) he saw more in terms of an experiment revealing a subject's perceptive processes, still some of the qualities inherent in his test rendered it particularly useful in the American behavioral science context. If, as Daston points out, the task of the historian is to recover the "slow process of accretion and absorption that accounts for the layered structure of the notion of objectivity," then projective techniques may mark a key moment in that slow process, which after all has been described less fully in the human sciences: an accretive shift and reconfiguration.[49]

After making a cameo appearance in Morgan and Murray's 1938 text, the term "projective methods" next appeared in a manifesto that psychologist Lawrence K. Frank, a man once described as the "procreative Johnny Appleseed of American social science," penned in 1939.[50] Looking around him at the proliferating array of tests with similar dynamics, the entrepreneurial Frank was the first to think of lumping them together and thus creating a veritable movement that would come to be called the projective test movement.[51] The "projective methods for personality study," Frank claimed, worked much like high-powered microscopes. They allowed researchers to see otherwise hard-to-glimpse phenomena such as how people organize their experience and "structuraliz[e] their life space." To look at a person's test results was to see revealed the keys to "that individual's private world of meanings, significance, patterns, and feelings." Up-and-coming social scientists learned with dawning excitement about the new possible powers this group of techniques granted. Suggesting that the key to projection was the bridging of inner and outer, Frank's piece used the phrase "private world" eight times, almost always italicized. Give a person a projective test and you will "induce" him to reveal himself—"to reveal his way of organizing experience by giving him a field (objects, materials, experiences) with relatively little structure" to project upon.[52]

The call to organize around projective techniques proved successful. Projectives took off as a movement and flourished in the 1940s, 1950s, and 1960s with a degree of fervor difficult to recapture. Alongside strong efforts to standardize projective tests and the energetic claims that tests were like experiments and thus shared the cachet of the laboratory, the projective movement just plain flourished like a dahlia garden in late summer. Varieties grew and variations proliferated. For children, there was the "Blacky by Blum," which serialized the adventures and misadventures of a black dog (one frame depicted the rabid Blacky grabbing another dog's collar, clearly stitched with the word "Mama"), and the Children's Apperception Test, which featured tigers and monkeys in existential or dangerous situations. For other age groups there was the Michigan Pictures Test, the Adolescent Fantasy Test, and the Senior Apperception Test. A series of strangely distorted pencil drawings of people falling off ladders and other activities going badly awry made up the Pain Apperception Test, a gauge of how a subject related emotionally to pain. Utilizing another sense dimension was the Auditory Apperception Test and a briefly explored "odor imagination test" involving some combination of violet perfume and Worcestershire sauce. Capitalizing on the capacity of visual representations to reveal their creator's preoccupations were the Make-a-Picture-Story test, the House-Tree-Person test, and the Draw-a-Man test. Niche-targeting proliferated for geographical areas: there was a version for the South Seas, one for Mexican Indians, one for West Africans, and one for Vietnamese; the Thompson TAT, with its vaguely Harlem-Renaissance style, targeted African Americans and Africans. A North Korean POW test explored the psyches of possibly brainwashed men. A welter of odd and surprising tests appeared during this heyday period: the Minister's Black Veil, for example, had participants complete a Nathaniel Hawthorne short story left off mid-narrative. The Rorschach, too, spawned a litter of specialized Rorschach tests, including the Inspection Rorschach, the Group Rorschach, the Multiple-Choice Rorschach, the Objective Rorschach, Baughman's Paired Comparison Method, and a range of methods for integrating the Rorschach with the therapy process.[53] Such supplements and modifications testified not to shortcomings in the legendary test, for they did not claim to displace the original, but to its fertility. And perhaps too they harked back to the very earliest meaning of projection, which came from alchemy: projection was transformation, the process that made gold from lesser metals, a sense employed metaphorically if prospectively accurately by Ben Jonson in his 1613 *New Inne:* "I feele the transmuatation o'my blood, as I were quite another creature, And all he speakes, it is projection."[54]

For such a successful test as the Rorschach, which remains perhaps the best-known psychological instrument in the world, it is odd how surprisingly common were dead ends, stumbling blocks, misappropriations, "creative misapprehensions," and ellipses of articulation. In Rorschach's own circles, his invention failed to generate much excitement for its first ten years—the test itself sat in a sort of limbo. Rorschach's precipitate death both inhibited and stimulated research, for his unexpected leave-taking left his test open—much like an inkblot—to free interpretation by others. His hand-inked blots, artfully designed to look as if they had been simply splatted on the page, bore at the heart of their original published form a momentous printer's error. And this error, due to the minute specifications for the cards' exact reprinting, then perpetuated itself in each future set of cards. Hogrefe Press's website describes somewhat wearily how the original printing press has been kept alive at great effort so that it could be used as the single instrument from which all Rorschach cards issue by a process carried out "on what can now only be regarded as ancient equipment," and if it is too humid or dry on the scheduled day, printing must be delayed "so as to maintain a virtually identical reproduction of the originals."[55] At the cards' heart, then, is an exactingly perpetuated mistake—albeit one that has proven generative.

Further contradictions, paradoxes, and lapses abound. Most prominent U.S. Rorschach workers did not realize that "Rorschach's inkblots are not actually inkblots" until 2000, when John Exner revealed this not-exactly-hidden though also never-much-discussed truth. Yet as diehard enthusiasts insist, the blots are not "mere[ly] accidental" but rather "special."[56] They have special qualities and elicit special reactions. This feature has always been at the core of the test's mystique, a mystique that has mixed liberally in the history of projectives with a quest for scientific authority. Finally, it was the Rorschach test's very indeterminateness (with regard to method) and intimacy (with regard to subjective processes) that allowed it to be recruited to establish a new sort of objectivity—about subjectivity—in postwar behavioral sciences.

So it was that in the United States the Rorschach initially found a home, presently circulated in a network, and eventually founded an empire when it joined the projective test movement. The fact that it flourished in the United States, I am arguing, had less to do with its promised special powers than with its hazy, inkblotty, indeterminate, error-ridden, ahistorical, and abandoned-at-sea qualities. In this light, it is interesting that the Rorschach test, though producing so much research and criticism—a recent bibliography of Rorschach-related literature runs to 492 pages and lists over five thousand publications, a number some suggest makes the test the most intensively studied creation in the field of psychology[57]—has rarely been the focus of historical analysis.[58]

Yet its history, along with the history of the projective test movement, highlights something about the nature of data, evidence, knowledge, and objectivity within the human sciences. The tests formed part of an effort in human documentation—an attempt to document the most human of human capacities (hence the seeking out of Nazi data, also known as "samples" of human nature). As the fate of the protocols shows, however confidently scientists may set forth, their efforts to "secure" reliable facts and finalize conclusions are subject to political and intellectual instability, undertows or riptides that at times destabilize the data, hold them in abeyance, and raise or lower the fortunes and reputations of the purveyors. The history of the tests reveals these fluctuations to a dramatic degree, and it is thus not surprising to find that Bert Kaplan, who for a time made it his mission to rescue forgotten or at-risk paper protocols, took an interest in the history of such tests. An undated handwritten note in Kaplan's files from Christiana Morgan informs Dr. Kaplan that Dr. Murray (who was one of his teachers from graduate school) "thinks it would be fun to talk to your group about the early beginnings of the TAT."[59] The group in question was likely a Harvard psychology course Kaplan taught at Harvard around 1949 or 1950, around the time he was beginning to formulate his idea for the data archive.

More recently, scholars have seen the history of psychological tests as a tale of increasing standardization, reduction, discipline, and a general tendency to become "less than" what they once were. For many Rorschach aficionados the attempt to transform a complex, multiperspectival experimental gauge into a simple standard instrument was a fundamental betrayal of the test. Historian Naamah Akavia, for example, writes of how looking at the Rorschach historically and studying its origins "readily reveals it to have been a much richer conceptual tool originally than the standardized projective test it has since become."[60] However, my goal in historicizing the test is quite different: it is to show that the process of standardizing was not simply one of boiling down, oversimplification, and loss of richness. The standardizing process was itself complex, at times bizarre, and though it did develop other areas than inventors such as Rorschach may himself have, in some ways it adhered rather closely to Rorschach's own "experimental" vision.[61]

This chapter has also been concerned with the paper trail these tests left. The paperwork of the test results (like the cards themselves) retained a kind of aura from its contact with Nazi war criminals, schoolchildren, Japanese book browsers, Navajo GIs, and many others who undertook the test. And as we have seen, the touch and feel of the paper records contributed to Klopfer's direct-from-the-source charisma and what one critic has controversially called his "guruhood."[62] Surprisingly—considering how sought after these troves of information were in their heyday—the records also became subject to a curious fragility and elusiveness. Eventually

Kaplan and his allegiance of data activists would retrieve some of them from off-site storage "graveyards" and careless accidents resulting in lost or precarious lodging. Other caches, in particular the Nüremberg Rorschach records, would remain in abeyance, a kind of remission, for decades, their absence serving as a reminder of the very phenomenon the "database of dreams" was intended to redress.

Without these qualities of projective test data constantly on view—as desirable and dahlia-like as they were fungible—it is doubtful Kaplan would have thought to attempt his own experiment in preserving data. First, however, the tests went around the world.

CHAPTER TWO

The Varieties of Not Belonging

A T some imprecise but remarkable moment around the twentieth century's midpoint, it became an obvious thing to do, when a social scientist met someone from far away or who was very different from himself, to sit down under the trees or over in the sand and delve into the ink-blot, House-Tree-Man, Stewart Emotional Response, or Draw-a-Person test. In certain cases, small rooms in missions or grass huts served as isolating structures to allow one-on-one testing, but generally the tests took place in "unconventional" settings under less than ideal circumstances—which might include having an entire extended family crowded around offering possible answers. Or the subject was engaged in washing some of his wife's silver in soapy water while taking the test, a circumstance not envisioned by Hermann Rorschach.[1] Still, social scientists systematically tabulated their results, and by the 1950s, the giving and storing of such tests was nearly standard procedure. As British anthropologist S. F. Nadel commented on the American-born trend, "A new kind of routine seems to be emerging whereby anthropologists, before setting out for the field, pack into their kitbag a set of Rorschach cards and T.A.T. much as they do cameras, a compass, or a copy of *Notes and Queries*."[2]

Before, during, and after World War II, such instruments spread from the field of professional psychology (where their inventors calibrated them to "X-ray" the inner struggles of a single human being) to anthropology and the behavioral sciences more generally, where they helped to capture the inner life and cultural data of a whole group or population—or so it was hoped. Instruments that first issued from Viennese drawing rooms, came of age in Swiss clinics, and later circulated in Greenwich Village salons—techniques designed, in brief, to plumb the psychological depths of the kinds of people who were very, very interested in their own psychological depths—traveled from clinic and doctor's office to field, island, tiny atoll, and Mato Grosso. They moved to a new set of scientific environments and very different sorts of people—in particular, the sorts of people who would neither have thought of asking to be tested nor would even have known what a psychological "self" was, much less a test to plumb it. Suddenly, what one expert termed "exotic cultures"—that is, "cultures clearly outside the main stream of Euro-American culture"—were the target of testing en masse.[3]

During the immediate postwar years, such psychometrics drew in wider and wider circles of culture-and-personality fieldworkers.[4] If placed in an imaginary collective archive (just such an archive would one day form the backbone of Bert Kaplan's Microcard experiment!), the combined results of their cross-cultural expeditions would add up to tens of thousands of pages of Rorschach, TAT, Bavelas Moral Ideology, Stewart Emotional Response, Draw-a-Person, and Sentence-Completion test results drawn from around the world. Most plentiful, however, were the Rorschach results, befitting the test's number one status among projective vehicles. These often appeared in the standardized forms that "Rorschach workers" (as they were called) adopted in the interest of regularizing procedures: Bruno Klopfer's six-page Record Blank, issued in 1942, which included a page of ten miniaturized inkblots on which the worker could add patterns, symbols, words, and lines to indicate where and what the subject had seen. The forms, when filled out properly, allowed any subsequent interpreter to "obtain a concise and simultaneous picture of what has been going on in the mind of the subject"—that is, almost to reconstruct the giving and taking of the test.[5]

A list of the Rorschach records later drawn together from data donors reads like a whirlwind world tour. From the Pacific islands came tests of Chamorro and Carolinian children, Palauan men and women, and Ifalukans of all kinds—in fact, every person, small and large, mad and sane, who lived on the small atoll of Ifaluk underwent testing by anthropologist Melford Spiro. From the Caribbean came the Rorschach results of a wide variety of Haitians, a sample of Jamaican Chinese children, and Montserrat men and women of the British West Indies. If a researcher was interested in Melanesia, there was a sample of New Ireland protocols available. From Central and South America came the tests of a Mexican "Ejido" community and the Pilagá of the Argentine Rio de la Plata basin. There were tests of Americans of ethnic origin, including Japanese Nisei and Issei (both men and women) and Spanish American young men. Sets from Asia included the tests of Nepalese men and children, Pakistani men and boys, Ceylon university students, Sinhalese third graders, Indonesian Alorese subjects, high-caste Hindu and Muslim men from Delwar, Uadipur, India, and a smaller sample of Bhil men from the same area. From the Near East came the records of Lebanese Maronite villagers and those of certain mental patients at a Lebanese hospital, as well as "Arabs from Algiers and from an Oasis."[6] In keeping with the Kaplan archive's overall tendency, North American Indian groups contributed most concentratedly and systematically to the collection of projective tests, with Rorschach results hailing from Attawapiskat Cree, Sioux boys and girls, Mescalero Apache, Ojibwe (many different Ojibwe groups), Tuscarora, Ute, Zuni, Menominee, and Navajo. A significant portion of the Rorschach material was the fruit, originally, of large-scale wartime or postwar projects, sometimes gathered with the help of military or governmental funding—for example, via the Indian Personality Project or the Coordinated Index of Micronesian Anthropology.[7] Each set, tagged with a code for geographic region, promised to sample the inner lives or "private worlds" of the subjects contained therein. When used cross-culturally, they also gave hope of showing how much and which part of the personality was culturally patterned and which part was changing under the "stresses and strains" of encroaching modern life.

The practice of giving such tests became normal for approximately a decade of the mid-twentieth century in cross-cultural research, after which it eventually ceased to be so. Projective tests' use began to fall off in the early to mid-1960s and had by 1970 reached a crisis—as indicated by the title of Marguerite Hertz's presidential address to the Society for Projective Techniques that year, "Projective Techniques in Crisis"[8]—although stalwarts continued what might be called their "offshore administration" well into the decade. The Rorschach test saw use in the Vietnam War, when behavioral science teams based in Saigon between 1966 and 1968 fanned out into the countryside to test villagers and soldiers with the Rorschach, the TAT, and the Self-Anchoring Ladder in order to discern their hidden motivations, yet this dubious adventure occurred near the end of the Rorschach's vogue as a fieldwork aide (and may indeed have contributed to it).[9]

Much in the way that people today no longer carry cell phones the size of small watermelons if they can afford a late model iPhone, so did the Rorschach and other projective tests eventually lose their feeling of up-to-date "rightness." To push the analogy a bit: a mobile phone is made to be portable, and at first its portability was communicated even via dimensions that would later seem laughably cumbersome. Likewise in the case of projective tests, the values of portability, ease of use, convenience, and experimental promise converged for a time, after which they no longer did, and it was hard to see how the technology had once seemed viable. No obvious modification of procedures or elements rendered the tests technological dinosaurs, and yet they became so. The devices in their black leather carrying cases and neatly nested boxes that seemed in their heyday to signify unending horizons suddenly looked like relics. Here the question is: how did such tests once seem the ideal choice for travel? How did this extravagant administrative history come about, first the rise and then the decline of the use of projective tests as "technologies of mobility" in fieldwork?[10] Recall that according to the origin story of the American Rorschach lineage he founded, Bruno Klopfer carried the Rorschach in a suitcase under the nose of the Statue of Liberty to escape the tumult and rising anti-Semitism of Europe; the tale itself emphasizes portability, as well as a kind of derring-do.

A larger question arises about technologies, from cell phones to psycho-technologies and how they move in the world.[11] To say this essential question has been little examined in the case of the Rorschach test is an understatement; in the history of the projective test, the history of the traveling test is a near perfect absence. In part, this omission is due to the fact that from the point of view of the growing Rorschach empire, the use of the test in anthropological field studies of people sometimes mistakenly called "primitives" represented only one among many worthy applications, including (as Klopfer once listed them) the study of juvenile delinquents and adult criminals, stutterers, epileptics and alcoholics, cancer patients, and twins. Klopfer and Kelley's 1942 Rorschach manual devoted far more space to the possibilities of Rorschach testing of armed forces personnel in order to discern the unstable and those who were officer material. The historiography of such tests, not itself robust, has ignored their adventures abroad.[12]

Here I attempt to bring the sidebar to center stage and tell what happened when the tests' mobility and malleability became of the essence, when the push to standardize them and the push to spread their use were sometimes at odds, and when (despite these challenges) for an energetic group of pioneers the project was full of promise. Eventually, due to changes in social science and social life in the second half of the twentieth century, the tests' erstwhile promise turned to obsolescence, decline, and the "orphaning" of the data they produced.[13] To explore this tale is to look into the role adaptable technology played in the database-of-dreams experiment.

On one level, the question of how tests exactly began to be taken around the world to be used with "primitives" is a straightforward problem of dissemination. Who borrowed what test from whom; at what dinner party, seminar, or colloquium (never underestimate the power of a dinner party); and then how did the intellectual spark catch fire? The answer to who first thought of "going into the field" with Rorschach cards packed alongside mosquito netting appears surprisingly simple: at a New York seminar in the mid-1930s on "The Effects of Personality on Culture" sponsored by the National Research Council, the renowned senior anthropologist Ruth Benedict began talking about the Rorschach to a man who had never heard the name before. Her interlocutor was A. Irving Hallowell, who went by "Pete," a recent convert to anthropology with a background in social work and a growing interest in psycho-analysis. He was struck by this new name, *Rorschach*, and by the possibilities the eponymous test seemed to offer. "Well ... so I didn't talk to Ruth about this at the time but ... I decided to look into this. And I did [for a few years]," he recalled decades later in an interview.[14] Little published research about the Rorschach test then existed, no translations of *Psychodiagnostik* were available, and Hallowell had not yet heard of Bruno Klopfer, but he struck up a correspondence with the "tireless" Chicagoan Samuel Beck, who had published a few articles. By the time he was prepared to set off for Manitoba in the summer of 1938, Beck's *Introduction to the Rorschach Method* had come out, so Hallowell, with no formal Rorschach training, went forth "with his book under my arm into the field."[15] In this way, another image of physical transportation etched into the story of the Rorschach diaspora.

This was by no means Hallowell's first trip to the area, how-ever. Almost a decade before, on July 1, 1930, Hallowell had disem-barked from the steamship SS *Keenora* on the shore of Lake Winnipeg at Berens River, where William Berens, a chief of the Ojibwe people, was one of the first he encountered. Berens, whose grandfather, Bear, had taken the European-style surname from that of a Hudson Bay Company governor, and whose father, Jacob, was the first Berens River Ojibwe to convert to Christianity, liked to meet the *Keenora* during its weekly summer visits, and their first conversation ranged far, as far as cross-cousin marriages and terri-torial issues, setting a pattern for a long friendship and "virtual collabora[tion]" (as Hallowell later put it).[16] In addition, since Berens had a mother of mixed Scottish, Cree, and French Canadian heritage, he spoke English comfortably ("very intelligent—excel-lent English," Hallowell noted in his diary that day).[17] Although Hallowell had originally intended to study the Cree, further afield in Manitoba, the conversation with Berens crystallized his interest in the Ojibwe and particularly "the fact that there were still un-christianized Indians 250 miles up the river in the Pikangikum Band," to which Chief Berens agreed to accompany him.[18] Two years after their first meeting, Hallowell and Chief Berens traveled a hundred miles by canoe (including fifty portages) to encounter, as they believed, those "less contacted" Ojibwe bands, and Hallowell

concluded that they were, if not completely untouched by white influence, outside of normal temporal bounds. Time had stood still,
at least in that redoubt. The two made several trips in the following
years, culminating in the Rorschach voyages.

The site at the mouth of the Berens River where Hallowell had
disembarked was a crossroads not only for Ojibwe people, but also
for the wider regional socioeconomic order, which had been rapidly
evolving and changing during the last quarter of the nineteenth
century and first years of the twentieth. Lake Winnipeg, on whose
eastern shore lay the Berens River port, had been the site of the
area's first commercial fisheries in the mid-1870s, run by Icelandic
fishermen. Fish shipments went to Winnipeg on steam schooner
and from there were transported by rail to larger depots. Icehouses
began to be used to keep fish frozen to access American markets,
and many Ojibwe in the mouth-of-the-river communities had summer jobs at the fisheries, even as native fish stocks were almost completely wiped out by commercial overfishing. Eventually, around
1930, Chief Berens negotiated to permit Indian fisherman to have
commercial licenses. Nonetheless, as Hallowell recalled with chagrin many years later, local whites invariably called him "Willie,"
rather than Chief Berens or Mr. Berens, mimicking oppressive patterns of address found, for example, in the American South.[19] The
Ojbiwe themselves were the fruit of much mobility in lineage for
the past three hundred years and had mixed with distant American
Indian communities, as well as with Crees, and with fur traders of
British, French, Scottish, Syrian, and Algonkian ancestry.

A few years later, in 1938, with the Rorschach tucked under his
arm, Hallowell arrived once more at Berens River, eager to use this
new tool to explore the differences he had seen among Ojibwe in
his previous studies. That summer he administered (or in his word
"secured") a total of one hundred tests. Upriver among Little
Rapids, Pauingassi, Poplar Hill, Pikangikum, and other bands he
found a rate of 50 percent "well-adjusted," versus far fewer downriver in the Berens River bands (including groups from Poplar
River, Ft. Alexander, and Manigotagen River), where several subjects, in particular, he judged to be severely maladjusted. After com-
pleting the summer's work, he met with Bruno Klopfer, and the two
presented a forum at that winter's "Triple A" meeting (the annual

American Anthropological Association [AAA] convention), which drew a large and curious audience, though overall, as Hallowell recalled, the feeling was, "Who could take seriously playing with inkblots?"[20]

Undeterred, Hallowell went back to Manitoba in the summer of 1940 to supplement his initial cache, creating a run of 151 protocols covering men, women, boys, and girls in two Ojibwe areas (upriver and downriver). All together, the data held out the promise of discerning personality changes that might accompany the degree of contact with the dominant society. He subsequently donated the whole data set to Kaplan's data-warehousing project, of which Hallowell became the guiding light and official chairman.[21] Despite the differences he found, Hallowell felt the tests showed that there was a unified "Ojibwa type" of personality, and that this Ojibwa-ness was in fact under attack from white culture ("acculturative forces"). Still, it was, at least for the moment, persisting in many ways.[22]

After Hallowell's last Rorschach voyage upriver in 1940, he never again visited the area. The Rorschach formed a final punctuation mark to his decade's worth of visits, though he continued to publish works about the Ojibwe for the rest of his life. Many Ojibwe remembered him fondly and perennially expected him to come back. In the summer of 1952, a visiting anthropologist met Chief Berens's widow, who handed over a small tent and old cookstove that had been Hallowell's, stored awaiting his return; in 1966, another visiting anthropologist was mistaken for Hallowell and then wrote him saying, "You are well remembered . . . and apparently are still expected to return, for I was flattered by one . . . woman's insistence that *I* was the '*Mide-oogemah*' [Hallowell's honorary title] from the States!"[23] Chief Berens himself wrote Hallowell periodically, thanking him for books or Christmas cards sent and emphasizing his regard, as in one letter: "You must not think I'm forgetting about you. But I never forget about you, but I had now body [*sic*] to write a letter for me."[24] Yet other priorities prevented or deterred Hallowell from coming back. In the summer of 1946, instead of visiting Berens River, he led a team of five graduate students to administer Rorschachs en masse at a more southerly locale among the Wisconsin Ojibwe of Lac du Flambeau

(known as the Chippewa), a group he deemed far more acculturated than the Manitobans. These data would fill out his Rorschach
sample with an additional contrasting case and complete his contribution to the collective Microcard archive, allowing future
scholars to test comparative hypotheses. By 1947, William Berens
was dying and wrote Hallowell asking him once more to visit: "I
am the oldest here now. But will be very glad to meet you again." It
seems that Hallowell did intend to (and Berens died not long afterward), but Hallowell was then immersed in the headline-garnering
murder trial of his adoptive son, Richard Kern Hallowell, who was
soon to be found guilty of killing two Philadelphia policemen.
Twenty years later, Richard, released from prison, would kill his
mother, Hallowell's first wife. (Historian George Stocking, a student of Hallowell's, has speculated about the unspoken influence
this ongoing personal drama may have had on his teacher's theorizing, in the form of an increasing concern for the interplay of
psychological, cultural, and biological patterns in human evolution,
but there is no room here to explore these questions further.) Half
a century after Hallowell's last visit, in November 1992, one of
Hallowell's interpreters and camp helpers, by then elderly, recalled
the anthropologist's visits and, aside from the question of his skill
as a dancer (affirmed), deemed his work important: "It wasn't anything that was useless. What he did was for a good purpose"—
especially his documenting of "the way the Indians lived before."[25]
For him and other Ojibwe, it was not the high-tech Rorschach but
Hallowell's pursuit of more "traditional" fieldwork—recording
dreams, rituals, and ways of life then losing purchase among the
young—that was of most value.

The picture of Hallowell as the first in the American lineage to
test non-Western, non-mainstream, non-literate people is slightly
complicated by Hallowell's own statement that perhaps he was not
the first. Such modesty and lack of a self-promotional instinct
made many esteem him as the "anthropologist's anthropologist"
during key postwar years of American anthropology. Hallowell
thought perhaps Jules Henry, who had also met Ruth Benedict
during the 1930s—in Henry's case at a Manhattan dinner party,
though with similar galvanizing effect—preceded him in the field

by approximately a year when he and his wife Zunia administered the Rorschach to twenty-six Pilagá Indian children and adults in the Argentine Gran Chaco sometime between September 1936 and November 1937. Nonetheless, those who care about such things generally, if incorrectly, credit Hallowell with "first use," perhaps because he organized his career around promoting the Rorschach whereas Henry was more ambivalent about it and, indeed, having been one of the first to employ the tests among "primitives" (his term), was one of the first to question their use and ultimately reject them, offering the following resounding statement twenty years later: "Were I to go into the field tomorrow . . . I would not bring the Rorschach test."[26]

But there were pioneers who preceded the pioneers. Before these Americans, a few European Freudians had employed tests to take the measure, as the matter was phrased in one initial case, of the "mental peculiarities" of "strange and mysterious . . . foreign people."[27] Between 1933 and 1934 a pair of Swiss brothers, Manfred and Richard Bleuler—both of whom, as sons of the world-renowned psychiatrist Eugen Bleuler, "literally grew up at the renowned Burghölzli Hospital in Zürich, surrounded by schizophrenics and discussions of schizophrenia"[28]—were among the first to use the Rorschach test in scientific research, as opposed to healing purposes. The brothers administered the test to a group of twenty-nine Moroccan villagers, "simple country folk living in the vast plains of Chaouia of West Morocco." Although struck by pan-human commonalities between themselves and the people they studied (who, they noted, also loved their children, sat idly chatting after a long day's work, and could be admirably hospitable), the two wished to delve into the extremer ranges where Moroccans were undoubtedly exceedingly strange and, they felt, verging on inscrutable to a European observer. A gap in processing visual information, for example, hinted at unbridgeable mental differences, for one could note that Moroccans, although they had certainly been exposed to photographs and newspaper images, could not properly or easily "read" them, so they were equally content to view an image upside down or right side up. "You can offer a Moroccan any old photograph," they reported deadpan, "and he will readily believe it is a picture of his wife."

In common with later studies, the Bleulers appeared in their conclusions more concerned with how their test fared on its outing than their subjects themselves. Above all they concluded that the tests, in having themselves been tested, revealed they did work in a non-European setting and surely "The Rorschach Test is a valuable tool with which to gauge the character of a foreign people." One might also note that the tests essentially revealed precisely what the authors, having made a multi-year, close-if-somewhat-condescending study of the "mental peculiarities" of Moroccan peasantry in the Chaouia, were already convinced to be the case. Not only their newness to looking at two-dimensional pictures, but also some "deep-seated racial characteristics," "some trait in the national character" were responsible for the Moroccan villagers' undoubtedly odd results. These traits were linked to their "racial psychology," which in the end betrayed more than simply an ignorance of "how to look at a picture" in the ways the authors were used to, but a deep-seated resistance to Western ways—for Arabs, they observed, actually rejected European imagery when they could easily have secured more of it for themselves.[29]

Several other psychoanalysts undertook this kind of racial-psychology study in the 1930s using the Rorschach to confirm judgments of superiority and inferiority. It is important to note that this approach was quite foreign to the founders of and contributors to the Microcard database of dreams that would arise two decades, a continent, and a war later, for all of them eschewed racialist conclusions of any kind (not that they failed to be condescending at times). Still, precursor studies such as that of the Bleuler brothers in Morocco must be classed as a distant if related enterprise.

Pilagá, Ojibwe, or "peculiar" Moroccans aside, within a few years, by the end of World War II, a grand-scale effort launched. There was resistance, but for approximately a decade, it seemed that using sophisticated tests in faraway places was a vital and important thing to do for a science of the self and society, and "The Rorschach and Thematic Apperception Test became practically . . . standard stock-in-trade of many anthropologists," as the prominent psychological anthropologist Francis L. K. Hsu confirmed. Or putting it another way, as another eventual Microcard contributor, George Spindler, did, "Rorschaching [became] a fad."[30] On the heels of the pioneers came troops of Rorschach workers who often combined the ink-blot with other projective tests and who by some estimates had churned out around 150 studies hailing from over seventy-five societies by the time Kaplan crystallized his project to collect masses of preexisting data sets across the social sciences.

Projective proponents who moved around the world administering their tests were, in a sense, countering the advice of the prominent stay-at-home expert, Beck, who in 1949, at the high point of cross-cultural testing's vogue, invoked an inward-turning version of Frederick Jackson Turner's famous "Frontier Thesis." As Beck put it, "We are now living in the one, closed world in which there are no more wide open spaces," and in such an utterly closed world the adventurous scientist's only remaining gambit was to explore the "Frontier Within" via tests like the Rorschach. However, for Hallowell and galvanizing groups of other workers, the frontier within also still lay without. The world may have been losing its wildest places, but the tests would help understand precisely how this happened and what were the psychic results of people experiencing such changes. Cross-cultural testers wanted to observe the effects of the changing outside world on the interior personality.

Indeed, the use of tests became a synecdoche for the growing field of culture and personality itself in the postwar years. As Kaplan commented, "The whole culture and personality area has somehow become prominently identified with these tests."[31] This was a big change undergirding the work of many of the "second generation" and galling to the first, including Margaret Mead, who complained of the tacitly accepted view among upstarts that "some work was done back in the 1920's and 1930's by the pioneers but that the *real work* in the field began when a few anthropologists borrowed projective tests for their own use."[32] Note that Mead's chronology here was slightly askew, as the first cross-cultural Rorschach users did set off in the 1930s, but she meant to highlight her own generation of pioneers as distinct from those she viewed as overly technocentric successors. Yet for these borrowers, the tests gained a status akin to totems, emblems of their own identity and their own sense of personal and professional possibilities.

By the late 1950s and early 1960s, criticism of projective tests, never absent, mounted in what Kaplan with a distinctly Kaplanian turn of phrase called a "somewhat violent" manner.[33] The purported objectivity of the tests, their vaunted ability to deliver workable "samples" of the inner life, and the claim that they operated in a value-free manner across cultures all came under attack. For some, this was an opportunity to double down. The field of culture and personality, itself often maligned as overly speculative, had staked a great deal on its ability to develop proper instruments, and the Rorschach and TAT were the most prominent candidates. Again, Hallowell addressed this point directly: "There was professional resistance on almost every front to investigations along these lines [of how culture affects personality and vice versa]. This was partly my reason for using the Rorschach Test; it would be an aid in accumulating relevant empirical data."[34] Tests, whatever their drawbacks, produced good-quality, relevant, reliable data stacks. Or let me qualify: the data were not perfect, and perhaps not even always good, but they were the best available under the reigning conditions (or so the tests' supporters felt).

In the face of vigorous anthropological criticism, projective tests flourished widely in the behavioral sciences, and Bruno Klopfer reported that some Europeans in the postwar period were even picking them back up. Swiss psychologist Gertrude Meili-Dworetzki, a student of Max Horkheimer and Theodor Adorno, was using the Rorschach to show how perception processes altered with age. Polish sociologist Tadeusz Grygier, a former Gulag prisoner who was based in London postwar, attempted to use the Rorschach and TAT to study displaced persons (DPs) in camps across Europe in an attempt to gauge the impact of oppression on human culture. He tested those who would agree to speak with him, though it must be said that not many Jewish DPs, just released from concentration camps, would work with a non-Jewish Polish researcher. Meanwhile, David Boder, the Jewish, Russian-born, émigré U.S. psychologist, also administered the TAT in these camps just after the war—nearly crossing paths with Grygier several times but experiencing slightly more success.[35]

By the mid-1950s, Klopfer reported experiencing "new stimulation" from the Rorschach's recursive return to its origins. A visit to Switzerland in the summer of 1954 excited him about the "exploratory-experimental approach many of our European colleagues seem to favor."[36] The suspicion among some that projective tests were merely a fluke or a "kind of 'war boom'" dropped away in the face of ample evidence of bibliographic heartiness, Klopfer exulted. For the first twenty-five years of the Rorschach's existence there had been 786 publications devoted to it (not bad), but in the ten years between 1945 and 1954, there were 1,899 (much better).[37] The *Rorschach Research Exchange*, a mimeographed newsletter that began feebly in 1936 under Klopfer with approximately nine subscribers, skyrocketed in popularity, broaching seemingly picayune debates over certain properties of the instrument or its methodological quirks or making bold attacks against rival camps. (After a succession of new names culminating in the *Journal of Personality Assessment*, it continues to flourish today.)

An abundant postwar literature devoted its energies in large part to standardizing efforts and dispersion effects in what was now sometimes referred to as the "test-and-measurement field."[38] In *Drawing Things Apart*, historian of science David Kaiser shows how the legendary physicist Richard Feynman's diagrams, from the late 1940s through the late 1960s, became a "new diagrammatic tool" that spread through the ranks of physicists, not only Nobel laureates, but also the rank and file of mid-century practitioners making everyday calculations, so that "physicists fashioned—and constantly refashioned—the diagrams into a calculational *tool*, a theoretical *practice*."[39] Crafting, deploying, and stabilizing their tools: this was also the work of the Rorschach cadres, and in particular those who wanted to take it far afield into "the field." Kaiser follows "unfolding variations within [physicists'] work," contrasting this approach to Bruno Latour's emphasis on "immutable mobiles" (1986) and the way consistency was achieved through reducing the variability of tools. In the case of the Rorschach diaspora both these forces— loosely, fealty and flexibility—were at work. The Rorschach and other tests were themselves "paper tools," in Ursula Klein's term. Chemical formulas on paper in mid-nineteenth-century chemistry (Berzelian expressions such as H_2O) allowed scientists, "at least for a while," to "t[ake] for granted" that they were true representations of the substance under investigation so that chemists could build models otherwise too difficult to express. Likewise, tests such as the Rorschach worked to stabilize an emerging field so that new questions could be asked and new resources utilized.[40]

Here, again, Hallowell proved a pioneer. The distinctive thing about Hallowell's Rorschach voyages was the almost obsessive methodological care with which he approached the giving of the tests themselves. He had an exacting vision of what was to be done. For him, the Rorschach movement was not a matter of simple application of the standard method but of working in different field sites with a "sense of the desirable mutability of scientific procedure," to borrow a phrase from the work of technology historian Amy Slaton.[41] Flexibility was an active and explicit goal. While "in principle," Hallowell admitted, standardization was a boon and was "basically important," still any final standardization should not be rushed and "must await considerable experimentation with the method itself."[42] Even if some decried the looseness of approaches as shocking—the redoubtable Marguerite Hertz, for one, spoke of the "deplorable lack of uniformity in the administration of the test" in different contexts[43]—Hallowell insisted that rigidity would result and that "rigid schematization in the administration or any other phase of the technique is of doubtful value."[44]

For Hallowell this was the key. Experimentation trumped rote replication, though it did not by any means replace the need for precision, as his much-commented-on exactitude and even-handed footnoting attest. Fifteen years after his test-giving voyages, when Hallowell donated his data sets to be archived within Kaplan's data bank, he stripped them down to include *only the protocols themselves.* Leaving aside his own interpretative schemes, he attached to the data an in-depth essay meditating on method, amounting to a "How To Give the Rorschach" manual, with an emphasis on what could be called a tinkering sensibility. For as Hallowell suggested, "The very flexibility of the Rorschach method as compared with other psychological techniques" was one of the intrinsic features recommending it for use among "primitive peoples."[45]

Among Rorschach stalwarts in U.S. expatriate and native-born psychological circles there seemed to be no minutiae too minute to be examined or discussed, and this detail orientation Hallowell extended to the challenges of the Algonkian wetlands. Throughout the essay introducing his Rorschach data, Hallowell paid thoroughgoing attention to the "how-to's" of what could be a frankly awkward situation, showing how the tests could be watered down, innovated with, experimented on, altered, or augmented without losing their adequacy. Difficulties uncharacteristic of giving the test in, say, Brooklyn, abounded. Some Ojibwe subjects did not want to take the test, some left in the middle, some said they would come but never showed up, and some appeared actively bored. Upriver, where people did not use timepieces or possess clocks or watches, one could not schedule hourly appointments. Downriver, rumors that the test was "hard" dissuaded some, especially women, from coming, for in Ojibwe society, where women were seen as less capable than men (according to Hallowell), they were sometimes laughed at for even contemplating trying a test that men had already declared to be challenging. In addition, most Ojibwe were not comfortable being alone with an outsider ("Any kind of isolation is foreign to them and even suspect"), yet this was the very first demand of the test situation. For this reason, Hallowell's use of interpreters may have had a secondary advantage, though he did not mention it: company. Likewise, among the Ojibwe who tended to follow "old native patterns," the only moment of complete isolation between men and women was "the sexual embrace," Hallowell observed, and thus finding oneself alone with the anthropologist in his hut might, for an Ojibwe woman, create psychological effects that "colored the background of the Rorschach situation when women were subjects." Daunting as such diversity of experience and subjects might have been, Hallowell took it as a challenge and demonstrated constant willingness to adapt his protocols.

Among Hallowell's most important modifications was the use of an interpreter—actually a series of interpreters, beginning with Chief Berens, extending to his son-in-law, and once attempting to use the children of a Syrian fur trader (ages seven, nine, and twelve respectively), an experiment that proved unsuccessful in getting the other children to be more forthcoming. The interpreter was a role not imagined by Rorschach and not condoned by many of his followers. Hallowell admitted that it was a radical innovation to make in the administrative procedure but a necessary one as his Ojibwe was not up to the job of examining in that tongue. (Most

Ojibwe were not fluent in English and were more comfortable in Ojibwe; only twenty-one protocols were in English exclusively.) Certain nuances may have disappeared in translation as a result, he admitted, yet when he took the step after his maiden 1938 trip of showing his protocols to Dr. Klopfer (in fact, this was the occasion of their first face-to-face meeting), the Rorschach eminence could "see no significant differences" between the Ojibwe records and others secured in English. This report of the Klopferian stamp of approval served as a mandate for continued cross-cultural Rorschach use, but it is also strange in light of the fact that the explicit purpose of using the Rorschach in the field was precisely to *discern significant differences*. What Klopfer meant, at least in Hallowell's retelling, however, was that there were no significant technical differences. This seeming confusion between method and meaning—or, one could say, syntax and semantics—dogged the projective testing movement—for example, in discussions of "good" and "bad" answers. Rorschach workers took pains to assure their subjects, who often worried about performing poorly or making mistakes, that there could be "no right or wrong" answers on the test. And yet, among themselves, they freely discussed right and wrong, good and bad responses—usually, again, in terms of technical quality. Good responses were those that made good data.

Counsels of perfection aside, if the desire was to use the Rorschach more widely among "primitive peoples," then interpreters were justified, Hallowell argued more broadly, "in view of the fact that most ethnologists never have completely mastered the languages of the native peoples on whom they have reported." Granted, the linguistically gifted Jules Henry could learn the Kaingang language in six months and Pilagá in eight and could therefore afford to insist that a shared language was the heart of the Rorschach. For his part, Hallowell granted that full command of language was vastly enriching, yet in the end it was a practical question of "whether the Rorschach method can provide any fruitful results at all with the use of an interpreter. In my experience it can," he declared. "Whether this will prove to be the case in other groups can only be decided by trying it out."

This "let's try it out" spirit characterized Hallowell's approach from the first moment of the test. He adopted a hybrid approach to the physical act of presenting the cards, taking a basic template from Hertz, combining it with tips from Klopfer, as well as his own knowledge of Ojibwe, and honing it in the field. In this way, he penned a working script: "I am going to show you some cards one after another. These cards have marks"—here his interpreters inserted an Ojibwa term, *ocipiegátewin*, which meant "picture"—"on them, something like you see on this paper (trial blot shown).[46] I want you to take each card in your hand (trial blot given to subject). Look at it carefully and point out what you see there with this stick. (Handing an orangewood stick to the subject.) Tell me everything that the marks on the card make you think of or what they look like. They may not look like anything you have seen but if they resemble something closely, mention whatever it is."[47] The native term *tab-sko*, meaning "something like," was often used by interpreters and by subjects musingly pointing with the orangewood stick before giving a response.

The use of the orangewood pointer was another innovation adopted for these conditions, for Hallowell found it made it easier to grasp the locations of their responses on the blot. Likely Hallowell got the idea from Samuel Beck, who mentioned another researcher then in the habit of handing a "small wooden pointer" to a subject at the start of the experiment, with no instructions accompanying it.[48] The pointer augmented the Ojibwe subject's expressive use of language, Hallowell felt. A fifty-five-year-old Ojibwe man named Moses used the pointer to explain how Card VIII was like an unknown and hard-to-pin-down animal, not a beaver, a little bit like a muskrat, a bit like a moose, and even more like a bear: "[More] like a bear than anything else. But different kind of hair than a bear. Cannot make out where they are putting their paws (running pointer up center). But they are learning or something. Something here in center—but can't see what—pointing up center and to white inside the blue."[49] The pointer made the inexpressible more expressible. Another subject, a young mother who nursed her intermittently fussy baby while taking the Rorschach, fanned herself frequently and used the pointer to tap her own arm now and again.[50] For Hallowell, too, the stick was an aid, allowing him sometimes to guess, from the movement of the pointer to a certain area on the card, the response a subject was about to make before he made it. "Thus the use of a pointer reduced the amount of interpretation demanded, saved time and facilitated accuracy in the location of responses."[51]

Throughout his set-up to the data set, Hallowell's attention to every angle of test administration shone through, as did care for the larger question of standardization of method—even as he innovated in both. Unlike most contributors to the database of dreams, he cited chapter and verse of administrative manuals and referred to differences of opinion, as for example in the question of the proper use of "encouragement." A premise of the Rorschach was that the subject should speak spontaneously and not be pushed. Yet at times the test giver might say a few encouraging words to prime the pump, as it were. This step was already a source of debate among different camps, but, in addition, Hallowell observed, "It is also a question of considerable practical importance that has to be faced in using the test among primitive peoples."[52] The Rorschach worker must remain alert, and if "productivity" dropped, indicating that maximum level of effort was flagging, the worker was justified in applying more encouragement even though it might seem to be approaching the "taboo" of stimulating subjects to give as many responses as possible. "I was amused when one of my interpreters suggested that I might offer a prize to the man who gave the most answers to the cards," Hallowell recalled, taking for granted the reader would be versed in the stipulation against pushing subjects. "He thought this would not only facilitate getting subjects, but that it would stimulate the Indians to give more answers than some of them were giving."[53] Still, a little encouragement could be ladled in: Rorschach himself permitted the examiner to urge at least one response to each card from a subject, and Beck suggested that a "rejection" should not be accepted unless and until a subject had been encouraged at least twice and looked at a blot for at least two minutes. On the other hand, Klopfer was against such interference "during the spontaneous reactions."[54] Again, Hallowell weighed all sides and found himself dictated to by practical considerations "and a desire to give the Rorschach technique a thorough trial among the subjects I had selected."[55] Ideals of perfect purity fell before pragmatics and concern for useful data. The instrument, so to speak, took some tilting and jiggering around to get useable answers, but this experimental attitude, when balanced with a good-faith adherence to the essential goal of the method, was in Hallowell's view the best approach. "While in dealing with my native subjects," Hallowell found, he got better results if he gave more time than Beck's recommended two minutes and if he sometimes might say, "Try a little harder, maybe you will manage to see something."

Verbal inhibition, diffidence of expression, worry over giving the "right" answers, especially in children (the smaller of whom often whispered their answers), could be overcome with, "Have you told me everything that has come into your head?" Generally Hallowell felt free to offer such prods, but the usual ones that stimulated Euro-American subjects such as "What do you see?" and "What could this be?" did not seem to have the desired effect of producing responsiveness in the field. Other types of remarks such as quips, sallies, jokes, and humorous asides—needless to say, not sanctioned by mainline Rorschachers—might be injected during the exam as a way of keeping or putting the subject at ease. On June 24, 1938, Hallowell even went so far as to suggest to nineteen-year-old Mary Ann Duck, who was having trouble answering, that she might see certain types of things such as "animals, persons, birds, etc."[56] Finally Hallowell noted that an important factor in further reassuring subjects was the fact that "the examiner was not a stranger to these people" but was well known, having had direct contact with many previously and having danced with them in their native ceremonies and made them gifts. "One of my Indian nicknames is 'good dancer,'" Hallowell mentioned, and his skill in dancing was confirmed a full fifty years later.[57]

The artificiality of this task, both on its own terms and even more so from the point of view of an Ojibwe schoolchild or fisherman, struck him. The act of musing over two-dimensional images was itself culturally specific. "To many of these Indians I am sure there seemed something 'phoney' about sitting quietly and looking at the blots one after the other. It was not only a totally unfamiliar task—as it is with everyone—but in terms of their experience it was particularly strange because there is not even the vague analogy of looking at pictures and they do not read."[58] In estimating blanket illiteracy Hallowell excluded the large portion of schoolchildren tested at the Little River Protestant and Catholic missionary schools, as well as others of his subjects who had been taught to read, because they only learned "after a fashion" and most had no books in their homes except hymnals or prayer books in the Cree syllabary—no daily newspapers or magazines. Most pictures seen by Indians at the mouth of the river, Hallowell reported, were advertisements of Eaton's mail-order catalogues (the Sears-Roebuck equivalent in Canada) or stray magazines. Experience in the field made it clear that even if perhaps an inkblot could be seen as culture free (more than, say, an expressionist picture of a factory), surely the situation was not. Along these lines, Jules Henry noted that a standard "impersonal test situation," in which a Pilagá test taker, even a child, was expected to sit alone in a room with an administrator he or she did not know for reasons he or she did not understand and obediently take a test that was unclear at best, might be entirely normal in Western cultures, yet "the native has no experience of this kind."[59]

Feedback and pushback from participants caused Hallowell (and other Rorschach researchers, to greater or lesser degrees) to make further modifications. There was a distinct quotient of resistance observed in some subjects. Often the tests were unpopular, a fact that their bearers sometimes reported and sometimes did not. In retrospect, reflected Hallowell's student Anthony Wallace in 2013 of his Rorschach studies in the 1940s, "Many Tuscaroras did not appreciate being studied in this way." Ulithian islanders in the Western Pacific displayed a "dyspeptic reaction to the cards" when anthropologist William Lessa administered them in the mid-1950s.[60] Other subjects poked fun, as when some Menominee nicknamed the Rorschach-centric George Spindler "Doc Psyche" and joked of a notorious local drinker that he "should have seen a bottle in them cards" if the test were any good.[61] When two anthropologists scrutinized the Rorschachs of Tepoztecan subjects in the late 1940s and found a distinct "lack of friendliness and co-operativeness of the people," the possibility arises that this unfriendliness may have been a response to the test itself.[62] Yet the test was not always deemed unpleasant, and some found it enlightening (as did an Alorese seeress, a Menominee peyote worshipper, and an elderly Manitoban medicine woman), while others found it beautiful, as did Wannatcos, an upriver Ojibwe man who exclaimed,

"Must have been a wonderful man who made these cards—so many animals but not just like those we see running around."[63]

Often Algonkian people saw dreams or dream-like images in the inkblots, though sometimes they were loathe to reveal details because it would dilute the dream's potency. In Ojibwe social life, dreams were of central importance, so important that when a person had a "big dream," he was under pains not to reveal it if he wanted it to work. A sixty-five-year-old hunter-trapper at Little Grand Rapids Indian Reserve saw Card V and said it reminded her of horns, legs, arms: "An Indian dreams this one." She deemed it a "pretty good" card and said, "[I] can't name it"—that is, she had never herself encountered the creature in a dream—"but I know it."[64] Card I caused a forty-plus-year-old man named Naman to narrate a dream from his boyhood in which he saw the same shape but with four legs rather than six. In the dream, he found himself on an island in the middle of a lake, where he saw the thing that lived there, entered its house (it talked like a human being), and answered its demand for eight kinds of sacrifice. In return, his children were protected—none had died in the succeeding years—and he himself was "never sick yet in my life."[65] Others saw visions they had once had fasting or revisited prophetic states when handling the cards. A Menominee peyote worshipper, Case 26, commented, "You know, this Rorschach . . . is something like peyote in a way. It looks into your mind. Sees the things that aren't out in the open. It is like that with peyote. At a meeting you get to know a man in a few hours better than you would get to know him in a lifetime otherwise. Everything about him is right there for you to see."[66] To him, it was recognizably a revelatory instrument.

The giving of the test intersected in a local economy of dream exchange. Ojibwe placed great emphasis on remembering and valuing all kinds of dreams from the garden variety to the vision dreams, in which the dreamer met his *bawaaganak* (dream visitor from which blessings come). Although most Ojibwe dreams were supposed to be shared with others and reflected on, certain visionary dreams—acting as "charters"—were not to be explicitly told but could be the subject of guesses and conjectures among kin and neighbors. For a people so dream-centric, the act of dealing in dreams was fraught with significance. Telling formative dreams to a visitor was a choice some made, as Jennifer Brown has described the Ojibwe dynamic of dream divulging. Fair Wind, a prominent Ojibwe priest during the years Hallowell visited, refrained for two decades from reciting his vision-fast dream; then at a Drum Dance in the 1930s, when he was blind and in his eighties, with Hallowell in attendance, he publicly declared it. And perhaps it was akin to a gift, for telling a strong dream rendered it powerless to protect the dreamer further if the dreamer still believed in its potency. (Likewise, a dream might be told if it had gone badly wrong. A man named Birch Tree told of a dying young man of his acquaintance who had dreamed too ambitiously: one night, he was able to see "every leaf in the whole world" and perished soon after, like the leaves that fall from the trees each year. This was to serve as a lesson that "it is better to dream of many things than too much of one thing.") Chief Berens, as a Methodist, declined the offerings of his dreams—which were full of creatures such as small magical beings who lived in rock cliffs and who vied to grant him special powers such as strength or the ability to dodge bullets—but was happy to pass them on to Hallowell. Another subject, when asked by Hallowell whether he had held back anything while taking the Rorschach test, said to him, as Hallowell noted parenthetically, "Yes—will tell me later. Refers to dream i.e., supernaturals. Asked him later about this but he would say nothing."[67]

Where people's responses to the test fell in the scale between delight and resentment often had to do with the preexisting relationship (or lack thereof) between the test giver and test taker. At other times, the test played into long-standing geopolitical rivalries: on the atoll of Ifaluk in the Marshall Islands, anthropologist Spiro offhandedly mentioned that similar tests (the Rorschach, TAT, and a few others) were being carried out in the Caroline Islands; the remark caused his interpreter, Tom, to conclude that if Ifalukans did well on the tests and outperformed the Carolinians, they "would be more high"; if not, they "would be more down." The next day, conveniently for Spiro, the chiefs called a meeting to encourage full-bore cooperation, and as a result he gained an almost complete data set. Noticing how technologies change via their travel and use, we see that the Rorschach was not simply imposed from on high (or not entirely so) but that there was a certain degree of negotiation. Test takers incorporated the tests into how they saw the world. Sometimes they saw recollected dream images in the inkblots ("I dreamed about a bad person ... looked just like this one [inkblot]"), and sometimes the test pictures later wove themselves into dreams. Attitudes from aversion to attraction, as well as other test-engendered behavior, fed back into the "Rorschach situation" to shape the developing protocols themselves. In turn, the test influenced on-the-ground relationships.[68]

Hallowell applied situational scrutiny to many additional topics, including whether to turn or rotate the cards, desirable seating arrangements, the use of missions or schools versus canvas tents or grass huts, the pros and cons of making trial blots, and other matters designed to regularize the tests and make them reliable. As is evident, Hallowell favored maintaining some slack in the application of the technique, reminiscent of sociologists Stefan Timmermans's and Marc Berg's insight into how technoscientific protocols are standardized: "Rather than being antagonistic to it, a certain looseness in the network may be the preferred (or only possible) way to achieve standardization."[69] All this—including Hallowell's diligent efforts to balance rigor with looseness, the contributions of his subjects, the debates among other workers, and the overarching project of testing the test in new and strange domains—tended to strengthen rather than undermine the technique—at least while the test was in its prime.

The question remains: what happened when the "primitive" Rorschach lapsed into obsolescence, when it finally became approximately as valueless as a melon-sized cell phone? It seems that once cross-cultural projective tests came under sustained attack and the political moment of postcolonialism arrived, the laxity and looseness that once buoyed the standardizing process, as Hallowell and others cultivated it, began instead to undermine the tests' authority and look like fatal flaws. (For example, the use of interpreters to administer a test that claimed to be "culture free," among myriad oddities and adjustments embraced in different locales, seemed anything but culture free.) At this point, decades' worth of carefully collected protocols, stored as scientific "samples" of complex behavioral personality structures from around the world, suddenly lost their theoretical framework and intellectual coherence. They became "orphan samples," to borrow a phrase Australian anthropologist Emma Kowal recently coined to describe a parallel case. Kowal traced the fate of thousands of blood samples taken from aboriginal groups over decades in the mid-twentieth century, samples that, when changes in ethical norms and scientific practices rendered them unusable and even embarrassing, entered a "state of dormancy," packed away in industrial freezers. In turn, the original researchers, hitherto celebrated in their respective medical and social scientific fields, began to experience "the agony of being viewed as unethical researchers," biopirates, or colonizers—or at the very least as naïve and methodologically bumbling.[70] This dynamic characterized too the data gathered by Rorschach workers, who have often expressed anger and confusion at the disregard in which their psychological studies came latterly to be held. After interviewing some seventy-five culture-and-personality veterans, anthropologist Peter Black found widespread resentment or sadness and some bemusement too, as in the comment, circa 1993, that the great age of Rorschach-driven fieldwork by now seemed "from another time, almost another planet."[71] And of course, the resultant hard-won psychocultural "samples," the product of so much invested effort, plummeted in value under successive new epistemic regimes, demonstrating anew the "alienability of samples," albeit behavioral rather than biological ones.[72]

This turnaround may help explain, too, why the workers themselves generally did not forsake their instruments in the face of fierce external criticism, at least not initially. They had always—at least the careful and experimentally inclined ones such as Hallowell, Kaplan, Spiro, and the Spindlers—incorporated play and tinkering. They had never ignored criticism but instead acted in its light and attempted to secure adequate, useable data in relatively large amounts. Their surprise at their chosen instrument's abrupt change in fortune, accompanying the decline in the tests' reputation, was marked and often painful. Criticism was nothing new, but disgrace was.

And finally the dynamic of testing the Rorschach in the field had a further effect, a paradoxical one. Methodological sensitivity pushed the test itself to the forefront in an unintended way. From the beginning, the Rorschach "situation" was explicitly designed to disappear and allow research to proceed seamlessly under controlled and fairly regular conditions, but it was often the case that the situation turned self-referential.[73] It became, in effect, a dollhouse-sized diorama of the social-scientific relationship itself, one that magnified a strange encounter involving a shared task and the exchange of assumptions, pleasantries, emotions, cash, or chocolates. In the process it pulled to the fore precisely the features of personal awkwardness, historical ironies, dreamy uncertainties, and social disparities it was intended to blanket. In this way, again paradoxically, the tests' data can be once again useful today—long after lapsing into anthropological obsolescence—as historical sources, if not precisely in the way their administrators intended.[74] The data, once promising, later much maligned, have a future.

If projective test results posed the question "What is a human life?—and how can it be captured?"—it seems an answer arrived from the field of data-storage technology: "Make it very, very small." In order to understand the technology that would turn Kaplan's bank of data into a unique double experiment, we must first consider the neglected history of the "micro."

CHAPTER THREE

The Storage of the Very, Very Small

W HEN Bert Kaplan and his cohort began to entertain the idea of rescuing all the accumulated projective data from around the globe—as well as related subjective data kept in haphazard stashes in offices and laboratories across the country—they entered the domain of the "micro": analog technologies designed to store multitudinous documents in teeny-tiny form. A brief history of photography's dalliance with very small renderings of information will prepare us for this eventual union of new format and new content. The fascination for rendering a mountain of data into nearly invisible units (each standard page the size of a fingerprint) was part of Kaplan's project. And rightly so. For any collection of data that aims to be comprehensive—any "total archive," that is—must work with the components of near-infinite expandability and collapsibility, with a mechanism of control, or the ability to maneuver between scales, built in.[1] The very small is inextricable from the very big.

For at least the past century and a half, the thrust of the "nano"—meaning "minute" or "very small" and derived from the Greek word for "dwarf"—accompanied the development of the latest in data-storage technologies, whether in microprocessing, archiving, or computing. In recent years, nanotechnology has emerged as a booming scientific-technological industry based on the ability to manipulate living materials on the scale of atoms.

Tiny technologies become increasingly powerful, stimulating what some call "nanovisions." Grandiose rhetoric abounds: innovations are said to bridge the human and the posthuman or then again to catapult beyond dichotomies to "the singularity." More startling than the rhetoric is the reality: the Harvard-based bioengineers George Church and Sriram Kosuri used stand-alone DNA itself as their data-storage vehicle and found they could lodge seven hundred terabytes on a single gram. The binary pairs of DNA nucleotides function as "bits," effectively 0's and 1's, so that they can treat "DNA as just another digital storage device."[2] Merging medium and message in an unprecedented way, Church and Kosuri published their own book about DNA data banks *on* DNA, including images, text, formatting, and all. Imagine, they say, holding entire libraries in vats. To publish you can spray it on walls.

Yet current nanotech, its arrival hailed since the 1980s via the inevitable pun as "something really big," finds its place within a longer history of large hopes for miniaturization. The roots of all this sweepingness lie in curious forgotten places. In particular, they lie in what can be called the fantasy of total information: the dream of condensing or shrinking immensities of text into small spaces, often advanced, as by a British observer in 1859, in the form of a striking vision of compression: "The whole archives of a nation might be packed away in a snuffbox." With the aid of microtechnologies, you could haul away all the world's knowledge in a van, inventor Vannevar Bush would claim almost a hundred years later. Expressions like this—in which the all is packed away in the small—pepper technological writings from the mid-nineteenth century on, featuring snuffboxes, matchboxes, small rooms, and an allocated thirty-five pages of the *Encyclopedia Britannica*. When Bert Kaplan, at almost the exact midpoint of the twentieth century, decided to turn his individual data-gathering efforts into a collective and nearly infinitely expansive enterprise, he shared in exactly this fantasy. Embracing "micro" storage for macro amounts of data, he thrilled to announce at its inception, "The contents of [our] new series are not even visible to the naked eye."[3]

The fantasy of total information could be made manageable via technological tininess. Key to this fantasy, and what made it different from miniaturization per se, was that the small must be

capable of becoming big again at will or whim. A starting place is
to ferret out early mechanical representations and micrographics
for clues. Look to the pin printers, pigeon breeders, and toy souvenir makers, for it turns out that seeking the small reveals a long
history. Early moments in the development of photography played
a vital role.

In the middle of the nineteenth century, a set of obsessive
designers built microphotographic devices, thus creating a shadow
history of the photograph, for the first microphotograph almost immediately followed the first photographs into existence, though
they remained sub rosa or at least occasional, always there but often
unseen as photography became more and more ubiquitous. A search
of this little-known history yields a series of characters that, drawing
on Tom Wolfe's observation, could not have been invented by the
most imaginative creator of fiction. A British multitasking engineer
named John Benjamin Dancer began in the 1830s to experiment
with shrinking photographs down to a shockingly diminutive size.
In his spare moments, the now largely forgotten Manchester resident also took time to invent a calcium spotlight—"lime lights"—
which he used with magic lantern projecting devices; the Victorian
"fairy fountain"; and also the buzzing forerunner of the modern
doorbell. His creation of the first microphotograph in 1839 occurred as a near simultaneous result of the birth of photography
itself, for 1839 was the year Louis Daguerre made instructions for
daguerreotyping public. After reading Daguerre's somewhat sketchy
descriptions—"unfortunately for my purpose ... [they] were crude
and obscure," he later recalled—Dancer, then twenty-seven, avidly
mimicked them over six weeks of trial and error, reporting "numerous failures, and accidentally [being] nearly suffocated by the vapour
of Iodine, before I obtained satisfactory results."[4] Nonetheless, he
mastered the technique and managed to turn himself into one of
the first English daguerreotypists.

With training since childhood in his father's optical instrument
shop and a tendency to tinker, Dancer perhaps not surprisingly
combined two instruments of interest to him. Later that same year,
1839, he installed a reversed microscope lens (with a 1.5″ focal
length) in a camera and was thus able to utilize the camera as
a means of reduction. The microphoto was the result. First on

record was a 20″-long document Dancer reduced to three milli-meters, a feat that achieved a 160:1 reduction ratio. However, be-cause the mini-daguerreotype the size of a small beetle rested on an opaque background, it could not be enlarged well under a mi-croscope and was legible only up to 20x magnification. He tried shrinking down other things, though it is interesting that a text was his first target. Sometimes using "lenses made from the eyeballs of freshly killed animals," Dancer continued to experiment mixing lenses, subjects, and techniques.[5]

Although his initial invention lacked brilliance and contrast, was on an opaque background, and was also unreproducible, Dancer was able to solve these problems by adapting in 1851 his friend Frederick Scott Archer's new invention of the wet collodion process. Collodion is a flammable, syrupy substance of pyroxidine dissolved in ether or alcohol. When dry, it makes a firm, flexible surface, thus allowing the production of the first photographic "negatives." This in turn allowed Dancer (by this time relocated in Manchester with an optical instruments shop and a young family) to experiment for a year in making the first collodion-based micro-photographs. He succeeded. The wet process brought out a star-tling luminosity in photographic subjects. These transparent tiny items, covered with sensitized collodion, were legible even when blown up and looked at through a 100x microscope. These micro-photos were smaller, too, more flea-like than beetle-like in size. In addition, whereas Dancer's earlier effort was non-reproducible (for it was a positive- not a negative-based method), the new process was reproducible.[6] By February 1852 he was making many such micro-images, each about one millimeter square.

In 1853 Dancer honored a recently deceased scientific col-league by making a microphotograph of the man's church plaque inscription, shrinking the 680 letters to one-sixteenth of a square inch size. Originally asked to memorialize the memorial by photo-graphing it, Dancer surprised his professional circle. Gazing into the microscope at a standard 3″ × 1″ slide bearing the dot of a photo (mounted in balsam beneath a cover glass), his colleagues could see their friend's encomium looming before them, perfectly legible too. Eminent Manchester geologist William Binney re-ported receiving it "with much gratification and surprise, having expected only a common, and not a micro-photograph." The microphoto was uncommon, and it spread through a vigorous Manchester scientific network of calico merchants, dye-stuff chemists, photographers, engineers, electricians, and others, who defined themselves variously on either side of amateur and professional; in the network "Dancer was the key figure."[7]

The scientific circle was most impressed and "the extreme interest aroused locally by these prints indicated a lucrative market to Dancer," who began supplying microfilm prints of cityscapes, the famous, and the Ten Commandments, among other subjects, to Manchester's novelty shops. This proved good business. In the 1850s, he presented copies of his increasingly feted miniaturized images, some 227 in all, to Queen Victoria and Prince Albert and to the Pope. Number 26 was the Lord's Prayer, illuminated, and number 27, the Lord's Prayer, plain. In this way, he joined a longer tradition of choosing the Bible—the great book with largest meaning, the good book with most lasting significance in time—for miniaturization. Another slide offered the "NATIONAL ANTHEM contained in the eye of a needle." The delight these afforded lay in part in the trope of containing something very large—emotionally powerful or even metaphysically uncontainable—in a format very small, but, in the case of the microphotograph, one that allowed the easy regaining of size. It took the reduction already entailed in printer's type or a normal photograph even further down while affording the means to bring it back up.[8]

By the late 1850s some went about independently inventing or copying the microphotographic technique. The *Manchester Guardian* in March 1859 congratulated a Mr. Amadio of Throgmorton Street, "whose portrait of Charles Dickens no larger than a pin's point was lately noticed."[9] Friends of Dancer took umbrage on his behalf: in his own hometown, his role in the making of microphotos was somehow uncredited. Shortly after Dancer made his opaque collection, Dr. Hugh Welch Diamond, a physician and amateur photographer who would go on to found the field of clinical photography, made the earliest microphotographic transparencies, and others entered the growing field.

"With photography . . . we enter something new and strange," wrote Walter Benjamin, but perhaps microphotography was stranger still. The method produced, as one commentator observed, "Photographic Curiosities." More precisely, two new capacities had now been achieved. First, the joining of collapsibility with expandability in nearly equal ratio was the key to their takeoff in popularity and perhaps the pleasure they occasioned. This key point is rarely emphasized. True, Dancer invented the microphoto in 1839, but few took note because it could not be blown back up to legible size and thus had little effect on the world of photography (aside from preserving Dancer's relative obscurity, a situation that contributed later on to rough-elbowed battles of priority, through the course of which Dancer, despite public debates urging his case during the 1850s, found himself denied official recognition for inventing microphotography for another century). With scaling, the invention took hold. The second boon was reproducibility. Using the wet collodion method, Dancer could now circulate his speck-like images because he could make many copies. They were cute curios, but they augured something more than mere adorability. They opened up the possibility of long-term storage, and the microphoto now suggested itself to several onlookers as a template for future data capture and archiving.[10]

Now visions of information storage began to arise among the scientifically and technologically inclined. In March 1853, the *Illustrated London News* became the first newspaper to undergo microphotographing; and even though the photographer was testing the lens resolution rather than attempting to compress information, it was becoming clear that the project had a future. Four months later, July 1853, John Stewart published a letter in the *Athaneum* to his brother-in-law, astronomer Sir John Herschel, in which he remarked on the possibilities of large-scale document storage: "Should your old idea of preserving public records in a concentrated form on microscopic negatives ever be adopted, the immediate positive reproduction on an enlarged readable scale, without the possibility of injury to the plate, will be of service."[11] A future in storage—libraries shrunk down to pocket size and blown up—came into public view only after Dancer added re-expandability to the microphoto.

Although born in Liverpool, Dancer was in a sense "native" to microscopy, for his father early on toured with him demonstrating

a solar-powered instrument that would show a human hair projected to six inches in diameter. Around the time he was inventing
the microphoto, Dancer also used a gas-illuminated microscope to
project a depiction of a flea that reached the size of a loaf of bread.
Dancer liked the way small things could suddenly loom very large.
Fleas, not incidentally, appear etymologically not simply as insignif-
icant things but as the embodiment of insignificance itself—or, as
the *Oxford English Dictionary* notes in its second definition, "as a
type of anything small or contemptible." An "ephemeron," in fact,
has the literal meaning of a short-lived insect. Yet in representing
so contemptible a thing as a flea, Dancer somehow reversed its
negation and made it important, even heroic. A forerunner flea is
worth briefly reflecting on. Two hundred years earlier, the Royal
Society's experimental curator, Robert Hooke, used his literary and
scientific sensation *Micrographia; or, Some Physiological Descriptions of
Minute Bodies Made by Magnifying Glasses, with Observations and
Inquiries There Upon* (1665) to present in writing and graphics a series of "some of the *least* of all *visible things*," among which featured,
as Observation LIII, an ordinary flea. Under the microscope's double lens the flea revealed whorls of exquisite complexities: "as for
the beauty of it, the *Microscope* manifests it to be all over adorn'd
with a curiously polish'd suit of *sable* Armour, neatly jointed, and
beset with multitudes of sharp pinns, shap'd almost like Porcupine's
Quills, or bright conical Steel-bodkins," Hooke wrote. By seeking
out and looking closely at an array of barely visible creatures, including silver book worms, louses, mites, flies that spin in the air,
the teeth of a snail, and "a small Creature hatch'd on a Vine," as
well as such phenomena as the putrefaction of molds and the flying
motions of the wings of flies, Hooke, the first "exponent of microscopy," defined and practiced a new way of seeing. Pursuing an
"*inlargement of the dominion,* of the Senses," Hooke looked through
the microscope's lens, using the instrument to "add[], as it were,
[an] artificial Organ[]" to natural sight. Perhaps it is no coincidence
that, as Steven Shapin points out, Hooke fashioned himself to be a
loquacious publicist for experiments and experimental results in the
Royal Society and beyond. Likewise his account of the microscope,
the instrument par excellence for inward-looking, is presented in a
distinctly extraverted fashion.[12]

Smallness itself represents interiority and secrets, as John Mack points out in *The Art of Small Things:* "Engagement with miniature worlds is a secret and often intensely personal activity."[13] Yet the point with regard to the technique Dancer inaugurated is that it was not only about the very small. Rather, it was about the constant possibility of collapsing and expanding on a scale. For Hooke, the flea's bodkins figured both as minuscule armor bedecking an ephemeron and also, via the microscope's view, something that created a large and imposing effect: a paradox. Dancer, who incorporated microscopes at both ends of his operation, took this concept further. In the realm of knowledge-seeking, the microphotograph represented a secret, private sphere but also the constant structural possibility of re-expansion into the shared and public realm of knowledge and delight.

Novelties and keepsakes fueled the growth of microphotography in the 1850s and 1860s, as the hunger for enchanting miniatures raged within the love of tourist goods and memorabilia. Why buy a full-size clock or crystal figurine when an adorably shrunken one will do? is a question still asked today and answered in dollars, krone, or yen in major tourist centers.[14] Dancer, by this time thriving financially, exhibited his tiny photos, each one "contained in the space of the eye of a needle," in 1857 in Paris and later in Florence and Rome, where they caused a sensation.[15] One spectator, a forty-year-old chemist and photographer who owned his own portrait studio, René Dagron, was moved to create a variation. He modified a small "Stanhope" lens from biconvex to plano-convex, making it flat on one side so that he could cement a microphotograph to the end. Now, instead of buying a souvenir slide, one could buy the slide with a built-in viewer. This was the first of many such viewing devices for microfilm. (Micro-readers would prove a problem well into the twentieth century, when "the limitations of reading machinery precluded an unmitigated success" for the data-storage technologies then debuting.)[16] Dagron equipped rings, watch fobs, watch-winding keys, penholders, small-scale ivory globes, and wooden toys with such lenses and micro-slides. Soon one could look into a peep-hole in a violin bow to see a miniature of Paganini perched inside (this innovation by an imitator of Dagron). His lab was employing a workforce of 150 and selling twelve thousand magnifiers a week. Ingenious tchotchkes along these lines, called Stanhopes or *bijoux photomicrographiques*—photomicrographic jewels—continue to be avidly collected as Victoriana. A 1864 booklet by Dagron, *Traité de photographie microscopique*, explained the process step by step to hobbyists, and in the same year Dagron devised his most powerful lens yet, through which a viewer could see, on a photograph the size of the head of a pin, the portraits of 450 men, "*députés de l'empire.*" Dagron prospered.[17]

Just as the microscopic photograph appeared destined for a permanent career of pleasure-mongering and frivolity, with the 1858 *Dictionary of Photography* labeling the process "somewhat trifling and childish" and other onlookers decrying the dead end of toys, it found more urgent use during the Franco-Prussian War.[18] In the autumn of 1870 Paris fell quickly, the Maginot Line proving unsupportable (as it would in a future war), and by September 18 the city found itself blocked behind Wilhelm I's "ring of steel" from receiving or sending normal mail or telegraphs, as well as food. Even as residents sacrificed Castor and Pollux, the celebrated elephants of Paris's zoo, and accustomed themselves to eating cats, dogs, and rats, citizens of the city made bold and experimental attempts to send out illicit texts. Postmen concealed coded letters in hollowed out coins and even in their skin, in tiny cuts, but only eight got through, and some of those caught were executed. After a series of further failures sending volunteers out through the sewers and subterranean catacombs and sending a boat to crawl along the bottom of the Seine, engineers hit upon the idea of balloons, which proved more successful and, incidentally, led to the founding of the first airmail service in the world. While 23,670 pounds of mail sailed out of Paris, the balloons could not carry any back in.

The air having proved the most amenable element thus far for reestablishing communication links, several elite pigeon clubs in Paris proposed sending their birds as emissaries to make the journey *into* Paris. The birds would travel out in balloons, rest and preen, then return to the city bearing mail, messages, and news. Since each bird could carry only around a one-gram parcel, a method needed to be found to increase the load of information without increasing the weight. "A number of persons, apparently simultaneously, thought of reducing the original, uncoded messages by photographic means."[19] Two professors, Joseph-Charles d'Almeida and Albert Fernique, set up a laboratory to work with the state-of-the-art equipment furnished by Dagron, who remained the outstanding microphotographer of Paris. Enthusiastic about the challenge, Dagron came up with the ideal printing surface—not paper but a dry stripping film of collodion, which was light, tough, flexible, and transparent. Here entered microfilm in a new role: pigeons could carry high-density microphotographic messages rolled up in hollowed-out goose quills. A pigeon handler sewed each quill with silk thread into the carriers' tails. Their wings, stamped with waterproof ink, specified their destination and other delivery information.

First, however, the microphotographic equipment had to be smuggled out of Paris. The likely dispatch spot was Tours (later another center), which had become a wartime communications hub, as it was the provincial city in Free France that lay closest to enemy lines. There, a central facility collected all of Free France's messages, and expert penmen began by reducing them to tiny code, which unfortunately was often illegible due to the imperative for tininess. On November 12, almost two months after the official start of the siege, Fernique, Dagron, and their assistants started out from Paris by balloon to land in occupied territory near Vitry-le-François. A deadly game of avoiding the Prussians ensued, but in a week the group at last made it to friendly territory with 1,300 pounds of heavy equipment in tow and "under the very noses of the enemy."

The new arrivals established a sort of mass-production center—or perhaps mass-reduction is more accurate, for it was there in the provinces that Dagron improved his method to allow assembly-line photographing of large plates of printed materials (the penmen were eventually dismissed) and, exposing them in a chemical bath, render them as "sheets."

Meanwhile, pigeon clubs pooled birds for heroic flights. Around four-fifths of a total 350–400 avian messengers died in the line of duty, either disoriented by the longer-than-usual flights, disabled by the cold weather, or brought down by specially trained falcons from Saxony—"Nineteenth Century Messerschmidts flashing down on liaison Piper Cubs," as one scholar envisioned it[20]—but the rest survived and returned to Paris replete with messages. One, Cher Ami, was immortalized in a civic statue designed by Frédéric Auguste Bartholdi (sculptor of the Statue of Liberty), only to be de-memorialized when the Germans took Paris in World War II and melted down the avian accolade.

How to get the information off the sheets? At first telegraphists with magnifying glasses set to decoding the contents, but later the Parisian recipients rigged a projective machine akin to a slide projector so that a corps of clerks could more easily read messages and transcribe them. When the hand-copied or printed duplicates arrived within the occupied area, workers stamped the regular telegraph forms "*Reçu par pigeon.*"[21] Dagron "in 1870 used a small oil-burning projector to read his microcopies," recalled a prominent librarian in 1936 of the historic accomplishment of the modern librarians' forerunner, invoking it as an inspiration to the Microcard movement that was then getting going.[22]

Some years later, the nanotech movement's founding document was to make a surprising link that on second thought may not seem surprising. On December 29, 1959, at Caltech, physicist Richard Feynman gave a talk titled "There's Plenty of Room at the Bottom," in which he launched the debut discussion of all that physicists and information scientists could eventually do in the "nano" fields. Although this visionary talk, later a tract, took a forward-looking gallop at the field, it began by looking back, posing a hypothetical problem with a real history. Feynman had been told, he recalled, that someone had years before already printed the Lord's Prayer on the head of a pin. (That someone was Dancer, of course.) This feat made Feynman think: could you print the entire *Encyclopedia Britannica* on the head of a pin—presumably a different pin? Yes, he answered, if you shrunk it down twenty-five thousand times. You could use an electronic microscope to read it then. In fact, you could surprise the Caltech librarian by telling her that "ten years from now, all of the information that she is struggling to keep track of—120,000 volumes, stacked from the floor to the ceiling, drawers full of cards, storage rooms full of the older books—[could] be kept on just one library card!"[23] As we have seen, this was not a new problem in 1959. The rhetorical use of shocking scalar reductions punctuated the previous hundred years. That many had pursued this line of fantasy and research before was in fact part of Feynman's argument, his jumping off point, and it is interesting to note that Dancer's research reemerged at this pivotal mid-twentieth-century moment. Likewise, note that Frederic Luther's investigations of microphotography's adventurous and illustrious history—he re-revealed Dancer's and Dagron's long-forgotten accomplishments in two well-placed articles in 1950—themselves played a historical role in the mid-twentieth-century self-definition of the American documentation movement. The work of Dancer, Dagron, and their collaborators marks the inception of "microfilm in its modern sense and on the modern scale," according to Luther.[24] Among latter-day aficionados of the "micro," European roots and a longer-durée history were suitably grounding.

Back to the outskirts of Paris, where, in January 1871, a single pigeon carried twenty-one microphotographic sheets, at one-twentieth of a gram per sheet, adding up to a total of sixty to eighty thousand messages. Perhaps it is best to leave the microphoto-graph—which was to become central to the database of dreams in less than a century—here suspended in the air, borne by swift wings, if only for a moment.

Microphotography was booming as it reemerged in the 1920s in, for example, the U.S. banking industry, where the "Check-O-Graph," a rotary microfilm camera that debuted in 1926 at New York's Empire Bank, allowed rapid copying and storage of canceled checks. The Check-O-Graph initially was a response to the grow-ing popularity of checking accounts and the concomitant problem of fraudulent personal checks. A vice president at the bank, George McCarthy, developed a camera to allow quick mechanical repro-duction of these items. Eventually McCarthy teamed up with Eastman Kodak's Research and Development lab to produce the Recordak, which aimed to fill the rapid-fire reproduction needs of all kinds of corporations, targeting especially department stores, transportation enterprises, and insurance companies. Commercial paper documents, produced on a large scale, needed to be retained. Micro-reproduction and -storage technologies gained new purpose within bureaucratic document-management systems.[25]

By this time, the "micro" in microphotography no longer facilitated tiny delights or wartime heroism but staunchly accompanied burdensome tasks such as the storage of necessary but unwieldy amounts of paperwork among corporations, the military, government, and banks. Sears Roebuck's mail order house was processing a hundred thousand orders per day, beginning at the turn of the century, and railroads were collecting and analyzing "literally tons of data" in the early decades of the century, all of which demanded office storage tools (vertical files), calculational aides (punch cards, for example), and copying capacity (mimeograph, typewriters, Photostats).[26] The availability of celluloid safety film, miniature Leica cameras, and cinematographic innovations spurred microphotography to become a more stable medium. Less in need of pigeon-borne quills as carriers, now eschewing gas-lighted lamps as projection devices, microphotography's new forms suited the emerging scale of the early twentieth-century corporate "octopus."

In its newest incarnations, microfilming addressed the tail of the paper trail: what to do with documents once someone generated them. It had the additional advantage of being beguilingly cheap. Miniaturizing size also produced miniaturized costs—an outcome echoed in social scientists' committees and action groups for whom "five cents per page" was a rallying cry. (By Kaplan's day this cry would become "half a cent per page.") Instead of pursuing more miraculous feats of reduction ratios, corporate labs and university publishers settled on modest reduction ratios "permitting greater rapidity in production . . . and greater ease in physical manipulation." Indeed, experts pronounced Dagron's legendary feats of minification a "lost art," one that fell into abeyance because no one cared to practice it at his level of exactitude any more.[27]

Libraries, too, experienced challenges for which micrographics offered relief. The research library, emerging in the late nineteenth century, saw its librarians drawn to machine solutions and embracing the "efficiency movement" of the early twentieth century. By the 1930s microfilming appeared as a technological sinecure to the problem of what a 1936 panel called the "two foes" of librarians: "brittleness and abrasion." Books were fragile things, subject to ravaging simply by sitting on the shelf. And the more books sitting on the shelf, the more space they required; brittleness and abrasion were thus related to the large problem of bulk or, more broadly, physicality. Shelf space was seen to be limited even then, and the "future of the library" opened as a topic for discussion and experimentation. Microfilm became the chosen modern means of preservation and access by the 1930s, touted especially by the documentation movement and even more especially by the American arm of that movement.[28] The technology appeared full of promise, and even if it would take two decades to realize that "Microfilm is not the universal panacea to problems of documentation that some of its first proponents thought it would be," it had proved itself, in the intervening years, an ever more pragmatic tool.[29] Librarians were not shy about taking it up.

Compare these North American developments for a moment with the continental European documentalists such as the Belgian polymath Paul Otlet and the French librarian Suzanne Briet, who embraced microfilm mainly in support of grand and idealistic universalizing projects, such as Otlet's Mundaneum, an ur-totalistic knowledge palace on a tree-lined street in Brussels.[30] Also they were more likely to pose poetic, lingering questions such as, "What is a document?" and were uninhibited in pursuing the further conundrum of under what circumstances a gazelle might be considered a document (Briet's answer: when in a zoo).[31] Science fiction writer H. G. Wells joined documentalists in 1937 at a world conference to hail a "World Brain" that would index all human knowledge via microphotography to create a "concentrated visual record" seen in a series of projection rooms. Wells's admittedly "colossal objective"? To dissolve human conflict through synthesis of all "human mentality."[32] In a prescient twist, Wells rendered the inwardness always attendant on miniaturization, from tiny prayer books to Hooke's micrographics, as a shared psychic domain. This is not to say that Otlet, Briet, Wells, and colleagues ignored the need for practical machinery but that they framed the project in pan-human terms.[33]

In the United States, documentation was less headily intellectual and more technologically oriented. First called "bibliography," it stuck to its more pragmatic roots, which lay in questions generated by the practices of library cataloging and business management. (President of the American Documentation Institute Vernon

Tate, for example, initiated the new run of the institute's journal not with a World Brain or Mundaneum but with a phlegmatic repetition of the standard international definition of documentation.) Major figures Tate, Eugene Power (see below), and Watson Davis were gadget-centered visionaries. University of Chicago librarian M. Llewellyn Davies compared the arrival of microtechnologies to that of the printing press. During this heyday for micropublishing, the *Journal of Documentary Reproduction* was born and from 1938 to 1942 regularly kept its readers abreast of developments. Microform became the leading-edge information-storage and -retrieval technology of its day: "The literature on documentation in the 1930s was as preoccupied with microfilm technology as it is now with computer technology and for the same reason, each being the most promising information retrieval technology of the time," observes historian of libraries Michael Buckland. Vannevar Bush, inventor of the imaginary Memex machine (he thought it up in the 1930s, though he did not publish his account until 1945—see chapter 9 below), actually built microfilm-based reading machines before the war and experimented with adding punchcards to microfilm to allow targeted retrieval.[34]

Machines for reading, projecting, collapsing, and expanding texts arrived in the 1930s from research and development labs to the marketplace. Devices such as the Fiskoscope were basically super-lorgnettes made to read special dedicated Fiske Reading Strips, long pieces of paper approximately the size of a current grocery receipt from a week's worth of groceries, covered with tiny print. Other clever devices innovated with different phases of what might have been, but was not yet called, the human-text interface. These mutable machines, many of them short-lived exotics, form part of what might be called, following Jon Agar, the "history of peripheral technology."[35]

Worried librarians embraced microphotography apace. Styles of microphotography formats multiplied in the United States, Great Britain, and other parts of Europe. Reading machines proliferated. Engineers experimented to make various microforms searchable. Bush's Comparator and Rapid Selector machines were only the most prominent of a number of failed attempts at searchability.[36]

Innovation arose when least expected, as in the case of the do-it-yourself micropublishing enthusiast Eugene B. Power, who set forth around 1930 to miniaturize all early English books through 1550, initially in his free time in the "rented back room of an Ann Arbor funeral home."[37] He experimented with formats and decided micro-photographed texts were best for his purposes. "This most famous of commercial microfilming projects now known as *Early English Books, 1475–1640* and based on the famous Pollard and Redgrave *Short-Title Catalogue* was completed after 40 years, and was the para-digm of a micropublishing product based on an available biblio-graphy," reported the *Microform Review*.[38] Power founded University Microfilms in Michigan in 1938 and with this inaugurated the large-scale micropublishing of dissertations, newspapers, and other research materials.

Like Dancer and other informatics pioneers, Power enjoyed tinkering. In addition to lending momentum to commercial and scholarly micropublishing, his efforts yielded an important techni-cal development. Searching for a satisfactory camera, he adapted the classic Draeger camera, designed by a U.S. Navy captain of the same name, a step that led to the adoption of 35mm microfilm as the recording medium of choice. A new "safety-base" film (cellu-lose acetate) became the standard over the decidedly combustible old cellulose nitrate film stock.[39] (Some years later, during the war, the sight of a bedridden soldier inspired Power to create a device that would project a microfilmed book onto the wall or ceiling of a hospital room. Dubbed Projected Books, Inc., it ran until 1970.) He coined concepts too. In these early interwar years, Power also began to speak of the possibility of "editions of one"—an idea only now popularly understood via the vogue for self-publishing. With textual miniaturization, he noted, the imposed limits of publishing were undone. You could use microfilm to store almost limitless amounts of information and publish anything in an edition not necessarily larger than one—and do so cheaply. Ushered or re-ushered in by Power and other pioneers, microfilm was the going means of preservation, access, and economical calculation.

World War II's outbreak focused efforts. Eastman Kodak in-vented the "Airgraph" in tandem with Imperial Airways (now British Airways): "Letters from the Forces Overseas in a fraction of the time thanks to AIRGRAPH," announced an ad run on October 17, 1942. It was not just time that was fractioned, but also space.

In the United States the microfilm reached national attention in the form of V-mail, a popular service by which twenty-seven bags' worth of soldiers' mail, microphotographed and spooled on 16mm microfilm strips, could be sent overseas in a single mailbag. On arrival overseas, on-site printers expanded the condensed letters to make prints three-fifths of their original size, on special stationery. Machines folded them and put them into specially tailored envelopes. V-mail operations thus hovered around—and in effect scaled up and down—the "three thresholds of legibility" documentalists identified: print read easily with the naked eye; reduced-size print barely legible with the unaided eye; and print invisible without a technological aid.[40]

A billion and a half letters arrived at the front in this manner between 1942 and 1945. Efficiency, privacy, morale: these values were served by the program. "V-MAIL IS SPEED MAIL," announced a supporting poster: "You Write, He'll Fight." In promotional photos, stacks of not-yet-shrunken mail dwarfed the micro-spools on which they would soon be stored. Soldiers equipped with the Schaffer V-Mail Kit could write back in like manner. The program was both a practical effort and a symbol—a symbol on the surface of commitment to the troops, but also, in a prescient way, a commitment to the charm of gadget-driven technological solutions. It hinted at a personal transformation of the human users of these intimate technologies, for in using them, as the media scholar Friedrich Kittler put it in another context, and as these promotions emphasized, "We are the subjects of gadgets and instruments of mechanical data processing."[41] Playful photos of epistolary mountains reduced to neat micro-reels once again evoked the ratio at the root of miniaturization's power: it could collapse space and, in effect, shrink time as well, with its access to automatic action and its rendering of the unstable human handwriting in the form of vast numberless accretions.

V-mail was in the public eye, but micro-methods also offered a solution to more urgent and strategic wartime emergencies. In the Asian and European war theaters, librarians and spies joined forces to protect endangered cultural treasures, national libraries, and more obscure ephemera as well. Spearheaded by a lively librarian organization called the Interdepartmental Committee for the Acquisition of Foreign Materials (IDC), a rescue mission under Wild Bill Donovan's Office of Strategic Services (OSS) started up. A corps of 150 agents and librarians set up a massive microfilming operation in neutral cities from Lisbon to Stockholm to New Delhi. Rigging together microfilm operations on the fly, interviewing refugees, interrogating prisoners of war, they sought out imperiled books and other forms of publications (such as rare single-run presses and underground newspapers). They funneled relevant information from on-the-ground texts—often literally on or in the ground, as they found stacks of books hidden by the enemy in caves and limestone pits, piled up haphazardly, or abandoned by fleeing Jewish families—and preserved them in tiny, tough format.[42]

In May 1941, the British Museum library calamitously lost a quarter of a million volumes in one night of air bombing. This loss dramatized the fact that collections could vanish while people slept or were herded into basements for air raids. Microfilm's role was to make textual rescue efficient and the materials lightweight. Eugene Power's firm, working with the OSS team on the project, and with the aid of a Rockefeller grant, photographed and microfilmed some 6 million pages of manuscripts in British depositories—making the irreplaceable in effect replaceable. (In 1978 the British ambassador to the United States knighted Power in Michigan on behalf of the queen in appreciative if somewhat inappropriate terms: "We in Britain," he said, "will long remember, with a deep sense of gratitude, his gift of Lebensraum for so many of our precious archives and libraries.")[43] Meanwhile, "the war put this technology to the test."[44] It started with a trickle in 1941, when Wild Bill Donovan thrilled to the sight of the first feet of microfilm arriving, and continued through the middle years of the war, when reels flooded official Washington. During the eight months from November 1942 to June 1943, the IDC microfilmed nearly eighty-two thousand published items, collected over twenty-three thousand original publications, and distributed nearly three-quarters of a million items to a variety of war agencies. This "massive microfilming effort" preserved many publications that would otherwise have disappeared from the human record, including obscure journals with small print runs, underground newspapers, and resistance pamphlets. Yet much as it was a savior, microfilm continued to disappoint because it was hard to catalog and search on the fly or, really, at all. OSS staffers came up with "ingenious solutions, including extensive subject indexing, abstracts, and full-text translations, and hired a small army of women and emigrés for this purpose." Indeed, the Library of Congress had hoped that the IDC would generally acquire publications in the humanities and sciences for its collections at a time when the European book trade was disabled and the fate of book stocks unknown.[45] Ultimately these war-orphaned texts ended up in the Library of Congress.

In 1943, the noir detective film *Sherlock Holmes in Washington* featured a search for a secret document stored on a cache of microfilm gone awry. Passed from a dying British secret agent to an American debutante, its tiny texts ended up stored in her "V for Victory" matchbook, where they sat, hidden in plain sight, producing anxiety in the audience every time someone lit up a cigarette. Other films of the day also displayed microfilm hidden near—but not within—the body, as critics Jonathan Auerbach and Lisa Gitelman point out. The format became a home for national secrets that were themselves secreted in homely spots, usually a lipstick case, matchbook, wristwatch, wallet, cigarette lighter, or the lining of a suit. It was a dramatic irony: "Concealed state documents are literally in the hands and under the noses of the unknowing evil spies in pursuit," like a mobile purloined letter sought by Inspector Dupin.[46] Microfilms were McGuffins, empty plot devices, as Hitchcock admitted (referring to his 1959 *North by Northwest*, where they perform a crucial plot function), but they were flexible McGuffins with unique qualities of being able rapidly to compress and expand in scale. In mid-century films, they appear in close proximity to people's bodies but never completely engulfed by them. Small and intimate, they yet contain huge amounts of information and are vastly important to affairs of state. Such a status resulted in (and stemmed from) an uncanny lack of stability. Compressed and miniaturized, documents were both secure and in jeopardy—as seen in the microdot, a nonfictional wartime technique that miniaturized documents and hid them "in obscure locations like under the period of a sentence."[47] Tininess was both protection and peril. The qualities of compression, scalability, and security-despite-riskiness had figured in the nineteenth-century microphoto but were now reconfigured and repurposed.

In the postwar years, micropublishing reached its heyday, and many different formats bloomed. Yet again, a war had transformed microphotography and its extensions. One driving element of the change was, effectively, the outsourcing of innovation financing to the state and the war effort. For the world of documentation, the story paralleled the development of automatic control technologies in U.S. industries, in which cheaper labor could be had only after factories integrated more "competent" tools that, with proper "set up," jigs, fixtures, and eventually near complete automation, could substitute for a skilled operator. In *Forces of Production*, David Noble describes how this kind of transition accelerated in a succession of wars: "In these cases where expensive technologies were introduced to make it possible to hire cheaper labor and to concentrate management control over production, the tab for the conversion was picked up by the State—the Ordnance Department in the early nineteenth century, the departments of the Army and Navy around World War I, and the Air Force in the second half of the twentieth century."[48] Ever cheaper, microfilming, like computers—though earlier and in an initially more accessible way—promised a vast expansion of capacity and a concomitant shrinking of labor (from clerical to cataloging) as well as storage costs.

The first new format to emerge after the war was the shiny Microcard, designed to revolutionize the research library and to "downsize" (to use an anachronistic but accurate term) the library worker. American-born, the Microcard technology constituted a "precomputer revolution" destined, after experiencing a vogue from 1950 to 1963, to lose out in the format wars to the European-born microfiche, which, its popularity among military and government users driven by its ease of duplication and disposability, then became a standard.[49] So ubiquitous did the " 'fiche" become (along with microfilm rolls used for newspapers), so unpleasant were most of the experiences it occasioned, that it seems retrospectively to have blanked out, for many users and even historians, the fact that high utopian hopes for the Microcard ever existed.

The brainchild of Fremont Rider, a visionary librarian and part-time inventor from Middletown, Connecticut, the Microcard was the solution to a problem Rider identified with some fanfare and exactitude in 1944. Information stored in libraries was increasing in bulk exponentially—sixteenfold per year was his calculated figure—and such increases meant that by the early 1970s, the mass of Yale University Library's books, for example, would require six thousand miles of shelves and eight acres of catalog files to hold them. (Rider's vision of an endless labyrinth of library shelves-upon-shelves evoked, perhaps, the "wand'ring mazes" where Milton's demons lost themselves in *Paradise Lost*.)[50]

Rider's response to this projected curatorial nightmare was to turn to the world of "micro-materials." Advances had been made but they were not thoroughgoing enough. A recent Readex Microprint Corporation publication of the "Church Catalog," as Rider pointed out, arrived on six leaves of 6″ × 9″ paper, printed on both sides, all enclosed in a linen-bound, slip-covered box that could sit upright on the shelf as if it were a book. So even though it had replaced twelve hundred pages with six and had thus "effected a more than 99% decrease in storage bulk," this savings in sheer page-space occupancy had been canceled out, from the point of view of the shelf, by the bulky box in which the microprint was stored. Rider, a working cataloger, was irritated. In an attempt to spur would-be "micro pioneers" (Rider's term) to be more pioneering, he created his own micro system to do so: bibliographic "cards" that could store, on their very own back sides, the entire text to which they referred. The library itself would be collapsed. Moreover, each card itself represented a double precipitate, of the bibliographic system and of the text itself.[51] Even the physical form of the book could be done away with eventually, he argued. For in Rider's system, at least as it was originally envisioned, the reference was also the referent.

As Rider saw it, scholars too could be, and perforce were, revolutionaries, if only because their need to have information at hand was stronger than others': "Having the text of his material conveniently near his elbow is his"—the scholar's—"sine qua non." Rider also envisioned the automation of library cataloging and the scaling back of the human labor element so that the modern research library would be efficiently packed, potentially much smaller, and less staffed. Unlike rival documentalist Robert Binkley, who responded to "a keen sense of living amid . . . a sea of documents" by becoming a decentralizer (via micro- and photo-offset technologies), Rider was a centralizer.[52]

As it happened, the Microcard never developed the storage-and-bibliographic merger Rider imagined, and in practice, two-sided printing offered too many difficulties. The Microcard simply functioned as a pre-electronic data storage vehicle. Each of his trademarked Microcards was 3″ × 5″ and made of a pearly, fine-grain, high-contrast paper. Step-and-repeat cameras automatically photographed and placed sequential pages of text, reduced to the size of a thumbprint, onto the same piece of sheet film. When finished, rows of contact prints taken from the sheet film marched across the Microcard.

Most libraries in America had them, and until the early 1960s, corporate and literary use grew as well, from "online" textual supplements for *Newsweek*, the *Saturday Review*, and *Science* to back records of the Air Force and the Atomic Energy Commission. This "ingenious proposed solution for the effects of [exponential] growth" for a time held tremendous promise.[53] "A future five-foot shelf," a writer for *Time* magazine observed excitedly in 1944 of the Microcard's potential, "may be no bulkier than a pack of playing cards."[54]

In 1960, an annual review of microtext progress announced the following item confirming the boom in the "micro" storage of texts and auguring the way micro-techniques would be recombined and woven into other technical systems: "US Army Rocket & Guided Missile Agency (ARGMA) has adopted a Microfilm-punched card system (aperture cards) at the Army's Redstone Arsenal (USA). This has enabled 100,000 drawings to be stored in only 4 sq. ft. of floor space and has cut clerical help by 40%. Savings in producing quantity requests for prints amounted to $53,000 the first month and as many as 10 days have been lopped off waiting periods."[55] Combining the microphotographed image with a punched card system moved the ever-adaptable technique into the realm of analog computing. The once-fanciful ratio of the Bible in a grain of sand appears here as one hundred thousand arsenal drawings in four square feet with cost and labor savings prominently on display. The micro-image appeared destined for this less fanciful future.

We have now traversed a long and discontinuous history of microphotography from its 1839 origin to the 1950 advent of the Microcard, a history that has veered, staggered, or sailed (depending on circumstances) from the apparently trivial to the evidently historic, most recently to land in some of the dullest stacks of the dullest paperwork ever known to mankind. (I'm thinking here of twentieth-century microfilmed collections of canceled checks and discarded engineering blueprints.)

For a long time, people dreamed of micrographic books both real and legendary, or sometimes both, as Benjamin Disraeli once remarked: "The Iliad of Homer in a nutshell, which Pliny says that Cicero once saw, it is pretended might have been a fact, however to some it appears impossible." Aside from such ancient rumored curiosities, miniature books' popularity peaked at the end of the manuscript era. Penned in tiny letters of gold or other inks on walnuts or grains of corn, such works trouble "the relation between materiality and meaning."[56] In their small size they called attention to the closed nature of the book, exploding and potentially expanding it. They marked the end of the written-by-hand book as state of the art. They announced the new while expressing nostalgia for the old. They destroyed even as they mourned. In contrast, microphotographic texts of the nineteenth and twentieth centuries were not written by hand nor produced via movable type but photographically made through scalar technologies, thus suggesting the limits of the human hand to make things nano.

What was different with microphotography was the zooming in and out capability built into the technique, including the viewing of it. Really, in this sense the microphoto was not simply a new kind of photograph, nor a different variety of miniature, but a set of acts and relationships: gaze into this Stanhope lens secreted in a violin bow, or view these pigeon-borne messages by means of a gas-lit projective machine, or scroll through thousands of early English documents on an electronic reader. Each required and helped bring about new social, cultural, and intellectual relationships.

Fantasies of total information, as manifested through the new possibilities micro-devices, micro-pioneers, and micro-movements engendered, changed in three ways from the nineteenth to the twentieth centuries. First, scalar became *scalability*. Once a surprise and a thrill, movement from the very small to the very large and back again was speeded up and normalized. Librarians, social scientists, and engineers built microphotographs and their retrieval and reading devices into systems. Entrepreneurs recruited them into offices, banks, and governmental record keeping. Tiny images on cards or film rolls came to inhabit their own rooms in libraries and corporations. Second, *compressibility* intensified. Recall that Feynman invoked a 1:25,000 ratio via electron microscope. This was not the reduction ratio commonly used in libraries or by the Microcard data archive's creators, as we will see, but the possibility was there. "What I want to talk about is the problem of manipulating and controlling things on a small scale," Feynman asserted, and the tight yoking of the two has become ever tighter. Third was the emergence of *storability*, not as a utopian inkling but as a practical solution to existential uncertainty. "The records of our time are written in dust," warned one documentalist in a talk at the First World Congress of Libraries and Bibliography in Rome in 1929. This consciousness would grow.[57]

On the other hand, microphotography predicted but was incapable of delivering some of the changes the digitization of data would soon bring about on a huge scale. Microphotography offered the easy up-and-down of scalable data storage, but as we will see, it stumbled in its incapacity to offer complex operations (though many, including Vannevar Bush, tried to enable them in this way), resulting eventually in fierce declarations such as this one in 1960: "Microforms have come to be one giant headache for library administrators, bibliographers and researchers."[58] With the turn to digital memory devices, according to historians of computing, an earlier capacity simply to aggregate data became a new ability to work complexly with them. Yet this capacity was already being cultivated during the early years of microphotography to the last years of the Microcard. Around 1870–1914, "massive new waves of information classification and standardization took place—international classifications were developed for diseases, work, criminal physiognomy, and so forth: facts could be split apart, sorted into pigeonholes, and reassembled in new ways. It is a direct outcome of this work at the turn of the twentieth century that we get the emergence of the database as a central cultural form." Here Geoffrey Bowker suggests that the database as a "central cultural form" emerged earlier than its technological realization in 1960 as the Database Management System or DBMS.[59] In a parallel manner, this chapter has traced the emergence of a fantasy of total information that predates digitization or the Internet.

Micro-techniques chipped away at the book's dominion and, long before the digital, encouraged fantasies of machines filling snuffboxes full of microcosmic knowledge. Likewise, within a few years of its debut, a group of U.S. psychologists and anthropologists would choose the Microcard format to store the particular kinds of data known as "subjective materials," including Rorschach, TAT, and House-Tree-Person results, alongside dreams and the stories of people's lives, fulfilling a promise of "looking within" that came built into early microscopic devices. This most unlikely unfolding of events remains to be told: how a massive data repository, including the dreams of Hopi grandmothers, the musings of Southwest Pacific islanders about moral ideology, and the psychological test answers of Iroquois boys and girls, ended up experimentally preserved via this burgeoning format.

CHAPTER FOUR

Data Mining in Zuni

ZUNI Pueblo, early summer 1947. An awkward meeting was under way.

Bert Kaplan, then twenty-eight years old and recently discharged from military service at Okinawa, had arrived without introduction in the pueblo, and although he had managed to secure a meeting with the governor of Zuni in his house, it was not going very well (at least from the point of view of the young social scientist). A graduate student from Harvard's experimental Department of Social Relations, he was in town to give the Rorschach, the TAT, the sentence-completion, and a couple of other tests to young Zuni males, especially those who had just fought in the war. He planned to write a comparative dissertation on the returning vets in four different American groups, and Zuni was his first stop. He was angling for access.

As the governor listened impassively, Kaplan explained "that I came from a large University in the East, that I wished to help the white men better understand the Zuni way of life and how the Zuni people thought, and that I had some tests that would help me do this."[1] In saying this, Kaplan engaged in an expository mode common to those who aimed to draw people into their psychometric testing projects, one that involved a certain amount of dissembling and also the condensing of a long line of occasionally tenuous logical connections. "White men"—scientists and administrators—would craft a better understanding of "how the Zuni people thought" from such studies. Eventually, they would also pass on the broad scientific and humane benefits of such knowledge, in the form of a more powerful social science. Kaplan was counting on the Zuni governor's faith in this process and in scientific progress itself.[2]

Unfortunately, the governor "did not appear to be enthusiastic" about his line of appeal, as Kaplan reported. He then switched to pragmatic considerations: "I further appealed to ... the probability that this study would help his people." Here, there was a certain built-in ambiguity. What did "help his people" mean exactly, and how direct was the help going to be? According to the logic in which Kaplan trained, getting test results from Zuni would help the Zuni in the broad sense that all scientific progress benefitted humankind, and the Zuni, as undeniable representatives of humankind, would also benefit via a kind of trickle-down approach to basic science. Many of the great philanthropic foundations of the day—Rockefeller, Carnegie, Ford—supported the social sciences with just such an understanding: the development of the social sciences as fully as possible would, they believed, in turn support mankind's betterment, as the Rockefeller Foundation's motto stipulated: "*The purpose of the Rockefeller foundation is in general the betterment of mankind.*"[3] In a more targeted way, psychological tests might help by telling the Bureau of Indian Affairs (BIA) or Indian Service officers what the Zuni—officially defined as of 1947 as wards of the state—were really thinking and thus smoothing problems attendant on what was sometimes euphemistically called "administration."[4]

A few days later, when Kaplan got the chance to meet some Zuni residents, he began introducing his tests, via an interpreter, with a version of the "these-tests-will-help-your-people" rationale. Kaplan typically introduced the storytelling tests this way, as he reported in his data set's introduction: "I come from a big university back East. We're doing a study of the different ways that people think. You know that the people here at Zuni think differently from the Navaho, or white people or Mexicans. Well, this test I have will help us in understanding how the Zuni people think." The tests benefitted the Zuni "because if white people understand you better they can do more to help you."[5] Delivered by an interpreter, this last point was usually received without comment, although it occa-sioned once, during a desultory Sentence-Completion Test that summer between Kaplan and a young man, the latter's interjection, "How does this help the Zuni people?"[6] This was a good question, in fact, for while Kaplan on some level believed his work would help the Zuni in an abstract future, he also knew that the direct aid his words evoked was not likely to pour forth. At the same time, a pitch of some sort was necessary. University-sponsored research was not "respectable" to the Zuni, Kaplan felt, not a good motivator, so it was better at times to dissemble about its nature: "Some explanation couched in terms and places and behavior that were part of Zuni ex-perience would have been preferable even if it meant deviation from a truthful account of our aims," he reflected in retrospect.[7]

A year or two later, Kaplan had an easier time at a boarding school in Lawrence, Kansas, where he administered the TAT to thirteen young Hopi volunteers: "After the fifth or sixth card, the Hopi students started asking questions, all inquiring essentially, 'How does this help you to understand the Hopi culture?'" Kaplan replied that the Hopi culture was different from other cultures and that by seeing "how folks use their imaginations and how they tell stories, we feel we can learn more about them"—an explanation that was acceptable to the students.[8] However, in Zuni, skepticism about testing was rampant, if rarely expressed.

Whether or not Kaplan fully believed his own account, he oc-casionally reproached the Zuni for doing so: "The Zuni approach to the testing was to consider how it would affect them. The test must either bring some advantage for them, or do them some harm. *They found it difficult to believe that it might be indifferent to their welfare,*" Kaplan observed in an uncharacteristically depreca-tory nugget.[9] He had used and would continue to play out the same lure, yet it was not simply a lure—it was sincere enthusiasm for what he believed the most advanced social scientific research could do. Logical contradictions like these were symptoms of the visible-yet-invisible moral heart of ethnographic inquiry. Along these lines, Clifford Geertz, who conducted fieldwork in Indonesia just after working on the Ramah project (he was a graduate student in the same department as Kaplan, only signing on a few years later), recalled, "Our view of ourselves as paladins of an improved, 'cutting edge' social science, our assumption that our work would benefit not just ourselves but our subjects," these factors served to obscure an "intercultural drama" that—looking back at it in 1995, at least—"vividly . . . reflects, in its unselfconscious, almost parodical way, what has widely come to be seen in the decades since to be the moral crux of ethnographical inquiry."[10] In the postwar years when Kaplan, Geertz, and many others went "into the field" to meet and study people they had never met before, the centrality of the encounter itself—including such details as exactly how it took place, where, under which conditions and auspices, who was paid what, who cajoled whom, or who was hoodwinked—was not yet apparent.

Part of the reason was that the unrecognized "drama" of working with human subjects meant staging one's relationship to the people from whom one wanted information. A burlesque of claims, gestures, song-and-dance routines, and unsaid things played continually on the surface of negotiations like these. Kaplan, for example, knew very well that the immediate effect of gathering Rorschach tests from Zuni men and women would likely not be an increase in the prosperity or health of Zuni people and dissembled to some extent about the nature of the tests as they were understood to function (for example, he neglected to mention their "X-ray"-like capabilities to penetrate subjects' most private thoughts and to reveal conflicts and psychical dramas unknown even to the informants themselves). Later he would frankly characterize his meeting with the Zuni governor as "my first strategic step." Yet Kaplan, it seems, was only partly aware of the dynamics of staging and strategy in which he engaged and was forthright in wanting to gather the data for reasons that in fact did descend from Enlightenment imperatives, newly remade for a peculiarly optimistic postwar world.[11]

Still, the governor's initial luke-warmth omened ill for Kaplan's work there. In that first meeting, he was unable to secure any token of enthusiasm aside from bare-bones politeness: "I did not actually receive his blessing, but he did say with the appearance of washing his hands of the whole matter, that if anyone wanted to take the test he could and it was no business of his. I offered him the honor of having his sons take the test first, but he declined politely saying that his sons were away sheepherding."[12] Pausing here only to observe that it takes a certain chutzpah, when face to face with a man manifestly uninterested in having his pueblo's residents serve as the scientific subjects of probing experimental tests, to offer him "the honor" of employing his sons first in that capacity, we proceed to the rapid end of the meeting. Undaunted, Kaplan embarked on his work in Zuni under conditions that, although they had begun inauspiciously, fulfilled themselves in unanticipated ways.

That summer Kaplan was one of two researchers who had come to the pueblo to study the mental and social lives of returning veterans. The other was from Albuquerque, New Mexico, and although they pretended not to know each other, perhaps it was clear that they did. John Adair, a Tennessee native and the first graduate student to enter the University of New Mexico's anthropology program, was spending a year studying the changes in Zuni that had taken place since Armistice Day. Adair took down six autobiographies, all later incorporated into the database of dreams; sometimes tried, though usually unsuccessfully, to give the TAT (Kaplan was better at it); and participated in Zuni life to the extent possible. Adair came in June 1947 and set up lodging in the village, with his pregnant wife and their first child to join him some weeks later.

Up the road, staying in rooms at the BIA headquarters, was Kaplan. He was there to run tests and do little else, planning to share his psychometric data with Adair and additional Harvard graduate students as well as his adviser, Kluckhohn. After one month at Zuni and a month each at nearby Navajo, Mormon, and Spanish American settlements, Kaplan planned to return home, the reason he and Adair pretended not to know each other: they did not want any resentment accruing from the Rorschachs and other tests to attach to the longer-term fieldwork. "We agreed not to divulge our connection . . . at least at first, so that any dislike of the tests would not jeopardize Adair's work," Kaplan recalled later in a statement, interesting not least because it indicates the two anticipated at least some resistance to testing. Balancing this awareness was a sense of brisk purpose. Granted, such tests were unpleasant procedures, but, like vaccinations or medical samples, they must be done. Why did Kaplan devote himself to testing above all? The specific reasons are not clear but likely centered around his strong conviction that tests offered a new methodological horizon. In this he joined a lineage of largely test-centric fieldworkers, those who wanted to assay this promising technology in the field as much as they wanted to find specific results. Others in this emerging tradition would include contributors to the database of dreams—Jules Henry, George and Louise Spindler, Edward Bruner, Anthony F. C. Wallace, Erica Bourgignon, and Melford Spiro. They could be called techno-modern fieldworkers, in a lineage of tool-centric anthropological-psychological collaborative studies extending back to the 1898 Cambridge Anthropological Expedition to Torres Straits.[13]

Lacking the governor's blessing, devoid of contacts in town, and friends only with a colleague he could barely acknowledge in the street, Kaplan strategized about how to get what he hoped would be a full projective series cataloging the young men who had fought, as well as non-fighters and older men for comparative purposes. Pressed for time, he noted the "necessity for working fast if an adequate sample was to be collected"—in this invocation of the language of "sample" and "adequacy" reinforcing the strict scientific spirit of his mission's research design.[14] He made the choice of securing lodgings at the Indian Service post in Black Rock, about four miles away, rather than in town among Zuni people, a decision that, although it may have been convenient and succeeded in deemphasizing his connection with Adair, had the further effect of estranging him from daily ordinary contact with people and associating him inextricably with the mostly despised Indian Service. He had hoped that the governor would "take me under his wing" but realized from the tone of the meeting that this hope had been unrealistic, and in the end he characterized the behavior of the governor as typically Zuni: "The Governor seemed to be afraid either to give us his support or to deny us the privilege of working in Zuni. His noncommittal, fence straddling position is characteristically Zuni."[15] Here, setting aside the unspoken conflicts his arrival foregrounded, Kaplan referred instead to the then widespread popular portrait of Zuni people anthropologists had circulated throughout the first half of the twentieth century, not least through the vehicle of Ruth Benedict's 1934 best-selling book prominently featuring the Zuni, *Patterns of Culture*.

In *Patterns of Culture*, Benedict argued that cultures were basically configurations and that each culture one came across was not just a motley assemblage of traits and preferred behaviors but a goal-oriented whole. One of her metaphors for the process was that each culture in effect chose a road and traveled down it.[16] Benedict did not explain who did the choosing of roads or the mode of traveling down them, exactly, a lacuna that was and continues to be one of the flaws in culturalist theories like hers, for they seem to rely on a sort of unmoved mover who cleverly lays out the "configuration" each culture will take—a process sociologist Pierre Bourdieu later described as "like a train laying its own tracks."[17] Impossible, yes, but it is *as if* it were so. In order to support this hypothesis—which drew on Gestalt psychological and German Romantic influences—Benedict portrayed four different cultures by way of Nietzschean contrasts: the wise and balanced "Apollonian" configuration appeared in Zuni and other Pueblo cultures' embrace of ritual and restraint, whereas the extremist "Dionysian" emerged in the ecstatic tendencies of the Plains Indians. She further argued that the "Treacherous" style of the Dobu of New Guinea and the "Paranoid" patterns of the (no longer culturally operative) Kwakiutl of the Northwest Coast were also mostly Dionysian.

Benedict's fieldwork in Zuni was confined to two summers' worth of interviews with kindly old men who spoke loudly, as Benedict was almost entirely deaf. "Mrs. Benedict generally worked with old men," according to the firsthand account of a Zuni-Cherokee woman who knew her. "She did not go out. . . . She would pick out someone and would bring him home and write down everything he said."[18] Note that Benedict did not aspire to "sample" the population, as Kaplan soon would. A few good informants would do. Her book popularized a view of the Zuni as mild and wise, rather like the men she interviewed, both of which (the qualities and the men) she idealized to some degree, remarking after leaving that she felt as if she had "stepped off the earth onto a timeless platform outside today," a real-life but out-of-time city on a hill.[19] "The Zuni are a ceremonious people, a people who value sobriety and inoffensiveness above all other virtues," was Benedict's official take, one from which Kaplan both conveniently drew and vigorously dissented.[20] When Kaplan called the governor's unwillingness to either support or kick him out for "fence straddling," he was in essence summoning the Benedictian Zuni portrait—even as he took pains to disagree with the thrust of what he found inapplicable, for, as he groused, "Zuni was far from being the peaceful, cooperative paradise which was suggested by Benedict's 'Patterns of Culture.'" Kaplan's dissent is not surprising.

Even today, it is hard to overestimate how important and influential Benedict's book was, not least because of its almost unrivaled elegance of presentation. In a recent assessment of the Boasian tradition (Benedict was a prime student of Franz Boas), Paul Rabinow remarks that the anthropological world view in these years was a redemptive one, in which understanding "cultural values" as primordial was a way to fight the rising totalitarianism of the twentieth century. Only later, with the rise of an apolitical symbolic anthropology, did anthropology in his view entail the "bracketing of truth and seriousness."[21] Despite decades of critique and the limitations of her fieldwork style, Benedict's ideas inform much contemporary understanding about what cultures are and how they work.

Increasingly desperate to secure males willing to take his tests and short of time, Kaplan called on the assistance of the Black Rock Indian agent, who had several Zuni men working for him at the local BIA office. Since the "Assimilation Era" of the 1880s, the bureau's agents had held sway over infrastructure (supplies, allotments, leasing) and operations (of schools, justice, most government functions). After 1934, the Indian Reorganization Act attempted to rectify the bureau's shortcomings by strengthening actual tribal-run governments. At Zuni this conflict was playing out around the time of Kaplan's stay. Even the Zuni staff of the bureau, although accustomed to working with whites, proved extremely reluctant participants and complied "only under the pressure of strong urging." Two convalescents at the nearby hospital provided two more sets of protocols and were "not nearly so reluctant," one can practically hear Kaplan sighing with relief. After three days, realizing conditions in Zuni would be tricky and a certain caution should be exercised, Kaplan had a lucky break. This came when he struck up a sort of friendship and "advisory" relationship with one of the Zuni men, "R____," who worked at Black Rock and who was ambitious, friendly, and interested in going to college. This acquaintance introduced him to a few of his friends, who in turn introduced him to relatives and further friends. The mission, John Adair, and the school provided some other subjects so that soon Kaplan had two or three to test per day. Indeed, it became clear that Kaplan had something of a gift for creatively procuring test takers in new places; when later that summer he drove down U.S. Route 66 to the Spanish American migrant-worker town of Grants, where few had any interest in talking to an Anglo scientist of any kind and "the testing situation was a some-what delicate one," he set up shop in a booth in one of the town's fifteen bars and, stationing himself behind a bottle of beer, paid the bartender to send subjects his way. The possibly intoxicated yield would eventually appear in the 1956 Microcard data repository as "Rorschachs of 24 Spanish-American Young Men."

At the same time as his test administration rate came up to speed, Kaplan became aware that he was triggering gossip among the Zuni, who suspected him of being a spy out to steal their secrets. He thought this a bit outrageous: "The most frequent suspicion seemed to be that I was somehow after Zuni religious secrets. This was an exceedingly sore point for the Zuni, although as a result of the thorough work of previous anthropologists there were very few secrets left."[22] In a kind of projected Catch-22, Kaplan found his hosts guilty of the paranoid suspicion that he as a social scientist was trying to capture their secrets.

In a sense, Zuni fear about white officials and deputized scientists out to steal their secrets was, from a historical and sweeping perspective, entirely reasonable. During the late-nineteenth century, when American anthropological interest in Indians was intensifying, the Zuni (along with the Hopi and other Pueblo Indians) became cynosures, for they were cultures that appeared miraculously "intact" since Spanish colonization. Seeming to live in a land that time forgot, they were desirable subjects to study. Historians sought to understand their "cultural persistence well after conquest," while anthropologists treated them as living "storehouses of knowledge."[23] "The Zunis and Hopis were appealing because they appeared less influenced by European cultures than other groups and did not put up hostile resistance to U.S. military conquest of the region," observes Leah Dilworth, while George Stocking

writes that the southwest somewhat later continued to attract anthropologists as "an area where culture seemed still vibrantly 'alive
and well.'"[24] Here convenience and safety draped themselves in abstract intellectual goals. Perched in their hilltop or mesa pueblos,
the Zuni were nicely physically circumscribed, disinclined to roam
about, and unlike the Plains tribes such as the Kiowa-Apache
or Comanche, whose empire extended into the pueblo country of
soon-to-be New Mexico at its height and whose embodiment of
classic images of "Indian violence" circulated well after its decline
in actual incidence, were not likely to be perceived as threatening.[25]

During the BIA's forced-assimilation programs from the 1880s
through the 1920s, few tribes maintained the ability to sustain
themselves without federal aid, and thus few were able to avoid, under humiliating circumstances, their forcible education into white
ways via the breakup of ceremonies, family structures, subsistence
methods, and linguistic traditions. Nor were Zuni exempt from
technological change. Small acts led to great alterations, amounting
to what one scholar describes as "agricultural hegemony," starting
with the 1856 gift to the Zuni from Fort Defiance's commander of
four steel plows and culminating in the 1908 completion of the
Black Rock Dam, which moved the area toward "progressive," "scientific," "rational," and even "patriotic" irrigation of a kind that was
transforming the West as a whole in these years—although this particular dam mostly didn't quite work.[26]

"Make them leave the old village of Zuni and live upon their
farms": this was the essential way to update Zuni, according to a
1901 call by Indian School Service superintendent Ralph Collins.
If agriculture was modernized, people would have to modernize
themselves, too, and Protestant missionary schools set up shop accordingly. The overriding concern of federal policy was "to introduce American technology to the Zuni."[27] Yet Zuni did not adopt
new technologies wholesale and refused to leave their hilltop village entirely. They defied irrigation reform in the wake of the oft-
malfunctioning Black Rock Dam and refused to observe the BIA's
neat plots and divisions of land, instead tearing up fence posts and
markers and persisting in planting "waffle gardens." Through resistance, flexibility, and selective adoption and through the almost
fanatical guarding of its religious secrets, persisting even today,

the pueblo kept up a degree of cohesion not seen on many other reservations.

Paradoxically, because of their success in resisting being overrun by "white ways," the Zuni attracted attention. Stationary, replete with esoteric spiritual knowledge, and seemingly unchanging, Zuni Pueblo attracted decades' worth of fieldworkers, who systematically sought to possess what the storehouse of knowledge held. By the time Kaplan arrived, despite fierce resistance to secret-breaching, not much about Kachinas and other forms of worship remained undocumented or unpublished, whether in professional journals, travel bulletins, or, around the time Kaplan set up lodging there, a chatty *New Yorker* profile by Edmund Wilson. Yet even as the Zuni's secret rites of passage appeared in the full daylight of print, the outsider-insider dynamic of secret-gathering and information lust did not diminish. In fact, it grew.

The first exposure of secrets had occurred almost three-quarters of a century before, following Frank Hamilton Cushing's 1879 infiltration of Zuni, where he lived for four and a half years, despite the fact that the Smithsonian-sponsored mission, the first of the just-founded Bureau of American Ethnology, was intended to last only three or four months. Cushing's expeditionary marching orders from the Smithsonian were to find "some typical tribe" of the Pueblo variety and study it. As Cushing recalled of his send-off, the secretary of the Smithsonian Institution, Professor Spencer Fullerton Baird, summoned him to his office to bestow these instructions: "Make your own choice of field, and use your own methods; only, *get the information*. . . . Write me frequently. I'm in a hurry this evening."[28] And grasping his umbrella (as Cushing reported), the professor exited. On arrival in Zuni, the twenty-two-year-old Cushing, according to some accounts, barged into the governor's house with his bedroll and received the dry inquiry, perhaps an early example of Zuni fence-sitting, "How long will it be before you go back to Washington?"[29] Undaunted, in the years that followed he documented the most closely guarded rituals, even the Bow Priesthood ceremony—indeed he managed to get himself anointed as a Bow Priest under circumstances controversial even to this day. He got the information. As he wrote in 1881 to Baird:

> Being now a *Pithlun shiwani* or Priest of the Bow, I am se-
> cured in the privileges of this strictly exotic society, as well
> as entrance into any meeting of, though not for the present
> membership in, all the other secret, medicine, or sacred
> orders of the Tribe. I bent all my energies toward this su-
> preme order of the Zunis, for more than a year, and my
> success in gaining admission to it is the greatest of all the
> achievements of my life perhaps; for it breaks down the last
> shadow of objection to my gaining knowledge of the sacred
> rites, not only of this, but of the Maqui [Hopi] tribes, and
> others as well.[30]

This was not the first time Zuni had offered a white man the
honor of becoming a Bow Priest, but it was the first time it had
been accepted. As Zuni was then a still-functioning theocracy, it
was (likely) politically useful to have a well-connected white representative as Bow Priest because this status also admitted him to the
tribal governing council. Meanwhile, Cushing published much of
what he learned via Smithsonian bulletins and posts to *Century
Magazine*, even as he complained about irritating missionaries who
"cast mud" at him by telling the Zuni he was after "all their secret
affairs."[31] This catty dynamic between missionaries and anthropologists entering local struggles would repeat itself in Kaplan's day.
    While Cushing profoundly loved and romanticized the Zuni,
he also saw himself as an ambassador for the "progress of science."[32] He wanted to live as a Zuni but also remain the authorita-
tive expert on them, and in a sense he had it both ways. He
"experienced Zuni as a sort of magic amusement park full of pup-
pets without strings" where he freely experimented with his own
identity, and yet he was in a position to document its most magical
happenings in the language of then-current science; according to
at least one scholar, Cushing's work on Zuni philosophy inspired
Émile Durkheim's and Marcel Mauss's groundbreaking concept of
the total social fact.[33] By his short life's end he had delivered into
print some of the most persistently vivid portrayals of the Zunis'
rituals, to their inheritors' future chagrin. He also was a strong
voice defending Zuni land claims against a powerful U.S. Senate-
backed incursion. Today, while some see Cushing's subsequent

death-by-fishbone—he choked to death in Florida some years after leaving the southwest—as retribution for unduly intrusive revelations, others use Cushing's work as a reference guide to recapture intricacies of the old ways, many of which have been forgotten. Lorandina Schecha, a Zuni artist who makes fetishes based on designs she re-appropriates from Cushing's 1880s publications, recalled, "My grandpa used to say that Cushing was a good guy and a crook."[34]

Luckily or unluckily, the Zuni received more than their fair share of anthropological attention after Cushing left in 1883 and before Kaplan and Adair arrived in 1947. Years later, a British-Indian researcher, Trikoli Nath Pandey, alighted in Zuni in the mid-1960s to be posed the question, "Are you an anthropologist?" which, he claimed, was by then how most strangers were greeted.[35] Between 1890 and 1935, ethnographic efforts included, toward the earlier end of the range, those of Jesse Fewkes, F. W. Hodge, and the "remarkably diligen[t]" Mathilda Coxe Stevenson, an early exponent of the goal of total information gathering via ethnography.[36] In subsequent years, there arrived Benedict, her friend Ruth Bunzel, Elsie Clews Parsons, and Leslie Spier. Also the travel writer Erna Fergusson took her turn, as did (briefly) anthropology's *éminences grises* Alfred Kroeber and Franz Boas, the latter's face receiving much comment, bearing as it did the scars of duels fought to counter anti-Semitic slurs during Boas's youth in Germany. Taken together, these studies pretty much laid bare the gamut of sacred, highly ritualized, and jealously guarded practices, as well as people's anxieties about this laying bare. Dick Tumaka, who had dictated Zuni esoteric texts to Ruth Bunzel for her studies, said while dying, "It is true I will die because I have given away my religion."[37] When such secrets were revealed and necessarily decontextualized, were they lost? Or were they transformed, beginning a new life?

For a long time, the Zuni and other pueblo people exerted a sort of hydraulic control over internal sacred mysteries, which they used to propagate tribal knowledge. Elders systematically hid the secrets of Bow Priest, Katchina, and Shaláko ceremonies from children and then dramatically revealed them (to boys) on coming of age. Revealing the secret that Katchinas were real men—men they knew—and not gods, priests sealed initiates into silence by means of terrifying stories of vengeful and, this time, *real* Katchinas, who beheaded secret-revealing boys with a cry of "Bu-ix!" Girls also received a progressive pedagogy of once-guarded secrets step by step as they grew up. With the increasingly widespread publishing of Zuni mysteries, parents were sometimes in the odd position of having to claim to young people that there were *further secrets* beyond what the anthropologists had uncovered in order to keep them from "drifting away." The power of secrets lay not in the knowledge stripped bare but in the economy of their circulation—revealed only at a prescribed moment under ritualized conditions, after which the initiate was in on the secret and actively conscripted to keep it from younger children. From their very first published probings, anthropologists threw a wrench in this economy. As one Zuni GI told Adair, in a remark eventually stored in Kaplan's hi-tech data bank, "Those old people don't know that there aren't any secrets left. All of that stuff was sold long ago."[38] Those secrets are now widely available, for example, on the Internet Archive, where one can find the corn maiden dance, a dozen sacred songs, and beliefs about death and attitudes toward diabetes, along with multitudinous other materials.

Nonetheless, this same GI, under the pseudonym Miguel A., repeated the revelations, yet again, by telling about his own personal experiences in sacred ceremonies and the mechanisms of instilling secrecy. "Tell me everything you can remember, no matter how trivial and unimportant it seems," Adair urged him. So encouraged, Miguel recalled receiving his initiation into masked dances his first year in school, about the age of seven or eight. Having thus revealed his secrets as well as their propagatory mechanics, Miguel, urged to provide still further information, began to reminisce about the corduroys that became the style for boys at school (full at the top and tight at the bottom), how schoolgirls wore their hair (braided in back), what kind of stockings they wore (black), and how he and another boy fought over a "triangle bar" after recess.[39] Earlier researchers like Cushing had not interested themselves in such minute particulars of *how it felt from a singular point of view.* For them, one Zuni ritual was pretty much like another, one Zuni life significant because of the light it shed on all Zuni lives.

By the summer of 1947, at the start of Kaplan's data-gathering enterprise, what had changed was that every Zuni life became by definition of interest and significant in its lived details from a person's first childhood memory to his first sexual encounter. The influence of Freudian psychoanalysis, in this, is plain to see. It is written into the records.[40] But it is as if Freud's ideas—about the significance of childhood experience, sexuality, and everyday life—were present but effaced at the same time. It was enough merely to record the stories, to get everything down *"no matter how trivial and unimportant it seems,"* and in this way to equip the database of dreams, eventually, with the accreted intimacies of day-to-day life. These things were the new type of "secret" that social scientists wanted. What exactly did the world look like to a Zuni Indian rapidly losing (as it seemed) his place in a traditional, religiously oriented world and systematically stripped of old-style secrets? What did a personality look like that was being dismantled culturally (the term "liquidation" was sometimes used)[41] and changed into something else? When an economy of traditional secrets no longer circulated as it once had, what new sorts of secrets might arise? Such questions were a preoccupation of Kaplan's cohort of culture-and-personality scholars. For example, Kaplan's colleague Dorothy Eggan, working on the Hopi mesa not far away around the same time, was interviewing a middle-aged man who brought up the story of Hopi arrival from the underworld, the Emergence Myth. However, Eggan refused it politely, saying: "I know that story. It is very interesting. I know all about the Hopi coming here, but I want to know more about your family and what you remember growing up."[42] The "recorder" desired a person's first memory with all its potential gaps and misremembered parts rather than a tribal tale. Each person would have his or her own private experiences, but it was the whole, an immensity of empirical testimony gathered up in standardized form and format, that attracted social scientists. A lump sum of minutiae pulled piece by piece from the realm of everyday encounters: these were the desired data. All these would wait for future viewers—whoever they might be—to make something of them.

Back at Zuni the summer of his data-gathering stint, Kaplan found himself maligned by a "fantastic" rumor that held him to be not an anthropologist but a spy, this view fomented because one subject, catching a glimpse of the Rorschach test kit's packaging, recognized German words and drew rapid conclusions that Kaplan was reporting to lingering Hun agitators. Here Kaplan admitted the Zuni did have some reasons for worrying about Teutonic spies, as there had been at least two episodes of infiltration during the war, each more outlandish than the last. "I later learned that there had been a full blown spy scare in Zuni during the war, which seemed to involve a *real* German spy." When the war broke out, German consulate staff spent their last few months in the United States in New Mexico, and a large number of FBI agents came to the area to track them, visiting Zuni and leaving "a blaze of excitement and an acute awareness of spy rings," Kaplan noted. In a historical sense, too, Zuni people had good reason to suspect infiltration. There was, too, the sore point raised by the tale of the itinerant fish peddler. It seems that during the war a man went around the various pueblos ostensibly purveying fish but actually, by means of a high-powered camera concealed underneath his aquatic wares, taking photographs of sensitive military installations in the area. On the face of it, this story seemed unlikely, if only on the basis of the widely known dislike for seafood among Pueblo Indians, and Kaplan attributed it to a figment of the Zuni imagination. Consultation with "reliable sources outside Zuni," however, soon convinced him that it was true, although the spy's peregrinations had likely not included actual Zuni territory.[43]

Despite the setbacks, rumors, and ill will that dogged him, Kaplan managed to collect in one month fifty-three Rorschachs of men between the ages of eighteen and forty, half veterans and half not. His tenacity paid off data-wise. His data set was nicely representative of the population in terms of location, status, and acculturation, with the exception of priests, not many of whom agreed to be tested.

Much as social scientists armed with tests and techniques tried to gain entrée to the inner sanctum where ultimate Zuni secrets reposed, they could not or only briefly could. It was as if they operated in a dream that on waking broke, its illusion—*all is now revealed, at last!*—dispersed. Yet this shared quest to gain access to others' experiential reality, to grasp the subjective sense of a world seen from a distant point of view, attracted its undaunted seekers even in apparent defeat. Both Kaplan's and Adair's data sets, issuing from this doomed struggle, are today on file in the culture-and-personality archive. What is the meaning of these revealing records? What, after all, do they reveal? They hold many things, including secrets, some sacred, some profane, but perhaps above all they are records of a yearning that was never fulfilled, on one side, and of a kind of sorrow-mixed-with-resentment on the other.

Let's return to Miguel A., a subject of both of Kaplan's and Adair's, and the man urged to reveal his no-matter-how-trivial childhood memories and Katchina recollections. The story of his participation is a kind of parable for the curious inexhaustibility of the quest for secrets. Complaining that already other Zuni called him "newista" or "white-lover," Miguel was not initially enthusiastic about giving interviews, but Adair succeeded in getting him to participate, likely because Miguel's wife, apparently one of the least shy women in Zuni and someone eager to cultivate anthropological liaisons and cash flow, brooked no opposition.[44] Egged on, he agreed to tell his life story at a rate of a few dollars per day. Yet after most episodes accounting for his early years, he would ask doubtfully, "Is that the sort of thing you want me to tell you?" and in describing his childhood and the Zuni initiation into tribal dance cult, he was halting in delivery and affect, mannerisms that Adair found less than satisfactory.

Things changed when Miguel left town for sheep camp and allowed Adair to follow him. Quite literally—for the anthropologist shadowed the Zuni as he worked, aware of the oddity of what he was doing: "My technique of recording had to be adapted to the situation. The anthropologist following the Indian around, clipboard in hand, while he is herding sheep represents a rather [amusing] image, but it was the way to make the best of the situation."[45] Away from the close confines of Zuni Pueblo, they were chummier, and Adair's own Zuni-bred anxiety decreased: "I had not realized how tense I had become as a result of living constantly under surveillance," he recalled years later.

At sheep camp while working during the days, Miguel divulged many things, including how, after their first furlough from the army, new recruits marched together to the station: "Everyone was drunk—some even drinking right then as they were walking along. We got to the train, all of the Zunis sat together. I sat next to Wilson, my cousin. Across from us was that old bald-headed barber. . . . I could hardly recognize those Zunis when they got their uniforms on." Miguel then shipped to England, first to a village, Little Stockton, twelve miles from Bedford ("all those chimneys in those houses, all just alike in a row"), where he trained in aeronautic mechanics and rode an English bicycle, and then to London. He was much struck by how the black-clad British workers took their tea at ten o'clock and two o'clock, squatting on the cement with their funny old-style jugs. He told about a Scottish factory girl he dated whose grandmother gave him Scottish napkins and lace handkerchiefs to send home to his family and whose brother played marbles with him. They were friendly to "Yanks," as they had visited the United States and been kindly treated in the past, and it is interesting that Miguel here in a Scottish village had his first experience of being seen first as a Yank and only second as an Indian. He discovered that since African American soldiers had often introduced themselves as tribal Indians, Europeans tended not to believe he was Zuni. A girl he met at a fair assured him he couldn't possibly be, as he was "too light."

As the war came to an end, Miguel saw Bob Hope ("not much of a show") and, on V-E Day, women so joyful they threw off their clothes and jitterbugged naked in the street. With two hundred other men he took an airplane tour over the zig-zag trenches, the ruins of Cologne and Aachen ("skeletons of buildings"), and then flew over Auschwitz. He remembered how it looked from above: "We hit, what's the name of that famous concentration camp? We could see long grey buildings with iron fence around, people milling around in there." On return to New Mexico, Gallup seemed to him like a little village, and "all those Navaho and Indians around" seemed strange; after some time avoiding Zuni, he finally made his way back.

One night at their campfire, accompanied by whisky ablutions, Miguel finally spilled, in Adair's view, all his remaining secrets, those "deeply private beliefs" he held and the truth of his experience as a Zuni, an event that filled the anthropologist-friend with elation, Adair recalled decades later. Drunk as he was, he made a mental note to write everything down, but he failed to. "That night, as I crawled into my sleeping bag, I had an exhilarating sense of having gained a profound new insight into the mind and heart of [Miguel] and through him into the quality of life in the pueblo." However when he awoke the next morning, "All had vanished except the memory of the intense excitement I had experienced. Even this was a hollow thing." His disappointment was alloyed only by the hope that Miguel likewise would not remember how much he had divulged of himself. Back in Zuni proper some weeks after the exchange of these intimacies, Adair saw his informant in the trading store, where Miguel was with some friends. Adair greeted him, but Miguel looked through him "without a flicker of recognition."[46]

Even as secrets embody their own potential loss, posing the question of whether a secret is still a secret after it's revealed, Kaplan's and Adair's project ran on a strong countercurrent of gain, one that is familiar over the course of Zuni-outsider relations: *Look! Secrets are taking solid form.* As if suspended in amber, they sit in field notes, data sets, and eventually a data archive even as they apparently evaporate from Zuni daily life itself. The process is never complete: the thing that is possessed somehow never fulfills the promise of the thing that was desired.

This trickiness is why, according to scholars, the "long term history of secrets" over the past thousand years remains a matter of fierce debate, and it is not even clear whether, overall, secrecy has decreased or increased.[47] Yet most agree that the Cold War was a key turning point for secrets of many kinds, including scientific secrets. As the result of World War II's atomic weapons debut, the very nature of hidden knowledge altered for good. During the previous world war, military knowledge had a "finite lifetime"—it would not always be necessary to conceal the location of access points to a munitions factory or the hotel where General Pershing was staying. With new atomic bomb capabilities, the need for secrecy was unending, and "the eternal threat of even a crude nuclear weapon gave the 'born secret' doctrine a new meaning." Hatched into secrecy, such knowledge has a putative eternal life. This was a new way of conceptualizing secrecy that persisted even after the knowledge of how to build a nuclear bomb had spread: it marked the emergence of a new "ontology of secrets."[48] Aside from more direct policy effects of this new doctrine, it seems that even in remote areas (such as Zuni Pueblo) and within supposedly unrelated data-seeking enterprises (such as Kaplan's archive) the structural shift in the quality and half-life of secrets made itself felt.

In these years of anthropological-psychological research, secrets became a kind of scientific currency, as we have seen. Until around 1934, they fueled an intricate Zuni economy of knowledge. Gradually and sometimes abruptly over the next decade, a transition took place. Denuded of their esoteric quality and exposed in popular newspapers as well as anthropological journals, they became ordinary things; yet the fact that they were now rendered as *things* that could circulate in a neutral, non-Zuni network was significant. It gave them permanence while stripping them of their specific potency. Of course, the breaching-of-secrets process preceded Kaplan's and Adair's visits to Zuni and would continue after their departure, and indeed neither can be said to make a mark in the *Who's Who* of researchers attempting to penetrate Zuni ritual knowledge. But they both, and the second-generation culture-and-personality school (as discussed in the next chapter), did exemplify this one new thing: the fine-grained search for the data of the inner life, accompanied by the push to render it—technologically—in properly usable form. Recorded intimacies became permanently stored secrets. These in turn circulated. The process, ever more technologically abetted, ever more penetrating, continues. Zuni professionals now manage their tribe's sensitive knowledge (such as through the Zuni Heritage and Historic Preservation Office) and request that a visiting ethnographer's research design win tribal approval in advance. As Gwyneira Isaac describes it, "Anthropologists at Zuni can clearly be seen to be working for Zuni rather than on projects propelled by external interests"—for example, anthropologists work as expert witnesses in legal land claims cases. "From a Zuni perspective, these new methods not only recorded their knowledge but also absorbed it." In contrast, the conversion-of-secrets process "disregarded the Zunis' responsibilities as teachers."[49]

In the late 1970s the esteemed Chicago anthropologist Fred Eggan, who would contribute his collection of Tinguian and

Ilocano (Philippines) dreams to Kaplan's data collection and whose wife Dorothy was the foremost anthropological "dream" expostulant of the day, wrote of the Zuni that although they had been subject to "intensive investigation" over past decades, not everything there was to know about them had been uncovered. "There is much that remains unknown," he observed, "particularly in regard to the inner world of the Pueblos and the meaning and significance of various aspects of ritual and ceremony."[50] The hearts and minds of others as manifested in their "subjective materials" were a final frontier, if an ever-receding one. This statement highlights a paradox that might be silly did it not deliver, in the dynamics of its playing out, so many ongoing and often deleterious or disastrous consequences. The urge to know secrets to the extent or at the very edge of knowability, to know things one has already discovered but to know them further and deeper and again, to ferret out the "inner worlds" of those whom one has (for centuries) seen hounded for other, if related, reasons—this is the paradoxical dynamic that revealed itself, too, in Adair's apprehension of full omniscience. Likewise, it appears as part of the motivating force of Kaplan's archive. The more one thinks one knows, the closer one comes or believes one has come to perfect knowledge, the more fragile, dream-like, and perilous does that summum become.

By the time Adair's dream dissolved in campfire ashes, Kaplan was back in Cambridge writing up his dissertation. There in the hothouse environment of Harvard, where the postwar social sciences attained perhaps their most exotic fullness of foliage, he would finally have his "seminal idea," which allowed him to go beyond Zuni or any particular place.[51] He would become obsessed with a new way to target these inner worlds, a true "beyond" of knowledge. He would declare with hope and a measure of glee that perhaps all behavioral and psychological data—records that lay unattended, uncounted, unused, disvalued, or improperly stored—could one day find a home in a centralized, hi-tech clearinghouse of his own devising. All the difficulties Kaplan well knew, and would soon know better, attended this quest: the evanescence of dreams, the trickiness of data, and the evident fact that the whole enterprise curiously resembled an approach to and tilting at windmills.

CHAPTER FIVE

Possible Future Worlds

WITHIN a year of returning from the southwest, in 1948, Kaplan began to entertain a grand plan. The past summer, while going house to house and field to field in four cultures, he had tried to secure full data from a range of individuals (including recently returned veterans from World War II), but now he had a bigger idea: to assemble the same kinds of data from a "representative sample of the world's societies." Instead of sampling the limited number of cultures he had originally studied in a limited area, he imagined a worldwide sample. He realized that he was ideally placed to build a new type of encyclopedia—that is, to go about "collecting, preserving, and making generally available selected categories of primary research data in anthropology and psychology."[1] In an ambitious turn, he set about making an ur-data set of what it means to be human in all the different places humans had set out to live.

Kaplan's shift from Indian-veteran-centered problems to data-centered problems was a classic symptom of a condition historian Joel Isaac has dubbed the "Harvard complex." Harvard in those end-of-the-war, early-postwar years was a place of ferment and even revolution in the life of intellectuals. Isaac tells the story of the "interstitial academy" that since the 1780s had taken root at Harvard, although not in a planned and technocratic way but in a "piecewise" and "haphazard" fashion that had resulted, by the post–World War II period, in a practice-based approach to theorizing the human sciences, a "hands-on epistemology."[2] Isaac offers a counter-narrative to the standard view of postwar American social science as a place of conformist ideology urging degrees of increasing unfreedom—a view advanced by a generation of protestors and disaffected former students, such as C. Wright Mills, who attacked Harvard's brand of "Grand Theory." Society, according to the grand theories of Talcott Parsons (one of Bert Kaplan's teachers), operated to maintain equilibrium at any cost—and, for some critics, that meant even at the cost of freedom. In describing society's controlling procedures objectively, social scientists such as Parsons appeared also to advocate for them.

Yet according to historians undertaking new research on these problems, advancing the blind conformism of the "organization man" was hardly the goal of leading-edge postwar American social scientists, who tended to focus instead on methodological newness. As historian Jamie Cohen-Cole notes, their shared ideal was not to close minds or close society but to cultivate an "open mind"—in short, a set of virtues such as autonomy, creativity, rigor, objectivity ("cognitive virtues sibling to free thought") that made up mid-century American liberalism.[3] To be human, for this set of actors, increasingly was not just something taken for granted but a *project* and a lifelong endeavor. This sentiment found its perhaps most poignant expression in the "Man: A Course of Study" (MACOS) social studies curriculum devised for fifth-grade children some years later, the goal of which was to ask, "What is human about human beings? How did they get that way? How can they be made more so?" The unintended irony was that in putting humanness on the curriculum in this way, they raised the possibility that some were more human than others and the further likelihood that being human might require how-to's, blueprints, and "exercises and materials."[4] In other words, it put "the human" in question. If, in these years, the pursuit of human virtues through sophisticated tools and technologies sometimes led to moral contradictions and a destabilizing effect, then that, in turn, is also part of the events to be narrated here.

At Harvard immediately after the war, problems once confined to desktop, armchair, or laboratory now circulated among these scholarly places and policymaking and military-government

spheres. As a result, researchers in the social sciences, specialists in
the fields of human and social life, felt their collective time had
come. The first war had been the chemists' war, the second war
was the physicists', and the "next one," as future Nobel Prize–
winning economist Paul Samuelson announced, would belong
to the social scientists.[5] Experts in the realms of psychological warfare, coercive persuasion, attitude measurement, cultural influence,
and the workings of creativity—to take a few examples—saw their
academic stock rise.[6]

While the European and Japanese social sciences stagnated in
these years, underfunded in the wake of war and reckoning with a
catastrophic failure of the human moral compass, the booming
United States energized its activities in almost all these spheres,
with scientists taking on subjectivity wholesale. A science of understanding how human subjectivity worked, pursued through the
application of penetrating technologies (for example, the sensory
deprivation method, the lie detector test, or the ever-increasing battery of projective psychometrics) and creative hybrid machines (for
example, punched-cards run through IBM machines devoted to analyzing soldiers' or consumers' behavior), was being born. This shift
marked the surge of an American techno-empiricist approach over
Old World methods anchored in philosophy. Across the United
States, the migration of many Jewish European social researchers,
such as the Frankfurt School alumni, aided the transition, some
setting up in New York at the New School for Social Research
(Claude Lévi-Strauss), at Columbia's Bureau of Applied Social
Research (Paul Lazarsfeld, Herbert Marcuse, Paul Massing, Leo
Lowenthal, Friedrich Pollock, and Franz Neumann), at Berkeley's
psychology department (Else Frenkel-Brunswik), and elsewhere,
including Princeton and Stanford. Meanwhile, Japanese social
sciences adopted "postwar modernism" *(kindaigshugi)* and soon followed the American lead, aiming to create a new human type, *Homo
democritus*, and European institutes too moved toward modern value
creation.[7] Emigrant scholars joined together on the urgent imperative to craft an anti-totalitarian personality here—that is, in the
United States, where an incipient totalitarianism was also discernible.[8] "Never was there a place where freedom was so much an illusion," wrote the monastic Thomas Merton to his friend the poet

Czeslaw Milosz, another émigré from totalitarianism, bitterly describing America's faults in a letter: "You will find here no imagination, nothing but people counting, counting and counting, whether with giant machines, or on their stupid fingers. All they know how to do is count."[9] Yet some of the foremost social thinkers wanted to use giant machines and clever devices to stimulate and explore the imagination.

The charged environment extended to Harvard, even if war exiles generally did not. So it was that in the spring of 1945, four innovators from Harvard's social sciences voted to name a new department and its accompanying laboratory "the Department of Social Relations" (DSR, or Social Relations) and Laboratory of Social Relations (LSR). Pulling from four core fields and moving to dissolve the boundaries of their respective home departments, the new creation would combine sociology, social psychology, cultural anthropology, and clinical psychology.[10] A push to build a "basic social science" animated the four founders, and they saw themselves joining forces to meet the "urgent and increasing demands ... for the study of the 'human factor' in a technological and atomic age," as Harvard's inaugural announcement put it.[11]

What was the human factor? How was it to be studied? On one level, the human factor was a terminological solution for a persistent problem that arose in the twentieth century human sciences: how to build a system or theory that took human qualities into account—and, ideally, one that could also calculate with them. In this sense, a human factor suggested something like a living number or a machine-like human, something that could be the object of scientific manipulation even as it expressed the workings of the subjective realm. Today the field of human factors psychology focuses on ergonomics, workplace safety, human error, product design, human capability, and human-computer interaction. It is a multidisciplinary field maximizing usability and increasing efficiency wherever applied. In 1946, the human factor was a signpost for a more diffuse approach to research. At Social Relations it was a priority to study the human factor via the most sophisticated new methodologies, including data crunching. What they did not fully reckon with was that *to study* the human factor in this way was necessarily *to change* it. Representing was also intervening: this was the increasingly simultaneous way in which the exercise of scientific rationality worked.[12] Finding the human factor within immense fields of extracted data would itself have consequences, as we will see.

During the next two years, while Kaplan himself was a student there (like many, on the GI Bill), Social Relations became an innovative place for what key figures saw as a "marriage" between grand theorizing and social-scientific experimentation, a pairing others had imagined and toyed with in the past. The department would do the theorizing while the laboratory would carry out the empirical groundwork and make itself a "seedbed" of experimentation— hundreds of experimental studies in all, funded by Carnegie monies. A graduate student who was spending a semester at Columbia's Bureau of Applied Social Research wrote a letter to Talcott Parsons reporting that "Much interest [here] focuses on collaboration between yourself and Stouffer"—the head of the LSR—"and the questions asked, if elementary, still get to the point ('Do they really talk' etc.). . . . I have handed out much re-assurance on this score, so that they do expect some theory-research marriage to bloom in Cambridge."[13] Experimental ambitions were on the rise in tandem with the feeling that a shot at a unified theory of social life was possible.

Tremendous confidence held sway. Once denigrated fields such as social psychology and cultural anthropology entered a heady and infectious state of ascendancy during the late 1940s and early 1950s. "For a short period after World War II," one witness recalled, "the social sciences experienced remarkable growth in terms of numbers, funding, prestige and influence. Social scientists were cocky and confident. . . . Social scientists became needed and they wanted to be needed."[14] At Social Relations this feeling was perhaps more intense, abetted as it was by Harvard's perennial Harvard-centrism (If not here, where?) and the correspondingly high levels of support it was attracting to the new intellectual creation. From the Carnegie Corporation came a total of $335,000 in the first decade of Social Relations' existence, with $275,000 of that going to the laboratory side to support informal, interdisciplinary pilot studies and experiments in methodology. The Office of Naval Research sponsored studies in the psychology of perception under Bruner and others in sensory deprivation.[15]

One of the progeny of the theory-research marriage was Kaplan's experimental data series. Initially, during Kaplan's intensive summer of data-gathering in 1947, the only reasons he gave for expanding his testing procedures during the frustrations of Zuni had to do with time saving and some glimmer of an idea that "conceivably" a full set would be of interest in the future: "At Zuni much time was wasted when young men were not on hand but older men, women and children were. Although I was not directly interested in the latter"—results from women and children—"the additional tests might conceivably be useful in the future."[16] His direct interests at the time were in the attitudes of returning Indian veterans (whose Rorschach test results Evon Vogt would use in his dissertation and a subsequent Peabody Museum paper). Kaplan's psychological batteries in four cultures turned the stuff of personal experience into "raw," reusable data, and although he initially gathered these for the purpose of asking a specific question about Zuni and other American Indian assimilation, he became convinced that research questions could be better asked on a larger scale. His adviser, Kluckhohn, was one of the Social Relations founders and (according to some accounts) was vying with another founding figure, Parsons, to see whose enterprises could arrive first at a working version of basic social science.

Eager to push on with this unifying project, Kluckhohn greeted Kaplan on his return from the field with the observation that Kaplan was bearing with him what might be the most *personal* data set so far in existence, for it contained materials seen as directly lifted from the inner life, veritable X-rays of the self, as many referred to these projective test results.[17] Given this unique array of data and its probing nature, could more ambitious questions be asked of it? Could the data set itself be reconceived? Kaplan began expanding his prospects, eventually seizing on a plan to capture, if possible, the entire range of psychological, sociological, and anthropological empirical documents in the form of data sets psychologists, sociologists, and anthropologists had collected in previous decades. Promoting a capacious and reliable science of the varieties of the human being was his goal: "*We intend our research to be one step toward a future science which will give biological, psychological and cultural forces* their proper places without the need to regard any one of these as a function of the other."[18] He began to conceive of a massive "file": not just a free-standing file in a cabinet but a raw-data "repository" that would be made "generally available" to researchers.

Kaplan envisaged a science that dealt with what it meant to be a human being of one kind or another, one that could, one day, answer fundamental questions. As he wrote to Kluckhohn,

> The development of the file of personality materials should facilitate research in the field of culture and personality. These materials, collected at much expense, are of virtually no use to science since they are in the possession of a small number of workers, most of whom have no intention of further using them. They certainly cannot at present be used for cross cultural comparisons except on a very minor scale. We may expect that as time passes the job of reclamation will become more and more difficult. Finally the file will be a repository which will receive and organise for cross-cultural comparison, the much richer and more sophisticated materials which will undoubtedly be collected in the future. The file could conceivably play an important role in stimulating research by providing rich and plentiful empirical materials in a field noted for the brilliance and sophistication of its theories but the paucity of its sound empirical research.[19]

These data already existed in back offices and filing cabinets but were in danger of remaining in back offices and filing cabinets or worse: being thrown away. What was necessary, Kaplan felt, was to build "channels through which the holders of these data can make them generally available to their colleagues."[20] He wanted to built a data pipeline and infrastructure to accommodate the "plentiful empirical materials" that were already in existence but threatened with the stasis of non-circulation and therefore in need of "reclamation."

In addition, the direct experience of administering psychometrics contributed to the way in which Kaplan's experimental data-collecting project first took shape, freshly arrived as he was from his intensive test-giving expedition. Hard-won knowledge about the challenges of getting such difficult-to-access data fed his desire to make them accessible to those not able or interested in taking the train to Hopiland or shipping off to the South Seas. "If, for example, someone wants to work with New Guinea T.A.T.s it is unlikely that he would consider making a special trip to New Guinea to get them, if he had easy access to a good sample collected by someone else," Kaplan would urge some years later in the *American Psychologist*, at that moment speaking more for psychologists or sociologists, who would tend to value data sets abstracted from fieldwork, than anthropologists, for whom the experiential component even of test-administration was more valued.[21]

In Cambridge that fall, Kaplan met with the DSR's gray eminences, including the TAT co-inventor and recent OSS officer Henry A. Murray and grand theorist Parsons, as well as Kluckhohn. Murray, the doyen of the "apperception" test (one of the four tests Kaplan brought to the southwest) was a natural source of advice to Kaplan, although the senior psychologist, with his Herman Melville literary obsession, self-mythologizing temperament, old-line Protestant family background, and recent wartime success in testing candidates for espionage and leadership mettle, did not seem to appreciate Kaplan's scholarly or personal style. (Kaplan also consulted with prominent social researchers outside of Harvard, including Margaret Mead, Otto Klineberg, and Ralph Linton.) But initially Kluckhohn was most influential, partly because he singled Kaplan out with his not inconsiderable charm. As Hermia Kaplan, Bert's wife, recalled: "I just fell in love with Clyde; most everybody loved Clyde." And in turn Kluckhohn thought a lot of Bert. (He also hired Hermia as his assistant despite her lack of typing skills.) The senior anthropologist had helped pave Kaplan's way at the Navajo reservation near Ramah, where the mere mention of his name had a near-magical effect. After alluding to his Kluckhohn connection at a town meeting, Kaplan found himself immediately swarmed by over a dozen people "in the most cordial fashion," behavior that was "refreshing in contrast with that of the Zunis."[22] A prolific and somewhat mysterious man—a man, as it turned out, with secrets—who trained in anthropology but exhibited in his career a powerful force-of-nature approach to many neighboring fields, Kluckhohn extended his reach from Social Relations during the early Cold War years to run the Harvard Russian Research Center, despite never having done a stitch of Soviet research. He had, to borrow a characterization from Dickens's Mr. Skimpole, "a strong will and immense power of business detail."²³ Hermia's fond estimation—"Kluckhohn was amazing. He was part of the upper echelons at Harvard. Clyde was a wonderful snob"—also begins to evoke the magnetism he generated within the Ivy League.

In welcoming Kaplan back from Ramah, Kluckhohn took a special interest in the young scholar's research, for he had been staying with and studying the Navajo on and off since he was nineteen and had even written novels about life there, the first called *To the Foot of the Rainbow: A Tale of Twenty-five Hundred Miles of Wandering on Horseback through the Southwest's Enchanted Land* (1927). Recently republished in the series Equestrian Travel Classics, the book features as a tale of a man who (according to breathless promotional language) "had a chance, a rare chance, to forget that he was born to ride a desk, not a New Mexican bronco. He had a chance, a rare chance, to turn his back on convention and schedules, wrist-watches and bills, misspent romance and a thousand other heart-aches."²⁴ Footage of a film made on one of Kluckhohn's trips shows skinny-dipping adventurers engaging in youthful hijinks (five recent college graduates, all to become professionally renowned in future years) and burros laden with gear in a dusty backcountry.

By 1950 or so, Kluckhohn had returned to more desk-bound activities, and having long since given up novel writing for scientific and administrative pursuits, he took Kaplan's project in hand. The data were excellent and Kaplan had secured them ingeniously, but the veteran angle was perhaps less compelling than the chance to test something fundamental about cultures and human personalities, he felt. As the war receded from immediate memory, returning American Indian GIs would likely soon resolve whatever initial conflicts with their own cultures had existed. But the postwar world would offer other challenges—how to understand cultural conflicts, most pressingly. As Kluckhohn wrote in his 1949 award-winning pop-science "manifesto" for anthropology's new role, *Mirror for Man*, the science of culture was "no longer just the science of the long-ago and far-away" but was a guide to the central challenges of modern life: "Of course, no branch of knowledge constitutes a cure-all for the ills of mankind," he admitted to an interviewer shortly after *Mirror*'s publication, but "anthropology provides a scientific basis for dealing with the central dilemma of the world today: how can people of different appearance, mutually unintelligible languages, and dissimilar ways of life get along peaceably together?"[25] Here after all were four sets of Rorschach tests extracted from four vastly different and at times conflicting cultures dwelling near each other, each set administered by the same individual. Kluckhohn saw that here was something unique and unheralded. The sets were *comparable*. This meant that a researcher could not only test a hypothesis in a clear and deliberate way, but also look for a common human substrate.

Within the Harvard force field, Kaplan made a turnabout. It was the first of several steps that would lead him to the creation of a data clearinghouse. Crediting Kluckhohn with "enabl[ing] me to leave behind many concepts of dubious value," Kaplan decided to take on what was perhaps the most commonly held but least commonly tested assumption in the field where psychology met anthropology, "the prevailing idea that there are wide personality differences between cultures."[26] Indeed, in going over the summer's Rorschach data of Zuni, Navajo, Mormon, and Spanish American cultures, he was finding more variations among individual personalities than expected. "My interest was therefore reoriented. . . . Critics had pointed to a certain looseness in both the theoretical and empirical work in this field"—the field of culture and personality or, as it was later known, psychological anthropology and cross-cultural psychiatry—"and it seemed within the realm of possibility that the influence of culture in creating modal personality trends has been greatly overestimated." Here Kaplan was criticizing, if rather mildly, the first generation of psychological anthropologists—Ruth Benedict prominent among them—who had characterized the "cultural patterns" they found as one pattern per culture and one culture per pattern. Indeed, Benedict worked with such a broad brush that it appeared at times that every Zuni person must somehow conform to the wise Apollonian type and every Kwakiutl be at root paranoid. When some of her peers, such as Edward Sapir, took issue with this broad approach, claiming she depicted "the passive enculturation of a personality essentially isomorphic with its magnified cultural counterpart,"[27] this did not stop her book from selling millions and her ideas from spreading among professional social scientists at a similarly wild rate.[28] At the root of culture and personality's movement lay this conviction: no matter whether you saw passive stamping or active co-creating of personality types at work, you likely believed that these *types* did exist in a dominant pattern.

This conviction was exactly what Kaplan had come to doubt. The discovery that the central assumption of his field could not be properly assumed would be, if Kaplan could make his proof persuasive, big news. After all, almost the foundational view of such studies was that group members would be similar in predictable ways. One influential theorist, psychiatrist Abram Kardiner, labeled this the "basic personality structure," a set of impulses and patterns (activated in child-rearing) that functioned almost as a recipe for self.[29] Most people in a culture or nation shared to some degree the same recipe, although even culture-and-personality studies deemed successful at identifying an existing modal personality within a single culture, such as Anthony Wallace's dissertation work with the Tuscorora a year or two after Kaplan's fieldwork, identified only a 37.2 percent rate of positive incidence.[30] Yet surely such basic cultural commonalities existed and manifested themselves in particular people, else Montaigne could not have written his essays nor Shakespeare his plays—and, for that matter, it would be hard to make half the jokes available to humanity, which usually rested on cultural generalizations recognized to be, if not uniformly true, at least anecdotally identifiable and experientially verifiable. Scientists only wished to make these commonsense insights more reliable and precise. French personalities would no doubt be more "French" and in ways that would ultimately be susceptible to scientifically *exact* representation through scientific tests and instruments.

Thus when Kaplan's study began, the going assumption was that one should be able to read an anonymous Rorschach or other document and say, "Here is a Zuni" or "Here is a Navaho, . . . a Bernese Swiss . . . or an Appenzeller Swiss." Similarly, one should be able to look at a culture and discover the kinds of personalities it would promote and also look at a single personality to see more deeply into the culture. One should even be able to scrutinize the culture that gave birth to Adolf Hitler and analyze him at a distance—know his secret weaknesses—via a close study of his milieu and its patterns, as well as any personal clues that could be gleaned. In 1942 and 1943, several of Kaplan's seniors in the culture-and-personality movement did just this with government sponsorship: Mead and Bateson crafted a culture-and-personality analysis of "Hitler's peculiar psychological makeup" for the army's Research Branch. Erik Erikson, steeped in culture and personality from the angle of a Freud-trained Danish Jewish psychologist, did likewise. And one of Kaplan's teachers, Murray, in 1943 presented the OSS with a portrait of Hitler's psyche, predicting correctly enough that he would commit suicide if faced with defeat.[31] Imprinting was the most common force in creating cultural regularities, and these regularities, surely, made themselves felt in the innermost core of the individual, the personality.

Kaplan took aim against this whole set of assumptions. Despite the fact that he held Kardiner's "basic personality structure" in high esteem as a concept, he felt its proponents tended to weight the "basic" part too heavily.[32] To be sure, certain commonalities did arise in his data. For example, three of the four cultures in his study—Zuni, Navajo, and Spanish American—could be said to be oral-dependent in a Freudian sense (although definitely not the Mormons), he felt. Still it was the non-commonalities, the amazing diversity registered *among* the people who made up each group, that Kaplan felt was more striking, perhaps because no one was looking for them.

As it happened, this contrarian idea was one of Kluckhohn's pet theories, "the problem of the heterogeneity of persons." Indeed, Kluckhohn had in 1945 taken a moment from his war duties at the Pentagon as co-chief of the Joint Morale Survey to make a complaint about then-existing anthropology. It had too often neglected to focus on what the individual was really doing or feeling at a particular moment—say, a Hopi who was attending a Snake Dance—in favor of generalities about what the culture was doing and how that (presumably) revealed a general cultural pattern: *this is what the Hopi do.* Could inner unity of experience be assumed? Wouldn't it be better to try to find out what a particular Hopi did or, even better, was experiencing? "In many monographs in the past," Kluckhohn complained, "statements were couched almost entirely in the general mode—'the Navaho do so and so' rather than 'I saw Navaho A, Navaho B, Navaho C, etc., do so and so' or 'Navahos X, Y, Z independently told me that they had seen Navaho A do so and so' or even 'I have spent eight months among the Navaho of such and such a region and under circumstances f, g, h I have never failed to see a Navaho behave as follows.'"[33] Kluckhohn attempted to address the problem in this case by adducing the evidence of Mr. Moustache, a Navajo elder who had dictated his life story one long day in 1938 to Kluckhohn. From this story Kluckhohn drew an analysis inspired by the "concrete sociology" circle at Harvard that gathered leading businessmen, sociologists, administrators, and interested others together in the belief that it was through concrete materials, as opposed to abstract theorizing, that true social facts and the precise mechanisms of "how cultural and social conditioning is carried on" could be known.[34]

It was also significant that this turn toward the concrete occurred during World War II, when men like Kluckhohn engaged in practical projects of an experimental nature and created, along the way, "vast new data bases of social information."[35] Nothing should be assumed about Navajo A, Hopi B, New Yorker C, or Mr. Moustache in particular, Kluckhohn felt. Instead, the social scientist must *find out for himself* and assemble models from there. There was a lesson, too, about data. Gather as many such "raw" documents as possible, was his urging, for each was an "authentic sample of concrete material."[36] Lest it seem that Kluckhohn was only a humble empiricist, however, it should be emphasized that his eagerness to gather up the details of A, B, and C under conditions of f, g, and h took place under a guiding theory, even a program, and was not lacking in ambition. Looking out from the cover of the January 29, 1949, *Saturday Review of Literature*, his presence in its pages conveyed that anthropologists now had "the beginnings of a science whose principles are applicable to any human situation whether it be international tensions [vide the UNESCO Paris conference to which Washington just sent him] or something on a smaller scale like the Navahos." Anthropology was akin to an all-purpose science that brought together vying ways of life and

apparently divergent values. Through anthropological assiduous-
ness, "competing ways of life" could be seen as actually unified, despite superficial differences, by "the principles that undergird each
culture." Such undergirding principles were nothing other than
humanness. In a sweeping claim that today may seem like sheer
chutzpah, Kluckhohn posed anthropology (spokesman: himself)
as the unifier not only of diverse cultures, but also of the social
sciences themselves, including geography, economics, sociology,
history, anthropology proper, and "even psychology."[37]

So it was not surprising that, finding Kaplan and his data cache
at his doorstep, Kluckhohn encouraged him in the "concrete" direction, as part of the larger unifying program.[38] Just after the war,
as the Cold War began, Kaplan turned anew to his data and rede-
fined their significance. Two things happened. First, he used his
dissertation to answer a question at the heart of culture and personality; meanwhile, as a second outcome, he began to think about
founding a raw data repository so that others in the present or future could use the pool of data he himself and a cadre of workers
were gathering—these would constitute the concrete, empirical
resources that would support an ever bigger social science.

Kaplan's paper for the Peabody Museum became a methodological
end run in the quantification of the personality—the elements most
easily quantified, that is. Despite admittedly "important drawbacks"
to this method, Kaplan wrote, it satisfied "a sense of the urgency of
the need for objective treatments."[39] "Objective treatments" of the
inscrutabilities of human beings, to be exact. Could personality and
what Bronisław Malinowski once called the "imponderabilia of
actual life," the subtle ways people saw and experienced the world,
be known statistically? Could scientists account for groups not only
as lumps of undifferentiated people, but also as very distinct, differ-
entiated beings: Navajo A, B, and C, for example? Kaplan thought
so. He framed the work, therefore, as an experiment.

In the experiment, "quantitative procedures will be applied to
the data." Unlike the full-pattern analyses other pioneers of psychological testing (Hallowell, Henry, Theodora Abel, DuBois) had
carried out in the late 1930s and early 1940s, Kaplan's Rorschach
results would be broken down into single variables, not interrelat-

ing variables. The prevalence of certain countable things—number of responses per card ("productivity"), for example, or types of animals seen in the cards—could now be ranked and compared. How many glimpsed grasshoppers in the plates? (One Zuni man only.) How many saw cats? (Six Navajo, eight Zuni, three Mormon, and one Spanish American.)

What he found in answer to the conundrum—Were people within cultural groups more alike or more different from each other?—was not one or the other result but both. He did find some Rorschach similarities among the members of each group: a typical Zuni-ness, Navajo-ness, Spanish American-ness, and Mormon-ness, in effect. "The significant differences between means, while not very great in number, are nevertheless very definite, and they indicate the presence of real differences, however small, between cultures," he wrote. But he also found a lot of variation within any one group: "the variability of individuals in any one culture" was found to be exceptionally high. For Kaplan, this latter fact outweighed the former. It was more significant, to him, that there were strong differences between this and that Zuni person than that one might also find some commonalities—by which he meant more shared traits than the number predicted to exist between any two random people. In other words, although Kaplan found data supporting similarities and variations among residents of each culture, he chose to emphasize the differences. A broader postwar American preoccupation with individuality and the fostering of creativity, both of which were seen as distinctive of U.S. society and a bulwark against Communist facelessness, might help account for this predilection, even if Kaplan was going against the grain of scholarly consensus.[40]

In Cambridge, Kaplan learned statistics painstakingly and painfully (according to his wife, his statistical turn gave him a hard time). Yet these labors paid off. He turned out an impeccable, statistically supported argument woven into a (to some) rather shocking claim: one could *not* assume there was a basic personality wedded to any particular culture—or, rather, what stood out were not necessarily the cultural uniformities so much as the picayune individual differences. Note that this claim still in some circles counts as significant today. Indeed, psychologist Gregory Meyer and his co-researchers at the University of Toledo were using

Rorschach results across cultures to establish parallel results as of 2012.[41] Not only did Kaplan's argument take on the sacred cow of his field, but also his methods upended it.

The coup de grace of Kaplan's Peabody paper was the "sorting experiments." Up to that point, as Kaplan admitted, one could emphasize either the differences or the similarities found. One could choose, metaphorically, to see either the old crone or the young maiden in the data—that is, the culture-borne similarities or the individual-borne differences. Surely the sorting experiments would clarify things. Could a trained expert working completely "blind"—that is, with test results stripped of external markings and names—place Rorschachs from four cultures correctly into four separate hoppers? Kaplan charged Dr. Marika Rickers-Ovsiankina of the University of Connecticut with carrying out an analysis of 116 tests. Her sorting did not succeed. Knowing nothing of the cultures involved, she was unable to make four correct typological groupings. However, in a second round featuring Dr. Alice Joseph, who was more familiar with the southwest and who was informed of the four groups included, though not the individual testees' names, the sorting was on the whole successful. At least her success-to-failure ratio was significantly greater than chance.

Kaplan's answer was . . . yes and no, again. Yet his conclusion was firm: researchers had for too long ignored variations within groups simply as inconveniences in the search for typical or modal person-alities, giving "only the hastiest acknowledgment of such variability before going on to the apparently more important considerations of the ways in which cultures create personality uniformities." It was all for the good, then, to insist on keeping the variations in view also, and the paper ended with a "plea for better balance."[42]

Kaplan's dissertation was the first to emerge from Social Relations and garnered a lot of attention. Next Kaplan took all his data and synthesized them as the paper for the Peabody Museum. "Working on the data . . . with the idea of demonstrating the effect of culture on personality, I found it was getting very difficult to find anything like typical personality patterns in the four groups of Rorschach tests, there being too much variation within the groups."[43] So he wrote what he saw or failed to see. He concluded that the energetic cohort of anthropologists and psychologists who were looking primarily for modal personalities were on the wrong track. He contradicted the received view. In doing so, he was displaying against-the-tide mettle related to his lack of careerism. Later he would worry about position and his ambitions, but during this period, seized by the strikingly contrarian patterns in the data, he found himself unafraid to pursue a controversial finding, as he recalled much later in life: "In 1949–50 I felt so sure of myself that I disdained doing anything for the sake of reputation."[44]

Thirty faculty members attended his thesis defense—a big turnout, due to the splashy findings and the fact that it was the inaugural dissertation of the DSR. Most were impressed, but the projective-test advocate, Murray, who had analyzed Hitler and built good results predicting spies' and officers' behavior during the war, distributed a circular in which he said (in Kaplan's synopsis) "the work was 'Wrong, unimportant and dangerous and besides I thought of it myself ten years ago.'"[45] Yet others saw Kaplan as a trailblazer. So impressed was Kluckhohn that he brought up Kaplan's results during his presidential address to the winter meeting of the American Anthropological Association, a tremendous coup for a graduating doctoral student. Despite the naysaying of a few, then, Kaplan enjoyed his new renown. The thesis and the subsequent Peabody paper made his career. It "made him famous."[46] He got hired at Harvard too, teaching for a year at Social Relations before making his way to the University of Kansas as assistant professor.

The Peabody paper also stands as the record of Kaplan's conversion to an ecumenical, data-driven vision. For the first time he offered his work as a public service: "It seems to us that if someone were to make available the Rorschach records which have been collected in perhaps a score of cultures, culture and personality research might be greatly stimulated."[47] He was beginning to see that the future of data lay in their agglomeration. "The thing about Bert and his project, to me, was that it always showed a respect, a humility," recalled dream researcher G. William Domhoff, a colleague of Kaplan's. "We've collected all this, but we've moved on, but these things might be useful to others. Even talking to others."[48] It took a particular kind of scholar to advance the collective possibilities of scholarship, and an unknown future, rather than his own personal interests. Kaplan was interested in extending the ability to see not by adducing better theories but by amalgamating data.

Having turned in a new direction, Kaplan launched within the next years a whole new enterprise: making very, very small images of very, very large data sets.

CHAPTER SIX

The Double Experiment

K APLAN embarked with vigor on his project of gathering a global set of subjective data sets. In its reach and aims, as well as its improvisatory character, the project was at least distant kin to the "global jukebox" of which, in the same years, ethnomusicologist and folklorist Alan Lomax dreamed (but, faced with technological roadblocks, never built).[1] First Kaplan began the task of soliciting the desired data from their gatherers, building on the area of his own research. Among specialists in culture and personality by the early 1950s, it was almost a matter of legend who had gone where and when. Anthropologists Jules and Zunia Henry had lived among the Pilagá Indians of the southern Mato Grosso in 1936 and '37 to collect psychological records, thus becoming one of the first teams to use the Rorschach and "doll play" techniques among non-Westerners. Kaplan approached seasoned travelers like the Henrys to "donate" their data. They promptly agreed, at least in principle, as did almost seventy other social scientists—including, to name just a few on the "yes" roster, Anthony F. C. Wallace (with his Tuscorora Rorschachs), David Schneider (with his collection of sentence completion tests, biographies, and dreams from the Micronesian island of Yap), and John Honigman (with his set of projective drawings by Great Whale River Eskimo children). In making this untoward or at the very least unusual request, Kaplan, a significantly less advanced figure in his field, was asking experienced scholars essentially to give away their data. With few exceptions they acquiesced. In most cases, the researcher had already made use of the data in some way and had already published and interpreted them at least preliminarily. (Some of the pledged data sets, as events would have it, never made it into the Kaplan database, either because funding ran dry or the collector experienced second thoughts or a lack of follow-though. A few examples of such nonpreserved data include that of W. W. Hill, from the University of New Mexico, who had a set of Navajo dreams; Margaret Lantis of the U.S. Public Health Service, who held Nunivak Island Eskimo test results, mosaics, and drawings; and Alan. Kerckhoff of the Air Force Personnel and Training Research Center who had gathered Chippewa children's TATs.) For the most part, however, compliance and carry-through proceeded at remarkably high rates during the early 1950s, considering that donating one's data was more of a selfless than a career-advancing act.

Kaplan's mission was based not so much on the question of whether these amassed results were definitive proof of one theory or another as whether there might not be more questions to ask of the data in the future (these were the "future values" his committee would soon invoke) and whether other scholars might wish to examine the constitutive data. There was also a greater problem: what was to be the fate of the data themselves, all sorts of "primary materials," as they were called, in their post-publication life? This was not a pressing concern among most mid-century researchers, and anthropologists were known for piling their field notes in trunks stowed in attics or boxes moldering in corners of research labs. They could see the advantages of saving data, but how and where to do so was not where they wished to put their time and effort. It was what one did with one's data that garnered attention and got one tenure.[2] Perhaps, too, there was a tinge of nihilism, an "Après moi, le déluge" attitude toward one's notes, as dramatized in the final scene of Barbara Pym's 1955 anthropological roman à clef, *Less Than Angels*, in which the bitter Africanist Alaric Lydgate showily burns his trunkful of notes, which had lain fallow for decades, to mark his liberation from the bondage they represented. If setting a match to field data was not common, though, withholding them certainly was. Nonetheless, researchers hired freelance secretaries or their students (typing services sometimes funded by Kaplan, who already had some support from the University of Kansas general research fund) and sent in stacks of their direct-from-the-field data, now neatly formatted on the page. Within a year or two, Kaplan reported, the collection added up to "perhaps 20,000 8 × 10 sheets of paper."[3] The sheets of paper held test results (including standardized Klopfer forms and quantitative analyses), life stories, various autobiographical interviews, and dreams as well as hallucinations.

That Kaplan's project, its backbone the projectives he and others had collected, would intersect with the work of A. Irving "Pete" Hallowell seems almost foreordained for "in many ways this story of the Rorschach"—its cross-cultural use in non-Western societies—"begins with anthropologist A. Irving Hallowell."[4] Despite the liminal status of cross-cultural Rorschach work within a growing Rorschach empire, Hallowell's rise had been as sure-footed there as it was in anthropological circles. By 1948 he had "reached the top of the pecking order in both the AAA and the RI," as a student playfully congratulated him in a letter—that is, he was president of the premier anthropological association in America as well as of the Rorschach Institute.[5] He had risen high at the NRC as well, and by the 1950s he had long experience heading the Anthropology and Psychology Division there. He had originally been invited to join the group twenty years earlier by Edward Sapir, who started it and who also initiated Hallowell, at the time a dyed-in-the-wool antiquarian with a salvage orientation and a persistent interest in bear ceremonials, into the more self-consciously subjectivist and up-to-date culture-and-personality concerns; Sapir "started me thinking about this psychological business."[6]

Not surprisingly, the projective pioneer was among the first whom Kaplan approached. Sometime in 1954, looking for guidance, he visited Hallowell in his office at the University of Pennsylvania to discuss what he then informally called his " 'personality materials' project" and was favorably received.[7] Hallowell began paving the way for an advisory committee to be created at the NRC and signed on as its first and lead member. His standing and enthusiasm helped, and Kaplan was grateful: "May I say that your warm reception of this project in Philadelphia was one of the really gratifying experiences I have had and has meant a great deal to me."[8] Kaplan, named executive secretary to the Hallowell committee, kept him abreast of his funding efforts as they leapfrogged from the relatively modest University of Kansas General Research Fund ($2,500–3,000) to the more prestigious and munificent National Institutes of Health and National Academy of Sciences (NAS, the home of the NRC). In turn, Hallowell, no figurehead, sent out queries to leading lights in the field to gauge responses and canvass for support.

A storehouse of data, others' and his own, was now in Kaplan's possession, and since financial support was beginning to arrive, the question was how to manage it. Initially Kaplan, a newly appointed assistant professor, did not want to go it alone (for despite the backing of Hallowell, he was indubitably the prime mover of the project), and initially he attempted to preserve and store the archive he had collected by subsuming it within an already existing and decidedly gung-ho data-processing enterprise, the Human Relations Area Files (HRAF). Since the mid-1930s, this effort, based at Yale, had sought to file on standard-sized index cards the key information extracted from myriad specialist texts concerning "every culture known to human kind"—or a representative 10 percent sample thereof. Although the dizzyingly ambitious Yale group, under anthropologists George Peter Murdock and Clellan Ford, innovated by building a sort of "disassembly line" for taking texts apart and filing them, they did not aim fully to miniaturize the data contained in "the Files," also known colloquially as Yale's Bank of Knowledge. Instead, they used miniaturization only as backup storage for the "second-tier" member institutions in the information network they built starting in 1948, gluing microfilm squares hand-cut from rolls onto index cards and by the mid-1950s microfilming their file system on microfiche cards that were then disseminated to less prestigious participants in monthly by-mail installations. (First-tier entities such as Harvard, the Sorbonne in Paris, and the U.S. Department of State received full-sized copies of the files, housed in full-sized Grade B Remington Rand cabinets, in these years.) Most important, the Yale group targeted the solid and visible parts of cultures—rituals, materials, rites, flora, fauna, handed-down customs of all sorts—in contrast to Kaplan's interest in grasping the netherworld of the subjective life and its "personality materials."[9]

On the one hand, it was not surprising that the Yale group was interested in accessioning Kaplan's domain, and "In the spring of 1952, I [received a] most encouraging reaction ... from Clellan Ford, who indicated that the collection of materials might very well become an adjunct of the Human Relations Area Files," wrote Kaplan hopefully.[10] Ford asserted that the HRAF had long been interested in adding personality materials, and after Kaplan visited New Haven to talk it over, Ford requested funds from the army and navy to support Kaplan for two years as principal investigator. However (Kaplan noted without losing hope at this point), the requests "were, as I understand it, laid aside temporarily in favor of other projects which they"—HRAF brass—"felt were more pressing."[11] That this temporary sidelining turned permanent was perhaps inevitable, for Kaplan was at this time a fledgling professor, wet behind the ears, and the HRAF was, by the early 1950s, hosting high-level foreign policy experts who recruited the system in order to target emerging Cold War geopolitical "hot spots." World information was at a premium in government and military circles—at least totalizing information about social and cultural life in Vietnam, Korea, and parts of Latin America—and Yale's files had what politicians and strategists wanted. Dreams and dream-like data were off the menu, despite initial interest. Dropped by the bigger enterprise, Kaplan turned to rethink his plans for networking and circulating his data collection. He then contacted the local Haloid Company representative in Topeka, Kansas, to learn more about Xerography as the potential medium for his data but was unimpressed with its advantages.[12]

Next Kaplan took a step that can in retrospect—in spite of the fact that his project was on some level a failure—be called visionary. A visionary failure, perhaps. Urged on in conversations with the University of Chicago anthropologist Sol Tax, he turned what was at the time a single experiment (in large-scale data management of subjective materials) into a double experiment, one that included format at its core. Tax, a radical in politics and a long-time expert concerning Fox Indian society, was himself a micro-pioneer. He had founded in 1945 the Microfilm Collection of Manuscripts on

Middle American Anthropology, a repository that eventually held everything from Catherine Pittman's 1951 *Modern Household Utensils of the Aztecs* to the 1949 *Chol Dictionary* and other highly specialized works delightfully demoted in size.[13] Tax felt that all anthropological records, including field notes, should be micro-recorded and made immediately available. "I've really been disappointed that the micro-filming of field notes has not been taken up with more enthusiasm," Tax commented sometime in 1954 or 1955 to Kaplan.[14] He encouraged Kaplan to take a close look at something called the Microcard, a leading form in micropublishing that gave promise of becoming the shared standard.

With the incorporation of Microcard technology into Kaplan's database project, what was originally a single experiment in capturing ever more subjective materials became "an experiment with two independent variables."[15] The second part was in the reproduction process, which, as Kaplan wrote to anthropologist John W. M. Whiting, head of Harvard's Laboratory of Human Development, "lends itself to easy coding and cataloging," along with being convenient to use and arousing less "psychological resistance" than standard microfilming.[16] Kaplan—an unknown from Brooklyn, a surprise success at Harvard, and soon to be an up-and-coming psychologist at the University of Kansas—now unexpectedly joined a rich and strange history: the Euro-American obsession with rendering increasingly tiny the possible dimensions of a text or image.

Inspired by Tax's success at Chicago, Kaplan, now in Kansas, turned to the microtechnology to make use of the wave of innovation that, he predicted, would within two or so decades allow "quick access to any information within the realm of psychology."[17] Initially he chose the Microcard arm of the University of Rochester Press and worked with the press on a system of coding and cataloging, but he ultimately found that the Microcard Corporation in LaCrosse, Wisconsin, was willing to fund a larger run in return for the guaranteed sale of at least thirty copies to libraries and academic departments. Overall, Microcard offered distinct advantages over microfilm (which was used largely for large runs of newspapers), for as Kaplan wrote to his friend Whiting, "I think this process is not to be compared with microfilm reproduction," as it was easier to navigate, code, catalog, and sell to potential users.[18]

Within a few years, Kaplan and the committee he soon brought together built a project characterized equally by sheer ambition (for it aimed to be a universal holder of social scientific data) and a hodge-podge scavenger's sensibility (for he targeted "a class of materials that is ordinarily confined to file cabinets, storage boxes in attics and basements, and, regretfully, occasionally to wastebaskets"). Kaplan's goal was to transform what others saw as trash into what he abidingly saw as treasure—the discards and leftovers, the neglected documents once earnestly pursued and subsequently forgotten: "I refer to the original research data of psychological studies, the whole body of empirical observations upon which psychology is based," he wrote in the *American Psychologist*.[19] He aimed to rescue all the at-risk data that lay strewn around in offices and unvalued in dusty files. Shrunk down to teeny-tiny size yet legible through the operations of a desktop Readex machine and saved on Microcards, such a repository would be available to researchers anywhere by "quick access" methods.

New Microcard pocket readers had just come on the market for $25 "so that work with microcards need no longer be confined to libraries," the group announced in a press release.[20] The portability of readers added to the project's sense of unlimited horizons. One did not need to go physically to Alor or the Gran Chaco to gain access to direct, inner-life data any more, and "the new publication will make it physically possible for a great many additional workers to do research" in the field of personality and culture.[21] A congratulatory note about the project came from Ray Birdwhistell around this time and indicated that the polymath ecological anthropologist was perhaps the project's ideal appreciator. In the 1950s and 1960s, Birdwhistell had a hand in many ambitious archival projects, ranging from collections of "microcultural events at Ten Zoos"—families feeding elephants, filmed via his "kinesics" method for super-close analysis—to dance steps archived from all over the world.[22] Birdwhistell congratulated Hallowell, and by extension Kaplan, on "your exciting microcarding project," in which he was "of course interested." His department in Buffalo would surely want a copy. "We have needed this badly," he signed off.[23] To some, the value of a "database of dreams" was self-evident and long in coming.

The data compilation could act as a remote-viewing technology—remote viewing the province of the self. In particular, Kaplan was very taken with the hand-held reader, which was portable and would allow, if not deep absorption in the data at any moment, at least quick reference while on the go. Research on readers was ongoing at the Microcard Corporation's West Salem, Wisconsin, labs, and there was even preliminary success making micro-spectacles to be worn as super-magnifying glasses called "Spectacle Readers," outfitted with glass, plastic, or multi-element lenses (though "some difficult optical problems remain to be solved," according to a 1960 assessment).[24] Lodging Kaplan's stores of data, collected from colleagues spanning the globe, in such an innovative format, the site of tremendous growth during these years, added an exciting dimension to the project.

Moreover, in a sort of reinforcing feedback loop, adopting the then cutting-edge technology of Microcard, hand-reader, and Readex machine confirmed and strengthened the group's commitment to experimentation in format. The members made a point in their minutes to make "explicit recognition that techniques of record keeping were deeply connected with the main concerns of the committee."[25] Choice of technological format became central to the project itself, as well as to the committee's thinking. From the outset these data enthusiasts discussed how new tools (such as the electronic calculator) made possible "new analyses of data collected in the past and now put aside."[26] The Microcard made the past's preservation possible for future use.

In December 1955 during the annual American Anthropological Association meeting, Kaplan and Hallowell convened a wide-ranging group of social scientists and technologists who met in a Boston hotel room to map out a future path for salvaging and amalgamating scientific data. Calling themselves the "Committee on Primary Records," under the aegis of the NRC, they concerned themselves with posterity. Present that day were the foremost psychologists and anthropologists of the postwar behavioral sciences who were interested in data-gathering innovations—a testament to Kaplan's prescience. Among the anthropologists were several "pioneers of data": in addition to Hallowell, there was Melford Spiro,

fresh from collecting a full data set of the southwestern Pacific island of Ifaluk, and, of course, Clyde Kluckhohn, "prophet" of the
new anthropology and Kaplan's mentor. Also attending was anthropologist John Whiting, a specialist known for calibrating the
statistical distribution of certain child-training practices—breast-
feeding, weaning, diapering, disciplining—and charting how these
influenced cultural development.[27] Among the attending psychologists were two "dustbowl empiricists," Midwesterners who favored
a sturdy, no-theory-needed approach to the collection of facts on
the ground: Wulf Brogden from the University of Wisconsin, a
student of John B. Watson, and Roger Barker, founder of the
Oskaloosa Midwest Field Station, which had collected for decades
the behavioral data of school-age children at play, at their desks,
and in just about every other circumstance. Barker's data encom-
passed a range of phenomena that included "what actually happened at a scout meeting, . . . what any boy actually did from the
time he woke up in the morning till he went to sleep at night"—
the gathering of which Barker characterized as "exact studies of
common phenomena." His field station aimed at the very least to
"pile up data points for unspecified future study" from Midwestern
towns such as Savoy, Pesotum, Laclede, and Odin (rather like a
meteorological station with weather data), and in this sense he was
a kindred spirit to Kaplan, who also looked ahead to as-yet-
undefined uses of data that were both ordinary and extraordinary.[28]
An executive from the NRC, Peter Finch, whose background was
in psychological experimentation, also took part.

A key aim of the Primary Records group was to explore "new
techniques and ideas in the field of documentation."[29] To this end,
Thompson Webb, Jr., director of the University of Wisconsin Press
and also the head of the Microcard Corporation, joined Kaplan's
committee in 1955. Webb pushed Microcards as a scholarly boon
that could supplement fields such as classics (by producing concor-
dances and word indexes in runs for small audiences) or biology (by
producing species lists or supplemental experimental data), all at a
viable price. Webb was an early adopter of the micro-opaque as a
scientific and scholarly publishing solution to the constraints of
space and time. Among innovators in information storage, a community that included Webb, the prospect (as a prominent library

science colleague put it) of "defrosting . . . frozen asset[s]"—that is, using micropublishing to free up materials otherwise untapped by mainstream publishing, making them "available as never before"— was appealing. Along these lines, Webb in a 1955 essay decried the "present lack of standardization" in the size of all the different micro-formats, each of which necessitated a dedicated reader and particularized storage solution, and thus he counted himself one of many who hoped the Microcard-plus-Readex machine might be- come that standard. The Kaplan committee's other constituents, their horizons gaining scope, hoped so too.[30]

The members met at an odd moment during the Cold War years when the past and the future seemed linked by just-emerging technologies. They meant to rescue past "human documents" by means of future-looking, new-fangled technology. Noting that among the group's ideas for potential data stores (floated at a later meeting) were Donald Hebb's Canadian isolation studies—the minute-by-minute notes of people confined for hours in sensory deprivation chambers as they went into trances and altered states of consciousness—one can pause to wonder—I did—how exactly records like these would figure in the group's ambitions. Such flor- idly hallucinatory states included vivid details of a temporarily in- duced psychosis, and it was not clear to me at first why a group dedicated to gathering the materials of non-literate cultural groups would also want such politically and scientifically sensitive materi- als. The answer, as I would discover, was that the group's ambitions extended far beyond culture-and-personality materials, although its leaders saw these as a good place to start. In effect, data collec- tors could start with the data of these groups they defined as decid- edly outside of mainstream modern life. Although, as it turned out, Hebb's notes were not included within the Kaplan database, and the closest materials to them in spirit, akin to data of "altered states," were the peyote records of Menominee Indians, still social scientists in these years concerned themselves with what may be called the "operationalization of subjectivity," the problem of how to take the inner life of another person as an object of study and so thoroughly understand it that a properly equipped expert might, in the most extreme cases, succeed in remotely controlling it. Even as strategists rethought war itself as having shifted from a "battle to conquer geography to a battle to persuade hostile minds," scientists who specialized in persuasion required reliable data about exactly what "hostile minds" and other types of (perhaps potentially hostile) minds were thinking.[31] There was a large-scale push to explore and thoroughly penetrate the last preserve of subjectivity itself—in all its varieties and all its elusiveness. In other words, if the goal was to understand subjectivity as a set of functions and complex (potentially controllable) interactions, then all varieties, layers, levels, throughputs, and the odd bywaters and surpassing strangenesses of human subjective experience should be charted and the resultant data kept. Hebb's sensory-deprivation experiments were inherently interesting because they posed the question of the stability of human functioning under extreme conditions; as data, they might prove useful to future studies.

The committee members wanted the most extreme and wide-ranging as well as the most "normal" data that tracked people's lived experiences. They wanted data that showed how a personality was shaped by its cultural surroundings and familial imperatives, its triggers, its checks and balances. They wanted data that reflected people who were rapidly and often painfully being "modernized." Personality materials, also called subjective materials, were those data, and despite patiently bearing these anodyne labels, they were somewhat incendiary.

In this spirit of understated subversion, the committee, while aiming at all data, began by fulfilling Kaplan's plan, calling it a pilot project "to assemble primary records on personality material that had been collected by research workers in societies 'other than our own,'" as the official minutes put it.[32] Primary records were the raw, uninterpreted stuff of empirical research, prepublished data sets. "Societies other than our own" was likely a delicate way of not saying something roundly insulting such as "natives" or "primitives" or "far-off peoples." There was no convenient term, really, to describe these other people who were the targets of study: they were interesting because they were different from the typical, urban-dwelling, anxiety-tending postwar social scientist, but they were not so different as to be seen as primitive. In fact, greater familiarity revealed that they—for example, Zuni dancers who consulted anthropological texts to remember exactly how their

dances went, or Hopi women who made traditional Piki bread and
served it with coffee and jam, or the Spanish American day laborers
who drank with Kaplan at the bar on Route 66—were not after all
so different from the presumptive "us," and with each passing year
they became even less so.

Under project leader Kaplan, members would meet every few
months to hash out exactly how to collect the collections of others—how to harness, in other words, the decades' worth of data-
gathering efforts in which fieldworkers had engaged, generally
with no thought as to the future of their hard-won sets. Kaplan's
committee was like a data-rescue squad finding "primary data" in
their fields of psychology and anthropology "which otherwise
might be lost or destroyed."[33] Within the first few meetings, they
enlarged their scope massively. Endangered species (pandas), endangered cultures (Maori), and endangered ways of life (family
farms): all these are familiar tropes that characterize what Claude
Lévi-Strauss would call a "world on the wane." Yet here the committee members were reckoning with the need to preserve something not yet commonly seen as a problem: endangered data.[34] In
an odd twist, it was not only 'far-off' ways of life that were disappearing in the wake of modern changes, but also the data that documented this process were themselves in danger of disappearing.
Kaplan and others noted that scholars were not taking care to
mind their raw findings. Such inattention would within several
decades become a pressing scientific issue at the forefront of
concerns about curating and managing vast amounts of information—but for now, as of 1955, Kaplan's committee wrangled with
an as yet unsung preoccupation.

When the committee met six months later, in May, at the La Salle
Hotel in Chicago, it had two things to celebrate. Word had it from
the National Science Foundation's Mrs. Helen Brownson, that a
healthy $20,000 was on its way to beef up its work and extend its
imperative to collect, retain, and circulate more kinds of "primary
data."[35] Looking ahead, the committee described its remit: (1) Find
out what is required for the retention and circulation of primary
data; (2) Inform scientists about what new techniques have to offer;
(3) Identify specific bodies of material that would be advantageous

to preserve or circulate. Accessioning "specific bodies of material" across psychology, sociology, and even ethology was the new frontier, the larger mission. Meanwhile, Kaplan announced that the first installment of the culture-and-personality materials (the pilot project, that is) was almost ready to go to press. It included twenty-six sets of data, representing two of the three kinds of records eventually collected in the Microcard archive. There were life histories (for example, Elizabeth Colson's "Autobiographies of Three Pomo Women" and John Honigman's life story of a Pathan [Pakistani] young man, including some discussion of his dreams); there were test results (Erika Bourgignon's "Rorschachs of 75 Haitian Children, Aged 7–15, and 42 Adults"); and there were combined grab-bag collections, such as that of the Science Museum of St. Paul, Minnesota, from Ruth Sawtell Wallis, who had agreed to contribute Micmac and Eastern Dakota life histories, Draw-a-Person tests, play protocols, dreams, fears, and children's stories. However, as of yet there were no pure dream collections.

Dreams were on the agenda, however. At one point, the discussion turned to the broad question of which kinds of data to target, hard or soft. Dr. Barker favored embracing the hard-to-embrace: "Much of the discussion," he observed of the group, "centered around materials such as statistical tables and data of archaeology and physical anthropology in which problems of preservation were less difficult than areas where more ephemeral psychological and social data were involved."[36] Why not aim at these tricky elusive materials—of which dreams were the sine qua non? The committee broke into general discussion, sometimes passionate, when one member (the minutes do not record who it was exactly) interposed that they need not make a final decision "among areas" but use the pilot project to "capture the interest of workers in many different fields and dramatize the possibilities of the new techniques for preservation and distribution."[37] He suggested, for example, a series in primatology based on observations of primate behavior at centers such as the Yerkes Primate Research Station in Orange Park, Florida.

The above comments were the first recorded outbreak in the official minutes of a debate that would arise periodically within Kaplan's group. The dilemma was whether to focus on content or technology—loosely, the message or the medium. To put it another way, the choice was whether to collect riveting amounts and kinds of data—for example, the dreams of American Indians or, speculating more broadly on further-reaching collections that would follow the pilot, the records of psychoanalytic sessions in New York, the five-hundred-page autobiography of a female heroin addict, and the real-time behavioral data of apes at the Yerkes Primate Research Station—or to focus on making an attention-grabbing spectacle of new technology by showing what could be done with the microtechnologies then becoming available. Some months later, on November 9, 1956, Professor Brogden, in particular, insisted, "Our present emphasis should be on experimenting with methods and techniques rather than actually developing a large scale archival program."[38] The committee agreed provisionally. They made an announcement: nothing along these lines—applying Microcarding to the "problem of the dissemination" of original psychological materials—had been done before.[39] Method and technique were, for now, the essence of the project.

Yet the committee also was unwilling to jettison the appeal of collecting some new, hitherto unsecured types of data, and Kaplan in particular found himself more strongly drawn toward data from the realm of the intriguingly ephemeral. Granted the experiment with technology was at the core of the efforts, yet the committee could also afford to experiment with kinds of data. Why not focus, Kaplan felt, on the things that flash by in the blink of an eye? In contrast, the more phlegmatic Mel Spiro emphasized that unaltered and unexpurgated field notes themselves would provide ideal materials "as first-hand description of real life phenomena." Accordingly, an archive of ethnographic field notes came in for intensive discussion; despite the fact that field notes were often higgledy-piggledy in organization, Barker and Spiro felt that they were replete with valuable "first-hand descriptions of real life phenomena."[40] "First-hand" and "real life," the "stuff" of life that of its nature disappeared as soon as it occurred—the passing moment, the unenduring pivot point of time—these bred the kinds of phenomena that ideally would be captured in these sorts of archives.

The "first-hand" was for Kaplan only the beginning. Aside from building a dedicated dreams archive, the committee could ask, "What other kinds of things flashed by? Could those things be captured?" Kaplan expressed interest in rescuing the data detailing Japanese civilians' experiences of the atomic bomb attacks, collected in civilian surveys of Hiroshima and Nagasaki. Ten days after the August 1945 bombings, U.S. strategic bombing command personnel set out in jeeps across the rubble to find eyewitnesses who would detail the "physical, economic, and morale effects of the atomic bombs . . . in order to arrive at a more precise definition of the present capabilities and limitations of this radically new weapon of destruction."[41] Psychiatrist and physician Alexander Leighton and other social scientists entered ground zero in Hiroshima's city center to conduct interviews with survivors. These interviews in their raw form were a powerful testament to an experience otherwise expressed in dull statistics or triumphalist headlines. Even though these firsthand records never appeared as part of the database for reasons we will see below, and one can imagine the dream-like, *Hiroshima Mon Amour* quality they would have contributed to Kaplan's already haunted archive, Leighton's wife Dorothy would go on to donate her data sets (containing myriad Zuni and Hopi inkblot responses) to the project.

As funding from the National Science Foundation and National Institute of Mental Health arrived to supplement the financial support the NRC had already extended, members enjoyed perhaps the high point of confidence in their project and sense of its potential unboundedness. Furthering the "increased accessibility of data" was their aim.[42] Sources and modes of access were opening up. Buoyancy extended throughout 1956, and the Committee on Primary Records announced it was considering expanding its purview to a "broader function in collecting . . . data," foreseeing further foundation support under Kaplan as executive secretary.[43]

Committee members ranged further beyond the pilot. Could their archive also include a series of "verbatim transcripts and tape recordings of psychoanalytic interviews"? Or then again, normal peoples' diaries and letters (as Mass Observation in Britain had done)? On the more dire but still ephemeral side of data drops, could they accession the records of the Committee on Disaster Studies, a set of field notes and firsthand accounts of airplane crashes, blizzards, earthquakes, epidemics, explosions, fires, false alerts, floods, hurricanes, mine disasters, tornadoes, toxicological substances, volcanic eruptions, and World War II bombings?[44] Or perhaps an archive of American utopias could be conceived? (It is not clear which records the committee referred to here, but an archive of utopian data does sound intriguing.)

Even so, it wasn't clear among stalwarts which priority would hold sway: to go for ever-softer data or to dramatize these futuristic data-storage possibilities. What everyone agreed on was that the mission was important: to revolutionize the storage of social-scientific data sets that were facing extinction. Above all, Kaplan wished that all the effort expended in amassing unique data not be wasted. Toward this end, the committee decided to do both things: go for the most ephemeral data and go for the most spectacular filing system. Equally clearinghouse and cheerleader, its members tasked Kaplan, as executive secretary, to "take the initiative in helping workers in certain areas, i.e. dreams, to organize themselves to get archives formed."[45] Also, with the new cash influx in hand, Kaplan volunteered to spend the next academic year on a mission delving into researchers' attitudes about making their data available. (As the most junior of the set and the most committed to the project, he was the obvious choice to go on the road.) Generous funding put Kaplan on full-time leave from the University of Kansas to travel from office to outpost to department, researching researchers' expert opinions about the storage and management of data.

Kaplan's year-long "polling" road trip amounted to a kind of fieldwork in data gathering. His diary recounts many conversations with the big men (and women) of the most data-rich fields in social and behavioral research. After intensively surveying and interviewing experts about their records of everything from primates to juvenile delinquents to housewives to the psyches of Swarthmore undergraduates, Kaplan reported the results to be "most encouraging." In February, visiting Harold Coolidge of the Pacific Science Board, he found him to be "highly enthusiastic and cooperative." Distressed for some time over the deterioration of Pacific records on land tenure and administrative matters, Coolidge "would like to see someone sent out with a microfilm camera to record data before they are lost."[46] Experimental psychologist Hans Wallach congratulated Kaplan and said he was "happy that someone is thinking of such things," for it was the "original protocols . . . one wants to see." A sense of both the precariousness and preciousness of such original records was widespread.

Northwestern's Robert Watson was "keenly interested" in archives of psychological materials, and his colleague William Hunt spoke beguilingly of the psychiatric records of 1,200 sailors he had from psychiatric hospitals during World War II. It is significant that Hunt did not feel ready to release them to Kaplan, as he hoped to conduct follow-up studies. Such attitudes of data "husbanding" would be occasionally expressed, Kaplan found. Once Hunt finished with his records, Kaplan reported he "[didn't] know what [would] happen to them" and imagined them landing in a graveyard for Defense Department data, the St. Louis Record Center. In the abstract, Hunt supported Kaplan's project but not in the details. Other researchers hinted of data gone awry, in distress, or inadequately provided for; Wolfgang Kohler claimed he had published all of the relevant ape data he had collected, that nothing of value remained, and that, anyway, "he would not be able to locate the data in any case since they have been lost." An Africanist reported collecting "several trunkloads" of Nigerian government anthropology but found no central archive where he could deposit them, though he still hoped to make them available to others. Others admitted they had fantasized about rescuing distressed data for years: dreams at the University of Pennsylvania Hospital, psychological tests at the Harvard Psychological Clinic, concept formation behavior in children at the University of Colorado Medical School, and "all the Rorschachs of people who commit suicide," which "are sent to Washington" to molder somewhere. One psychologist, however, declared he had outgrown such rescue fantasies, and "in retrospect I think this was an anal attitude and perhaps neurotic. Mostly people rightly want to collect their own. The people who want to use other people's data seem few." There was something potentially improper about using others' materials, almost like borrowing their socks, some felt.

One or two of the experts began to think in grand terms of the scope of possible resources Kaplan's project implied. As eminent psychologist Solomon Asch, formerly Kaplan's undergraduate teacher, told Kaplan, "My mind reels at the quantity of data which would be involved." Asch recalled he had to "sweat blood to extract [my data]" and implied that this compound effort should not go in vain. Another researcher continued the visceral descriptions by painting her child research institute as "reeking with good data" and arguing that therefore it was a moral imperative to protect them. Selection and standards would have to come into play: "We should not be flooded," continued Asch. In addition to social researchers, Kaplan consulted with prominent librarian Verner Clapp, who advised him, in selecting and recording data, "to be as forward looking as possible and to try to keep in mind the *potential* usefulness of data rather than be oriented only to the uses that were clearly envisaged at the present time." Kaplan began to embrace a future orientation and spread this attitude to the committee when he reported on his mission.

The thrust of Kaplan's road trip was that a purely psychology-centered Microcard data bank of primary records was doable, and although the constraints, aims, and limits would still need to be worked out, Kaplan heard hopeful words from so many that "I think the prospects for such a series would be very good," as he reported in a letter to Hallowell.[47] Specific support for Microcard as a going format encouraged Kaplan—"definitely relevant" (said Joe B. Wheat of the Colorado Museum) and "very feasible" (responded Jesse Jennings of the University of Utah).[48] Beyond that, worlds of untended but important "primary records" from sociology to primatology were ripe to be saved.

Such sideline cheering aside, it was not for decades that social scientists would understand the need to hold on to data, their own and others'. Kaplan was well ahead of his time in seeing, along with Vannevar Bush, "that gadgetry is not necessarily trivial" to the creation of knowledge.[49] Alongside his committee's concern for data-storage-and-retrieval platforms, the focus on primary materials was dizzyingly prescient, a situation underlined by the fact that these issues are still urgent and unresolved. December 2013 saw a rousing debate among Oceanian anthropologists about how to preserve their field notes and data, and shared among them was the rueful fact that even in the twenty-first century, providing for one's primary field materials was not yet part of professional training or practice, as one New Guinea ethnographer noted: "Collections of primary materials have not, to date, been allowed to be considered a necessary part of scholarly outputs, but that is changing."[50] This comment serves as an index of just how timely the "database of dreams" in fact was—so timely, as we will see, that few initially grasped its significance.

Micrographics, from their origins in microphotography to their burgeoning twentieth-century forms, inspired a continual reckoning in scale and ratio, as we saw in chapter 3 above. Scaling up and down at will was the essence of the form, as was the touting of mind-blowing ratio calculations (the entire Bible in the size of a walnut!). One distinguishing trait of Kaplan's archive was that it exploited these historical qualities and evoked the fantasy of total information. A second distinguishing trait lay in the contrast between the expansive, unheralded, wide-ranging data contained within and the recessive and pointillist force of its reduction. The initial series of cards (dubbed a "volume," perhaps to evoke for potentially nervous users more traditional forms of publishing, though "stack of data on cards" was closer to the truth) sold for $35.00, while the following run's brochure showed some added typographical flair: "Just Published," it read in large fancy letters, and Kaplan's name was now bigger. The price had gone up, too, to $58.00, due perhaps to the much greater number of documents, but also to the undoubted momentum of the project.[51] Here again one can see the push and pull of strong forces: on the one hand, standardization (the cards fitting into existing forms and drawers, making the Microcard a standard), and on the other, ephemeralization (shrinking down, choosing the most elusive materials, rendering the not-quite-visible in a format legible only with the help of a machine). While descriptive headings in large type ran across the top of each card, the data themselves, originally in standard typeface or standardized forms on 8 × 10 paper, were shrunk down by a ratio of approximately 1:24 so as to be inscrutable. With Readex machines, the cards would assume legibility, and in the event of flood or war they might be swept away but could be perfectly reproduced from Microcard's backup files in Wisconsin and thus were in effect indestructible: "Should Microcards be damaged or destroyed by fire or flood, exact duplicates can be quickly prepared from negatives which the Microcard Corporation, upon request, keeps on file." Likewise the cards themselves—space savers, mini-photographs—also stood as emblems of toughness that "do not smear when handled" so that the "image remains sharp, clear and legible indefinitely." Thus although they were not readable by the naked eye, they were in another, possibly deeper sense permanently legible for an infinitude of time. The cost? "Nominal"—half a cent per page.[52] These factors constituted a revolution in psychological research, for microproduction offered "the revolutionary possibility of the publication of large amounts of material in small editions at very low cost."[53] Dwindling per-page calculations were a coin of this realm, but it was not a realm many professional psychologists or anthropologists cared about in the mid-1950s. For Kaplan it opened up "a new kind of flexibility" unconstrained by stodgy letter-press standards of what was worth setting in print. "Scholarly value can again be the main criterion of acceptability," he announced, with the advent of microcopies and micro-reading machines.[54]

But the promising new microtechnology did not act alone on the data the Primary Records Committee saved. It was only one of a nest of technologies that would ultimately work together as a sort of Rube Goldberg machine to enable intimate data to be extracted from sources around the world and circulated to scientists.

For such a feat, one had to be able to turn a dream or a life into data.

CHAPTER SEVEN

"I Do Not Want Secrets. . . . I Only Want Your Dreams"

THE primary dream collector involved in the Microcard archive was a woman named Dorothy Eggan, who began to record the dreams of Hopi informants in the 1930s. Dreams offered access to the "television qualities of the night life of the mind," she once wrote—as a sort of provocation to those who saw dreams in a more traditional light—and social scientists ignored these nightly broadcasts at their peril. "I do not want secrets. . . . I only want your dreams," she demanded of one of her subjects, a fifty-year-old Hopi man recorded in the Microcard archive.[1] It sounded like a very bold program, as if, somehow, dreams were no longer secrets dredged up or roads to hard-to-reach unconscious desires but to be treated as something less intimate, less flimsy, and more shared.

Eggan's papers contain—alongside stuffed folders of letters from Hopi friends, piles of stenographers' notebooks full of records, and hand-drawn sketches of Hopi cosmography—a perfectly typed-up rendering of all her most important Hopi dreams. This was her husband's posthumous memorial to her, buried in the papers after her suicide. He had her remaining dream materials (outside of the ones circulated in the database of dreams) extracted from the scribbled-in notebooks and readied in preparation for someone to finish the project she had never managed to write, a companion volume to the 1942 *Sun Chief: The Autobiography of a*

Hopi Indian—a shadow autobiography told in dreams. In turn, this constituent material slated for an unfinished project itself has scarcely been touched or referred to in succeeding years. Yet Eggan emerged from the archival ruins of this invisible memorial in an exorbitant light.

Raised in Dover Hill, Indiana, and subsequently in the obscurer parts of southern Michigan, Dorothy Eggan dropped out of college at nineteen or twenty due to her sister's death and father's illness. She married young, getting introduced to the southwest when she and her architect husband moved to Santa Fe in the 1920s, and, "lucky as always where people are concerned,"[2] she befriended several Hopis, including the soon-to-be-renowned artist Fred Kabotie, who was still a high school student who was wandering around the Santa Fe Museum of Art when he and Eggan encountered each other over a painting. Kabotie and (later) his wife Alice became not "informants" but friends, a distinction they upheld over four decades, and a notable one because it did not hold for many of Eggan's main dream sources, who vacillated between informant (subject) and friend (non-subject).

In the 1930s, Dorothy and her husband returned to the Midwest due to hard economic times, and her husband entered as a graduate student of archeology at the University of Chicago, while she took on clerical jobs around the university and migrated to social circles with anthropologists, including Arthur Radcliffe-Brown, Robert Redfield, and Fay Cooper-Cole. Ultimately, she divorced and got married again, to Fred Eggan, a well-known expert on Hopi and Philippine "acculturation." Although she continued throughout her life to lack a college degree, much less a PhD, and would sometimes irritably chastise those who insisted on addressing her as "Dr. Eggan"—"I am *not* 'Dr.'—I told you that before," she admonished Bert Kaplan early in their correspondence—she amassed over three decades, alongside the dream data and careful records, fan mail from some of the most eminent practitioners in the United States, such as sociologist David Riesman, who praised her scholarship, which in his view "combines accuracy with eloquence—a rare achievement." Or, as Clyde Kluckhohn wrote her, "You must realize, of course, that I am far from being alone in valuing your work. I know of few, if any, papers in recent years which have had as much influence among a wide and discriminating audience as yours."[3] In eight lapidary essays written between 1942 and her early death in 1965, she sidestepped a focus on deep, Freudian, or symbolic meaning of dreams in order to grasp their televisual, *projective* quality. Championing manifest (surface) over latent (or hidden) content as the key site of analysis, she advanced a self-consciously experimental approach. She also wrote about the shared communal qualities of Hopi life glimpsed as people underwent adaptation to "white ways," a transition she saw as an inevitable if never total process.

Eggan was a kind of archeologist of data sets even before she met Bert Kaplan, at least in spirit, for she encouraged a wide array of researchers to access the "wealth of dream materials buried in many anthropological files."[4] Anthropological files were themselves sites of excavation. She saw such data sets not as personal hermeneutic repositories to draw from and then lay aside (as most anthropologists were trained to do in the virtuosic research tradition), nor as sites for deep excavation into personal intricacies, as most psychoanalysts preferred.[5] "Psychoanalysts are interested"—Eggan observed of her dream collection—"but *they* want me to try to do something which *I* don't want to do," she protested in a letter to Kaplan, continuing, "They can do it if they wish after it is made available to them."[6] Beyond serving her own immediate interests, her dream materials were to function as a would-be public clearing-house of data. After the "stuff" of dreams went up (or out?) in the Microcard network, other researchers were free to use it. This was a data-centric vision very few social scientists at the time attempted to pursue—among them Kaplan, the Yale-trained Murdockian group including John Whiting, and Claude Lévi-Strauss, each in his own idiosyncratic way her kin.[7]

Eggan treated her own data sets to different kinds of analytical approaches. She sequentially numbered dreams that had been collected in the 1930s and 1940s, and these dreams eventually fed the Microcard archive in the 1950s and early 1960s. Each of her subjects, from five Hopi families, generated a series of dreams marked in the order they dreamed them. She took these numbered dreams and created sorting lists (looking for thematic threads) and "series" (temporally arranged). Further arraying dreams in charts, she discerned patterns through the "massing of items." Such devices as lists, series, and synoptic charts, she argued, were useful for "massing evidence in readily available form."[8]

The word "massing" comes up fairly often in Eggan's writings. It was not simply the telling, singular dream such as "Irma's injection" in Freud's *Interpretation of Dreams* (1902), but also the piling up of dreams as documents that moved her and led to significant insights. Step by step she made a system for turning dreams into data to be worked with. The goal—not just Eggan's, but a sort of shared social scientific imperative among culture-and-personality scholars—was to create a new kind of scientific subject whose subjective space was understood as projective and shareable and therefore amenable to projecting and sharing techniques. Data about the inner life of such subjects would issue forth. Eggan wanted to understand Hopi ways of being in order to understand what it meant to be human.

By 1939 Eggan was living in rooms in the Hopi pueblo of Old Orayvi with her anthropologist husband, Fred Eggan. She had befriended several Hopis and, growing closer over time, feeling their differences both lessen and grow more remarked, decided to study their dreams. Eggan found the dreams of preliterate people to be especially revealing. One might find in an autobiography or life story the view a person wants to give of himself (to the anthropologist), "But in dreams, where preliterate people particularly are off guard, in others' accounts . . . and in data dug out by various psychological techniques, the picture begins to be more rounded." She was speaking of Don Talayesva, who was Eggan's main dream source, but the claim applied broadly in her view: dreams, as well as psychological tests and other targeted materials, were special, providing "field data" that allowed one to understand someone fully, not just to the extent he or she was willing to allow. They gave access. Anthropologists and social scientists more generally could use dreams for a new, previously unimagined purpose, as "social scientific documents"—that is, as data ("document" being a near synonym for "data" in mid-twentieth-century social scientific circles).[9] In search of dream documents Eggan, like most anthropologists, paid informants an hourly amount of around twenty-five cents, a remuneration that was a significant motivation for Hopi often short of cash, especially, as we will see, since consorting with anthropologists became more politically tricky over time.[10] Her friends tended to be the more educated Hopi, who spoke English fluently, and with others she used an interpreter.

As Eggan wrote in "Collection of the Data," remarks prefacing her first Microcard contribution, she benefitted from the fact that in a society that valued group effort and considered "talking about one-self ... exceedingly bad manners," the chance to adumbrate one's dreams with a sympathetic interlocutor was welcome, and people usually enjoyed the exchange. As we will see, though, after 1942 and the publication of the controversial and occasionally downright embarrassing *Sun Chief*, the sexually and ritually explicit life history of Don Talayesva (who donated 341 of his dreams and would go on to achieve the unlikely distinction of most thoroughly recorded human in history),[11] such exchanges often came accompanied by the informant's fear of being discovered talking to an anthropologist and "after publication of *Sun Chief* ... all informants increased their demands for anonymity, and a few would no longer work as informants."[12] The coffee-klatch atmosphere of the early dream-collecting sessions could be seen to dissipate after 1942, and Eggan's informants then began to have an air of running scared. Eggan herself took pains to "anonymize" her informants by blanking out incidental names (for example, those of acquaintances or small children), stripping away notable features (sometimes up to two-thirds of the details of certain dreams, which would have easily given away the dreamer), devising three- or four-letter non-Hopi names as substitutes for real names, and hiding clan affiliation. Here is an example, with all names given as pseudonyms and any identifying elements removed:

> Joab: My wife get after me for working with you. People will be mad, she say, if they find out. I tell her we need money for my son's house. She say we can't trust *bahanas* [white people]. They get rich off of what we tell them, and then people know who talk, and we get trouble. I tell her you promise no one ever know who talk to you. You promise again?
> Recorder: I promise. Unless you or Daf, or your wife tells it, no one will ever know. [Daf: the interpreter and daughter of Joab's older sister.]

Joab: Well what did we talk about last time and what do you want me to talk about today?
Recorder: Tell me more about your family and what you did when you were young.[13]

The subject also gave evidence of enjoying himself, telling Eggan, "I often think of you people"—that is, Dorothy and her husband Fred—"and am always thinking over my dreams."[14] This man's great fear of being discovered was offset by his desire to build a house for his son, for which he needed money, and he punctuated his recitation with demands for assurance of fealty (likely never suspecting his recollections would be on file at the Library of Congress within two decades).

Above all Eggan noted the Hopis' "determined 'Hopiness' in spite of an equally determined effort on the part of whites to change them." No matter how much internal strife, gossip, or social up-heaval was to be seen in Hopi—"what seemed to me to be almost universal bickering and tension among themselves"—there re-mained a striking adherence, in Eggan's view, to an unflappable quality that seemed to spring inextinguishably from their own cul-ture.[15] Yet there was a paradox, not always named by Eggan but ev-ident in her work, and the first of a layered set of paradoxes that arose in the process of turning dreams into data. How did this en-during "Hopiness" square with the inexorable, if unevenly arrived at, "Westernization" she also observed? Her data set (Series I) de-fined as "Hopi" only those who were not yet Westernized, but she went on to include a whole set of seventy-four Christian dreams, including a subset from a woman, "Gar," who spent ten uninter-rupted years at boarding school, married a half Spanish American, half non–Hopi Indian and who disdained Hopi life. Likewise, in successive dreams one of her "traditional" Hopi subjects is found filling his borrowed car with gas, encountering an evil spirit (a "mixture of Hopi witchcraft belief and Christian avenging angel, via the Christian section of his family in another Village"), trying to buy a bed for three dollars, driving a school bus from the high school to Polacca (during which he is laughed at for having an old-fashioned hairstyle), and singing a song about butterflies while dancing with Dorothy's husband.[16] These matters intersperse in dreams with direct encounters of Spider Woman and Masau'u, the Hopi god of fire and death. Even self-consciously traditional Hopi did not dream "traditionally" in any imagined pure sense; they dreamed of the world in which they lived.

Architecturally, too, the dreams formed creative annexes reflecting the mixed nature of daily life. Another conservative Hopi, Chad, dreamed of wandering down a hallway, a feature not found in Hopi architecture but which he knew from Oraibi High School, and, as he reflected, "I have seen many nice houses of white people in which they have hallways."[17] His wife, Debbie, about sixty and "thoroughly 'old Hopi,'" dreamed (dream #2) of visiting a friend and climbing through an opening in the ceiling to find not a room but a beautiful space, where the "ground was like rubber sponge," and there was a pool of water with "green grass like the lawn in front of the *banaha* houses." She found herself flying into black clouds where she met her mother-in-law.[18] The paradox is this: even as "Hopiness" was the goal to be described (by Eggan's definition, at least), in order for it to be available for description, it needed to be at least partly ablated.

A second and related layer of paradox cropped up in a seemingly mundane research practice of Eggan's. In the course of her research she handed out to "non-literate" Hopi people stenographers' notebooks with brand names such as Progress, Satin Finish, Hy Tone, or Tumbler ("Turns Quick, Lies Flat, Stands Upright, Eye-Ease Paper") on which to write their dreams. Yet their ability to use these notebooks rested on the extent of their training in the very schools—schools to which they were often dragged at a young age by Navajo policemen or U.S. "Negro troops"—that imperiled "Hopiness" and in fact were designed, at least originally, to extinguish it. An artifact of this painful conflict was what allowed a younger generation to take the first step in Eggan's project to turn dreams into data by recording them. (Older Hopi couldn't write, and in those cases Eggan herself would transcribe their dreams, as well as the context and associations, with the help of an interpreter.) At times, notebooks themselves became "manifest content" in dreams. An example is Joab's dream #34:

34. . . . I came back down from the housetop and went to my house. *Then an old, old lady came to me and asked me all kinds of questions, and she had a book and a pencil* and she wrote down everything I said. I looked at her and started to laugh at her and said, "Give me your book and let me read what you wrote down. You can't write, for you are an old woman." [An old woman among the Hopi would not have been taught to write, Eggan noted.] I looked at this woman close and it was my mother.[19]

Just before this, his dream had Joab prophesying a return to old ways; now he saw himself fielding "all kinds of questions" whose answers were taken down in a book by a Hopi woman who could not *by definition* write—a switch that turned the recorder of non-literates (Eggan) into a non-literate recorder (his mother). The dream scrambled its salient worries, for all of which the ability to write in a notebook was both a symptom and an aid.

In a sense Kaplan's data archive was a projection of such paradoxes on a worldwide scale, its contents describing how cultures and personality types seemed to disappear yet not disappear, to be *gone and not gone*. Binaries such as "primitive" vs. "modern," "non-literate" vs. "literate," or "them" and "us" broke down before the recorder's eyes, even as other distinctions arose. Cultures seemed to fall apart and yet to reassert themselves in new ways. Eggan set out to study closely what this process meant by means of the dream life and inner "subjective experience" of Hopi people.

This chapter traces the life cycle by which certain dreams turned into data and what happened afterward.

Eggan's and Kaplan's first encounter was via a soliciting letter in which he asked her whether she would like to add her data to a slated National Institute of Mental Health (NIMH)-sponsored Microcard project. She said yes, eventually contributing, in 1957 and 1962, two dream "series." Within a year, they were hosting each other for talks and visits, north to Chicago and south to Lawrence, Kansas, where Kaplan continued to work after winning his post as assistant professor. Preparing to go speak to the University of Kansas psychology and sociology departments, Eggan crowed, "The 'dream' of my life has been about getting psychologists interested in working with dreams" (that is, anthropological dreams). After Kaplan's reciprocal Chicago appearance, Eggan further thrilled, "Your visit here did the dream project more good than you will believe. Beyond publishing an occasional paper about it and talking to classes or seminars when needed, I've felt in a blind alley—hopeless about publishing the material in a manner which would make it as useful as it should be, and still acceptable to the informants. But your interest in even these abbreviated examples of the material gave me new hope. . . . Your encouragement has really been a catalyst." Her enthusiasm was infectious and the ingénue tone of the letter was much like her speaking voice, according to friends. Even as Kaplan admitted to a perennial reserve in epistolary style that he could only attempt to overcome in order to reciprocate her enthusiasm ("I enjoy your chatty letters, and hope you will excuse the business-like tone in mine" for "I do not think I have learned yet how to be friendly and informal through the mails"), he too expanded the scope of his ambitions.[20]

Eggan's enthusiasm *was* catching. In a meeting in November 1956 the Primary Records Committee "looked very favorably on the prospect of a Microcard series focusing specifically on dream materials," Kaplan reported to Eggan, and he would be setting off to discuss this with Calvin Hall, Leon Saul, and Montague Ullman that year, when he was slated to work full time for the NRC conducting his "roadtrip" survey of scientists working intensively with data. (Hall, Saul, and Ullman were psychologists sharing a "shallow," oft-statistical approach to dreams. All developed proto-experimental approaches to dream study in the 1940s and 1950s, and all focused on manifest content. Hall pioneered the collecting of large amounts of dreams, beginning mainly with those of American college students but eventually extending to dreamers in other cultures. His "comprehensive system" subjected dreams to scoring and statistical analysis applied to elements of settings, actors, objects, interactions, and emotions, and Hall's work continues as a growing data bank today. Ullman's too resulted in a worldwide network, especially concentrated in Scandinavia, of Ullman experiential dream groups.)[21]

Although Eggan had doubts about Kaplan's faith in interspersing dreams with other kinds of data, recalling in a letter, "I know you said, 'I doubt whether anyone will ever look at these dreams (in the Microcard series)'—and that made me wonder if you really want them," in fact Kaplan's committee was working toward a dedicated dream repository. It wanted more, not fewer, dreams. In so doing, it was upending a rationalist hierarchy of evidence in which the "soft" things like dreams—almost not even things at all—rose in desirability because of their trickiness and their access to the non-rational.[22] They served a threshold function especially valuable after World War II.

When Eggan met Kaplan in the mid-1950s, she had been collecting dreams for just over fifteen years. She had gathered dreams from "some twenty informants in five villages, along with native interpretations, life history data, and interviews, based on 'free association' and modified analytical techniques." In addition, she had assiduously secured the dreams of that somewhat unusual Hopi man, Talayesva, between 1939 and 1947, visiting him frequently, and although the war interrupted this pattern, she would have Don write his dreams down in notebooks, and Eggan would interview him subsequently, when resident in Hopi, to glean his thoughts and associations. As of 1949, she reported 295 dreams from this single subject in eight years. She continued to be in touch with Talayesva and to receive his dreams, often by mail, through the 1950s, resulting in a total of 341.[23] Why dreams? One place to start is that she had been in psychoanalysis for almost as long as her dream-collecting activities, and her first analyst was dream expert Thomas French, a fact that may have provided initial motivation, although it is perhaps an irony that she eschewed strict psychoanalytic readings of dreams. She prefaced her first Microcard dream series with the statement that her psychoanalysis at the time of dream collection was as yet incomplete but subsequently completed, as if showing her bona fides.[24]

Kaplan and Eggan hit it off, she feeling that the "creative minds of the sleeping Hopi" were an oblique but effective tool to establish a scientific approach to understanding how cultural forces and individual personalities interact with each other—"interactions between the entity which is a personality, and the entity which is a specific culture." He too was interested in the sleeping Hopis' dreams as a way to build up his data collection. For both Kaplan and Eggan, dreams were part of the data that served as the empirical support for a unified human science, one with the capacity to explain and predict the effects of culture on individual personalities. Indeed, dreams were special not in the way Freud had suggested—as direct access via what he called "a new psychic material" (that is, latent content) to the deepest unconscious—but in a technical way, as in-born, human-powered projective devices: "Dreaming is a built-in projective process," Eggan exulted in a letter to Kaplan fueled by the possibilities of their collaboration, but this fact had so far scared off most social scientists, cowed as they were by Freudian dicta: "So we ignore it [dreaming] and use inkblots just because Freud put a fence about it and no one really tries to jump over the fence!" Dreams were special evidence, procedurally just like Rorschach responses— or, rather, they were likewise part of a "projective process," as Eggan wrote in prominent articles for the *American Anthropologist* in 1949 and 1952. There she specified that a dream was a "projective phenomenon" for the dreamer by which he or she free-associates about matters that may be painful or difficult to think about. Further, dreams could be used as techniques, for, as Eggan noted, just as projective testing such as the Rorschach and the TAT had been made relatively standardized for fieldwork, so too might "comparable methods" for the gathering of dreams' manifest content. (This line linking dreams to projective techniques found new figurative life in 1980, when leading dream researcher J. Allan Hobson described the neuroscience of dreams as pointing to how "Dreaming sleep may thus be viewed as a physiological Rorschach test, self-administered four to five times a night.") For Eggan, Hopi dreams elicited this type of special projective evidence in spades. Likewise, Eggan's position outside the establishment—really, as an inside-outsider, having no advanced degree but married to a high-up Chicago anthropologist—seemed to make her particularly available for participation in Kaplan's experiment in gathering up such evidence.[25]

A third layer of paradox in this reckoning of the special value of non-acculturated people's dreams was that they subsequently served as proof of the pace and destructiveness of acculturation itself. Seen as somehow unadulterated, these dreams testified to how the exigencies of modern life intruded into the inner self. Qualities seen in "dreams from old, conservative Hopi"—such as clarity and a direct relationship to fantasy, myth, and problem-solving laid bare—gradually ebbed away. An example occurs in the dreams and stories of Don Talayesva, wherein the deities and figures of the Hopi pantheon actually interact in day-to-day or night-to-night experiences (Spider Woman hanging out by the spring or lingering by the corn fields or encountered on the Salt expedition), and the language Don used, despite his rather thorough education in a California boarding school, showed he did not find these to be "symbolic" encounters requiring interpretation but absolutely real. One night, he dreamed of a wind, and the next day he prevented his son from injury in a sudden storm. Seventy-year-old Yaw, an excellent weaver from a largely Christianized family, was recounting his dreams to Eggan when he offered in addition a real-life event, which appears as no. 18: "This is not hardly really a dream, but I saw it myself. I saw it on the road; all fire. It was Masau'u." In fear and trembling, he had encountered a deity whose existence he had come to doubt, and told the man who presently appeared in his ragged robe, " 'So you really are alive; I've heard of you and I believe you really are alive like we are. You go your way and I'll go mine.' ... Then I rubbed my eyes but I wasn't really dreaming. [Now] I dream of him sometimes and am scared."[26]

One of Don's school companions, Albert Yava, in contrast, referred in his autobiography, *Big Falling Snow*, to ritual and religious activities from a distinct and "objective" standpoint, as if he had absorbed the social-scientific eye—a point his biographer made. "Yava can see a ceremony externally as well as through the eyes of a participant. The experience he communicates is substantially different from Talayesva's which is largely internal, the experience of a participant cast by fate in a role. Talayesva describes supernatural events as aspects of everyday reality. Yava, though dedicated to the values of his group, finds it relatively easy to say 'they believe' or 'they' do this or do that." Although the two were chronological contemporaries, the latter "began to see himself from the outside. Don acted out tradition in his personal experiences, whereas Albert became narrator and chronicler." This change manifested itself too in the way dreams were dreamt, as if from the outside. Younger, Westernized Hopi displayed a difference, and "Their dreams, like our own, become almost unintelligible at the manifest level, and they have great difficulty in giving associations to them." Or so Eggan argued. Hopi were crossing a threshold, not all at the same time and not all in the same way, but nonetheless this was happening. It was this crossing rather than any "primitive" essence that made them interesting scientific objects, but the very objectification process belied the fact that the relentlessly intimate forces of modern life galvanized all people who encountered it, including, not least, the experts (rightfully sometimes called "middling modernists") themselves.[27] And perhaps this was just as the experts liked it, engaged as they were in their own projective pantomime and dreams of science.

As part of Eggan's collaboration with Kaplan and his micro-archiving operation, a large amount of Hopi people's dreams came into new hands and then disseminated. Dreams—as the Hopi dreamt them, as Eggan or her subjects transcribed them, and as Kaplan's committee micropublished them—left the village of their origin in the Arizona mesas during the 1950s and traveled as far as naval outposts and Swiss libraries, finally summoned up on Readex screens and tiny, pocket readers. By the early 1970s or so, the peak of their use had long passed, and they rested in microfilm collections scarcely consulted, mostly forgotten, but legible still, neglected artifacts, until their revival today. Via this process of punctuated travel, circulation, and rest, they became part of a unique experiment. Although the history of dream studies is replete with collections of significant dreams, prophetic personal dreams, elaborated dream interpretation, and even cross-cultural dream interpretation, no one before had gathered up dreams of non-literate people, collected them en masse, and made them available as "raw data" for present and future use. "I certainly agree with Mrs. Eggan that we have too long neglected the dream," remarked anthropologist and co-dream collector Melford Spiro at a 1958 conference.[28] He felt her concept of "manifest content" was itself in need of some interpretation, yet he hailed her commitment to such an approach, versus one that searches for ever-more-latent content in the quest for ever-deeper pockets of the unconscious.

Dream scholar Barbara Tedlock has argued more recently that Eggan was part of a trend: "Beginning during the 1940s, much of the anthropological work on dreaming had as its main goal the creation of a corpus of manifest dream contents that were then analyzed, counted, and tabulated in various ways for numerous reasons." This claim, although accurate in some ways, misleads by listing Eggan as one in a long line of manifest content gatherers. (The primary citation supporting Tedlock's claim about the emergence of manifest-content centrism, indeed, is Eggan's own seminal 1952 piece on exactly this topic!)[29] To the contrary, Eggan was the one who brought manifest-content analysis to the fore, for as she noted to herself in the margins of one of her own proposals, her recipe was to "discuss impossibility as well as *inadequacy for social scientific purposes* of 'symbolic interpretation.' Illustrate with dreams." Where Eggan led in collecting and manipulating dreams, others followed. It is perhaps a testament to the reach and influence of her writings—what one colleague described as their "Rembrandt" quality—that her original insight would be latterly attributed to a whole epoch. As she pointed out, "Trouble is, there have been no 'methods' worked out for the use of dreams cross-culturally; psychoanalysis still holds everyone's mind in the cellar so no one tries to look in the attic." So dismal was the methodological landscape that Eggan surveyed its scarcity as follows in a letter to David Schneider (the "Dave" in question): "There is *Freud, Freud's followers*, Dave, and me! If anyone . . . else has a 'method' I don't know it." She added parenthetically a reference to "Hall and Wolfe, of course, which I doubt that most social scientists will want to follow." Most anthropologists stuck to Freud, in her view. As it happened, David Schneider's engagement with dreams was more of a dalliance, to Eggan's disappointment. Meanwhile, when Bert Kaplan published Eggan's dream approach in his 1961 edited volume, he assured her that despite the delay in the book's appearance, her article was still at the forefront of the field: "Each time I read it I became more aware of its importance. I'm sure it's way ahead of most contemporary thinking about dreams." Another Microcard contributor, Edward Bruner, wondered at Eggan, "In what way do you look at the data which makes your papers somewhat unique and so excellent?"[30] More recently, dream scholar Waude Kracke claims influence by Eggan.

Countervailing trends were strong: a robust scholarly tradition then and now argued for the unsuitability of dreams when used in such empirically minded projects, as did Judaic scholar Elliot Wolfson in his recent *The Dream Interpreted within the Dream:* "I do side with those who detect in the dream a mythopoeic propensity that cannot be subsumed under the stamp of scientific explanation, no matter how broadly the criterion of empirical data is conceived."[31] Insistence on the irreducibility of dreams, during Eggan's time too, was a staunchly held scholarly view, especially among Euro-American psychoanalysts and psychologists, for whom dreams were special in their singularity and almost infinite interpretability, rather than en masse as data. Dreams were not suited to be reduced to data, almost everyone aside from Eggan believed, although she had a few allies in nearby fields. The preponderance of symbol-mindedness among psychoanalysts was one of the reasons Eggan was happy to meet Kaplan.

Here was something different, then, for the Microcard archive would hold dreams without the usual forms of elaboration, stripped of interpretation and thus apparently bereft of their "imaginal quality," their mythopoeic propensity, their depth. Rather, it would hold dreams treated to be as close to "raw" (or radically empirical) as possible: taken down in this or that outpost, recalled under odd conditions and unlikely auspices, and typed out but otherwise unanalyzed. "Original dream materials" were what he sought, Kaplan explained in a 1955 letter to Schneider, then in residence at the Center for Advanced Study in the Behavioral Sciences—further specifying in a follow-up that "I am not, at the present time, interested in the theoretical articles or in the analysis of dreams except insofar as they point to some existing set of dream materials."[32] Perhaps the only then extant resource that made similar "dream materials" easily available was the British-born Mass Observation collection, but these self-recorded dreams by trained citizen-observers simply resided in a file in Bolton or London, where, as they were not microcarded or otherwise reproduced for ease of circulation, they were not available across a network.[33] Similarly, Berlin-born Jewish journalist Charlotte Beradt assembled three hundred of her neighbors' nightmares between 1933 and 1939 (she asked them to tell her their dreams), a cache that served to support her view that average Germans convulsively dreamed into reality the coming terror as borne out by their nighttime visions. However,

Beradt closely guarded the dreams and would allow only her writings about them to circulate. On a bigger scale, some researchers in the United States did share an enthusiasm for dreams amassed in numbers, and in 1951 Calvin Hall splashily announced he had collected "more than 10,000 dreams thus far, not from mental patients but from essentially normal people," stacked-up "dream materials" that he had subjected to statistical analysis to discover at long last what exactly people dreamt about.[34] Yet Hall, like Beradt and others, did not prioritize making his dream data available as a resource to others, and, in fact, Kaplan hoped one day to include Hall's materials in his own database, beginning with the dreams of five hundred Harvard undergraduates, a proposal to which Hall was amenable.[35]

Kaplan's accessionary ambition was clear: he was working on a different order than most, interested as he was in *collections of collections, or data sets of data sets*. In this regard, Eggan was the perfect partner.

White European presence in Hopi land came in degrees of intensity and waves of incursion. After three hundred years of strife, rebellion, and subsequent lackadaisical letting-alone that characterized the Spanish relationship with the Pueblo Indians, the late 1800s and early 1900s saw a more intensive set of penetrations into the southwest. Sidestepping major military conflict, the Pueblo Indians nonetheless experienced great changes during the Western Wars, especially after 1850. By the turn of the century (and see chapter 1 on Zuni parallels), each Hopi characterized him- or herself by an allegiance, not so much to white culture per se but to a position taken in regard to whites. To say that "Friendlies" and "Hostiles" did not get along is to understate the nature of the dispute, for in it resided two philosophies and sets of pragmatics. The terminology came originally from Indian Office administrators, but when adopted in the late 1890s by Hopi themselves, it reinforced brewing disdain of each faction for the other, calcifying attitudes and ending with physical confrontation in the form of a fateful "push"-of-war on September 8, 1906. This event, the culmination of decades' worth of conflict resulting in a showdown that would fuel further decades' worth of conflict, took place after thirteen years of drought and fifty years of below-average rainfall, environmental conditions contributing to stressful relationships and exacerbating preexisting sociopolitical disputes. Divided by four lines in the sand, the two leaders battled, backed by their respective factions, and the Friendlies "won." Kicked out of Orayvi, 102 families relocated down the road at Hotevilla, marking the "Orayvi Split." Orayvi eventually broke into five factions. Some renegade leaders were banished to Alcatraz and to the Sherman boarding school, which also functioned as a de facto prison.[36]

The Orayvi Split still resonated in the dreams Eggan collected, not only through memories of hunger and suffering, but also in sensitivity to alignment with anthropologists' projects. One of Eggan's informants, in his fifties when she interviewed him, was a young boy at the time of the split and still dreamed about it, as in dream #19, in which he was shooting birds, surrounded by sunflowers as big as evergreen trees. But then "big strong men" came and fought them for the birds, taking them all. "All this fighting I dreamed. I think it meant the fighting being driven away from my old Village to this one at the time of the split." His memories of childhood were of starving and watching his mother carefully apportion food to the children, giving less to herself. Debbie, a Hopi grandmother who was also a child in the years leading up to the split, dreamed of a spooky ghost coming in the door to take her grandson, causing them to run into the small back room to hide, as she awoke screaming. Eggan noted, "Small back rooms are usually windowless and used for storage. During the days when the government was kidnapping children to send them away from the mesas to school, these rooms were also used to hide the children behind rows of stacked corn."[37] The split exacerbated an already existing anxiety about being two-hearted (Kahopi), each side accusing the other of betrayal of Hopi ways, for the Friendly accommodators were not particularly friendly to whites necessarily. (They did not want to adopt white customs wholesale but merely to acknowledge, by agreeing to accept gifts and send their children to schools, the reality of white presence and the power of white politico-military structures.)

In some ways, all of the dreams, and all of Eggan's stacking up, listing, charting, filing, and rendering of the dreams, are haunted by the split, which was itself a reflection not only of internal fissures and pressures in Hopi society, but also the pressures linking

Hopi fate more and more closely to the larger society surrounding them, for as Don Talayesva put it around 1938, "We might have been better off if the whites had never come ... but now we have learned to need them."[38] Eventually, almost all Hopi children went to school (eighty-two from resisting families), supplying to the government's schools their "quota of children" (in the words of the U.S. secretary of war to Congress).[39] Finally, the split arguably had as one of its consequences Eggan's own presence in Hopi land: her first Hopi friend, Fred Kabotie, was one of the eighty-two children from resisting families kicked out of Orayvi, with the result that he attended the Oraibi Day School and then Santa Fe High School (initially against his family's will). He received lots of encouragement from schoolteachers for his artistic talents and ultimately became one of the great twentieth-century painters of Hopi images. Kabotie never participated in Eggan's dream collecting activities, for (as noted above) he and his wife were always more friends than informants.

During World War II, Eggan captured the dreams of thirty-six-year-old Lars (not his real name), who had eight years of schooling. One of his dreams found him in Europe touring a bombed-out city. He assumed, as he put it in his Hopi-inflected English, "the city must be Paris because some building that I have seen in picture Life's Magazine in Paris were there." Presently he noticed the city was somehow now located in Keams Canyon Valley near Hopi (where the Keams Canyon Boarding School was located), and the houses were "all tumble down." In associations, Lars described how he liked to hear war news from all visitors and was interested in the war in Europe. Two others dreamed of German planes bombing Hopi villages, some with Navajo pilots glimpsed in the cockpit.[40] Along with their longtime hostilities, Hopi-Navajo relations worsened before, during, and after the split as a result of the government's using Navajo as policemen to commandeer Hopi children. Hopis were modernizing and globalizing, accompanied by abundant growing pains, traumatic events, confrontations, and conciliations occurring on and off the reservation, as their dreams showed.

By the 1940s and 1950s, Eggan was documenting dreams even as she was witnessing the partial revival of lost elements of Hopi life.

She scurried to get the dreams ready for the typist, to send to Bert Kaplan, between madcap trips to the southwest—for example, one with Fred to Arizona for ten days in 1957 "to see an archaic ceremony which was being given for the third time in the memory of living Hopi and which may never be given again."[41] Likewise, in one of her data sets Chad (in dream #6) meets the recently deceased Flute Chief, a Rain Clan man, after whose death his nephew discontinued the Flute Dance that August. "I was worrying over the dances when I had this dream. We are giving the dances up and it means bad for us." In the next dream he is in Old Orayvi, but it is "not exactly like it," and it is "like the houses at a ruin."[42]

Eggan's question was: "To what extent then, may the 'private world' of the sleeping mind—including the emotions involved in its action—be equipped or limited, *or even used* by a specific society to maintain equilibrium?"[43] Data from non-literate dreams could be used to answer it. Find the functions of dreams, therefore, was Eggan's instruction to herself and others. As Eggan frequently remarked, they had a television quality best seen in the dreams of the relatively primitive ("non-literates" or "pre-literates"). But, of course, only for a little while longer.

Eggan's lifelong quest was to make the most of the "potential information so widely expressed in dreams." She used all her "data from non-literate dreams"—namely, some 650 dreams she collected over two decades from three generations of Hopis—to "explore more actively the potential of the dream in extending our knowledge of the moulding influences of society and culture." As with the work of others in the culture-and-personality movement, the goal was to capture something called *"humanness."* Humanness was both transcendently universal and immanently particular in its manifestations. "Dreams as a universal production of human mental activity … expose … the essence of what Robert Redfield has called the 'developed specific,' as well as the 'developed universal,' of humanness," she wrote. More candidly she confided in a letter about her universalistic hopes and of exactly what brand they were: "I'm so sure that if enough became actively interested in working hard on dream collections that something very startling would emerge both as regards 'basic,' 'modal'—and maybe even *universal* personality (and here I don't mean a universal Freudian symbol system whether this exists or not), as well as with regards to culture concepts, that I feel completely frustrated because more able people don't work on dream collections." And this "basic personality" was to be known through the path of data, in piled-up amounts. "I believe that an adequate sampling of dreams from enough cultures could give us a glimpse of . . . a possible 'modal personality of mankind.'" Here the enthusiasm for the method—the TAT, Rorschach, or dream record—is inextricable from the enthusiasm for the goal: grasping or grappling with human essence, a common quality. ("For the study of dreams, like the Rorschach and the Thematic Apperception Tests, are an adventure in understanding and the three are, of course, close cousins," Eggan wrote.)[44]

To go at this question via the collection and study of "subjective materials" was at once to sidestep the non-essential frippery of custom and outward comportment and move directly to the *What-is-it-like?* of this or that human example, the experiential reality of someone else seeing the world and, in effect, seeing one back. The database of dreams is full of such moments, refractive and glancing, brief and fleeting, but pointed, at which the subject breaks the flow of data extraction to comment on it or on the "Recorder," as Eggan names herself, or to mention meeting her in a dream (as in dream #3): "I was dreaming I was going somewhere in a train and I was reading papers and some women were sitting by and asked me everything and asked me about my dreams and I told her and she started writing them down," recalled a young "modern" Hopi man who owned a store but was ambivalent about *bahana* values required to succeed materially in life. Dream #3 continued:

> So I asked her what she was going to do with it. She said it was nothing at all and that she was only interested in my dreams and she laughed, so I didn't tell her any more and I got off somewhere, but I don't know where. I was at the beach somewhere because I could see people wearing their bathing suit and I went looking around and found the old chief at Old Oraibi in a bathing suit. I was just laughing at him for he looked so funny, and he was sure talking to some white man in English. He was telling the white man

about war and what he thought of it, but I didn't care to hear it so I went away. Just when I came away I woke up.[45]

When Eggan asked him to give associations to his dream, he said that he wondered "why you always asking me about my dreams" but was in no doubt about the good money it was possible to make providing them. The chief at Old Orayvi, staunch in calling for a return to old ways, would frequently scold this young man for adopting Christianity, so it was odd to see him in a bathing suit at the beach talking to a white man in English about war; in the dream, the chief was guilty too of intimacy with white inquisitors.

In other instances reported in the data set, the Recorder could not be addressed safely, as in the case of Joab's wife, a very conservative Hopi of about fifty-five, another who was forced by U.S. soldiers as a child to go to school for two years, during which time she did not learn English. "She refused to talk directly with the Recorder, not because she was unfriendly, but because she was afraid of Hopi criticism if she acted as an informant." Her dreams were recorded by her daughter-in-law, "Daf." A pervasive atmosphere combining elements of fear, temptation, reproach, suspicion, and accord surrounded the extraction of dreams. In the forty-ninth dream of Tammy, an educated Hopi mother who came from a conservative family but married into a modern one, the dreamer went to a Snake Dance on the plaza, where the housetops were thick with people, and brought unnamed "white friends," whom she placed in front so they could see well. In associations to the dream, Tammy explained to Eggan, "I had been worrying over trying to find a place for you and Fred at the dance. Yes, people do criticize us for such things but we don't mind for you." In this, Tammy interpreted her own dream, seeing the anthropologists inhabiting it. Likewise, Chad's view of his dream circled around the temptations the ethnographer presented and his own self-reproach: "I have taken things from *Bahana* friends like you sometimes but I feel guilty afterwards and that is why my dream scolded me."[46]

The dreams people told Eggan, then, reflected the truth of their circumstances, not only the Orayvi Split and stresses, but also the situation before them—in which their dreams actively were being extracted for data. As Eggan emphasized, her pursuit was humanness, and in this dreams were especially useful. Their use lay beyond the fact that they were said to qualify as "universal" in a simple-minded sense—that is, everyone probably dreams, as it was surmised but not yet known as of the early 1950s. (Only in 1953 was the correlation between REM sleep and dreaming made by Eugene Aserinsky and Nathaniel Kleitman and followed in 1957 by William Dement and Kleitman's differentiation between the mental activity associated with REM and that associated with non-REM sleep.)[47] Instead, by "universal" Eggan meant that dreams manifested the dynamics of any particular person's living reality as it was endlessly being constructed and wrestled with out of conditions and circumstances. Some might say a novel or any other creative act offered such dynamics, but here the process was unself-conscious, scientifically graspable, quasi-statistical, and potentially global. Note, too, Eggan's frequent use of the word "potential" regarding dreams, as in the "potential information" they held: her work was to secure them with an eye toward as yet unknowable future methods of accessing this potential.

In terms of timing, the gathering and dissemination of Hopi dreams coincided with the emergence of television as first an experimental and then a ubiquitous household machine or information distributor. The debut demonstrations of "modern television machines" in the United States in 1936 (343-line systems displaying grainy and greenish-tinged images) had a combined audience of about one thousand. By the 1950s the advent of the large, bright screen marked a golden age of television saturation of mass audiences. In Germany, England, and across North America, the device embedded dreamlike sequences of televisual programming in the normal hours of waking life. The extraordinary became ordinary. Man's strangest dream was coming true in one's own home.[48]

Taking the clue Eggan left to her intentions—that she aimed to access via dreams the "television qualities of the night life of the mind"—what can we make of the rise of television as a new medium occurring at the same time as her dream extraction was going on? The buildup of television as a household norm altered the texture of family life and the "ordinary." It was not simply adding a sense domain (sight) to radio, although this is how it was conceptualized at its origins. It was a different way of seeing and being a psychological self, a fundamental shift David Foster Wallace remarked on: "Television, even the mundane little businesses of its production, have become our interior."[49] By the time Eggan was advancing her dream theories in full force, the great era of television in its mass-media format had arrived, as well as the increasing concern by expert psychologists with mind-image-transmission phenomena such as the Petsal Phenomenon (the recovery of unseen phenomena via dreams), subliminal messages (through film imagery), and brain-washing commands delivered via media.

Putting it another way, the dream began to be historicized and materialized in different fields, almost as if floodgates had opened. Early psychoanalysis (Freudian and otherwise) drew on a connection between dreams and films, circa 1906. Film scholar Lydia Marinelli argues in favor of historicizing the dream. She takes a close look at how filmmakers, while not referring to psychological theories of dreams, of which they were ignorant, forged congruous theories of dreams of their own. Once the dream becomes a historical object, one can speculate that early film worked by analogy with a different kind of dream than did early television. And Eggan, in harnessing the non-traditional dreams of Hopi people rapidly becoming modern, on the cusp of the emergence of modern scientific dream research, was herself adding a new element of possibility in the figuration of and mechanization of dreaming; the mid-century operationalization of subjectivity was coming in this way into clearer view.[50]

Eggan by the late 1950s was becoming a well-known methodological innovator. Now, through her data, a viewer in Lausanne could tap into a batch of Hopi dream images (text-based, of course). If man's strangest dream was coming true in one's own home, one's own dreams might be landing in other people's homes. It is perhaps no surprise that in 1963 the British psychologist and writer J. B. Priestley issued a call over television broadcasts for pre-cognitive dreams and received through the mail over seven hundred.[51] In different ways, dreams were now circulated, networked, broadcast, and projected *things*.

Unique among anthropological dream researchers, Eggan began to value cross-cultural dreams for their television qualities. Sometime around 1955, when her friendship with Bert Kaplan began, a strong shift occurred in the place of the imagination (source of dreams) in her work. Less and less should the imagination be plumbed; less and less did she speak of the Hopi adaptive "way of life" as her central subject. More and more she asserted the capabilities and capacities of dreams-as-broadcast-technologies for understanding psychocultural formations.[52] The Hopi were an example of this. Dreams (and materials like dreams) were to help found a true human science.

By this time, Eggan was gaining a name for herself and began to travel in new circles of experts who shared her instrumental view of dreams. She wrote letters—at first intimidated, later exhilarated— from a conference she attended at Royaumont, a Cistercian abbey situated just north of Paris, where she joined a lofty company of scholars to consider "The Dream and Human Societies." Eminences such as the erstwhile surrealist and polymath Roger Caillois, she found, agreed with her that dreams gave clues to whole ways of life and human meaning, and were not simply the upwelling of personal preoccupations. When Eggan found herself to be celebrated, toward the end of her career (abruptly terminated by imminent blindness and subsequent suicide in 1965), she was appreciated for this fresh modernizing and externalizing approach. At the 1958 American Psychological Association conference, she found herself in the company of experts on subliminal perception, the neurophysiology of memory, and sensory deprivation pursuing the topic of "Dreams and Related Problems of Consciousness." There, as if it were the precondition for admission, the moderator eschewed any concern with "content" at all—"which we will not go into this evening." Instead, the heavy-hitting panel treated dreams as "problems." As if in accord with a tropism, they turned away from content and bent toward the sun of the "mechanism that accounts for the occurrence of imagery, the experiencing of imagery in its externality."[53] They were technique and tool oriented.

Let me keep the focus on this 1958 event for another moment: its significance is in its alternative formulation of the significance of the dream life. What brings together the following things under the rubric of *dreams*: hallucinations brought on by sensory deprivation or resurrected unseen images in subliminal perception experiments? These are phenomena at the edge of conscious thought and perception and thus qualify as "problems of consciousness." Dreams are the ur-form of such things—ordinary bridges into the extraordinary—and it was just being discovered, as William Dement testified at the same conference, that dreams do happen to everyone, every night, and likely (he thought) in babies as young as six months. Dreams were the Rosetta Stone but not, as for Freud, sending one down to the hidden recesses of *Seele*, self or soul. No, they were palimpsests for understanding what could be called "not-self," the place at which the self begins to shade away into nothingness or something else. As tank-bound polio patients dreamed of being elsewhere and produced wishful hallucinations (a boy in a tank hallucinates he is in his own bed), such subjects become "unusually susceptible to indoctrination procedures." Like brainwashed POWs, they can be turned from one to another set of beliefs. Likewise, Kinescopic experiments performed at the edge of the "so-called visible," according to Dr. Max Pollock, when followed by a mandated session of dreaming, caused the lost images the next day to enter the realm of the visible.[54] This traffic across the threshold of visibility, and the ability of the subject to recall "unseen" things, was of prime interest. It was a mechanico-neuropsychological process that could be identified. Labeled the "Petsal Phenomenon," it pointed to the "problem of isolating the dream as a crucial variable" for the "recovery of unseen material."[55] If the public's fear that a Madison Avenue Big Brother could brainwash us "in the comforts of our living room through television" was overblown or not yet feasible, still (from the other side, as it were) it was a vision of possible remote control capabilities that galvanized experimentalists.[56]

Eggan's views on the functions of Hopi dreams may have seemed like odd company for these exegeses, but her discussion of their operationalization as data, their connection to fantasy life, and their materiality (emphasis on manifest content *as a method* rather than on the content itself) linked her work securely to these concerns. Hopi dreams were, in her hands, phenomena akin to the Bethesda hallucinations found in John Lilly's experiments (at the Neurophysiological Laboratory of NIMH, where Lilly placed patients in a large water tank or "Lilly pond" within a dark room so that they floated in a state of sensory deprivation so as to produce visions) and the Kinescopic recall of forgotten images. "Television qualities" were built in to the

Hopi culture of dreaming. At least, it had been so. During a question-answer session, Eggan explained: "During the twenty years I have been going out there [to Hopi] there have been an increasing number of people who simply can't handle life as it is and they had to be taken to institutions. Now there may be two things at work here. I am sure that one of the things is they can't accept Westernization calmly. . . . They break down half way. They half believe in the Hopi god and they half believe ours. You see they are mixed up. They want the old way but they know the new way is coming. So, they are disturbed for that reason."[57] Eggan's lesson was that as Hopi lost their "Hopiness" and embraced "our" values through what she called Westernization, the televisual quality of their "night life of the mind" was transferred and transformed. It was no longer internal but external. Hopi dreams would soon be like everyone else's dreams, but until they were, they should be collected, all the more so from sources themselves undergoing this disturbing transition from old way to new way. They provided the key to everyone's dreams. The key to "humanness" itself was in their minute particularities.

This chapter has described how almost a thousand dreams from Hopi people in Arizona became social-scientific data, just under half of them microcarded. The transformation involved work of different kinds: fieldwork, methodological work, and theoretical work. Some of this work comes up against the unavoidable ephemerality of dreams. Even the setting down of a dream in handwriting or typescript entailed a necessarily massive distortion. Eggan elicited associations to dreams, conducted many follow-up interviews, and handed out notebooks for dreamers to write in (if they wrote)—an odd paradox for a group defined as non-literate. She needed to take advantage of their schooling, on the one hand, and still to choose those she considered *not yet thoroughly* schooled in white ways, on the other.

It is hard to capture a dream in any medium. Recently, writer Michael Chabon characterized dreams as entities that are by their very nature uncatchable by any scientist or artist, even one as gifted as Maya Deren, whose 1943 film *Meshes of the Afternoon*, "in the flickering of its pseudonarrative, the ostinato of its imagery, the strange urgency of its tedium, comes closest and yet still rings false, camera-bound, hokum-haunted."[58] Dreams do defy rendering as things. (For another thing, full dreams are not easily recollected.)[59] Still there is something unparalleled and strange in finding the dream of an Iluocan or a Sioux warrior, or of a Hopi grandmother, tucked away, even in its daily-ness and the terse, sometimes annoyed form found in the database. Perhaps these get closer in some way to the specific quality of actual dreams, this "success" due in part to the haunting sense of knowing that *this* dream, for no particular reason except that someone was there to trade a cigarette for it or pay a trifle and then to catch and store it, would be preserved for seventy-five years and more. The same goes for memories.

CHAPTER EIGHT

Not Fade Away
(A History of the Life History)

KAPLAN's database was a slumbering giant. Rare forms of data rested there. They did not exactly live, and they did not exactly die. Made up not only of dreams and test results, the database held people's life stories, known by such terms as "life histories," "personal documents," "human documents," and "expressive autobiographies." In its recesses and through the mechanics of its collections, people's lives became scientific data.

A Menominee Indian man who lived on the tribe's reservation in the Wisconsin woods told his story in 1949 or 1950 to an anthropologist, George Spindler, who gave the data to Kaplan in 1957 to be held in the database's second run. The central incident of this man's tale was his memory of escaping as a boy from the Tama Indian Sanatorium in Iowa, a long way from home. He had lived there for a year and a half, ever since the Indian Authority doctor on his home reservation said he had tuberculosis. "Well, for as long as I remember I was a sickly boy," he recalled, and when the recommendation came that he be sent away, his father agreed. (His mother had died not long before.) After arriving at the asylum, the boy, known only as Case 9 in the Microcard archive, lived his life as the institution defined it. He never got permission to go home, never saw his father, never heard his native language spoken, and never got really better but never any worse. He saw some come in "too far gone," and go out on a pallet in the morning.

When he was about thirteen or fourteen, in 1929, an older boy suggested he and another friend run away together.[1] They left some sandwiches and their suitcases concealed along the road, and around suppertime, with their dress clothes under their coveralls, they took off empty handed. The one road along the hills led to the highway, but they headed across the fields instead and traveled all night. Eluding the head doctor in his Model T, who was looking for them, they crossed the Minnesota line, sleeping under a bridge. Toward evening the next day they gathered their courage and asked a restaurant owner if he would shelter them, introducing themselves as "runaway jacks." The owner said they could sleep at the fire department, where, after helping out with the sweeping up, they slept. (Asylum school had taught them a lot about dusting and cleaning, by the boy's account, so they weren't long in fixing things up.) That night, the restaurant owner gave them coffee, fried eggs, and toast; in the morning, breakfast; then, coming out of the kitchen to say goodbye, he handed them a paper sack with sandwiches. The next night, still in Minnesota, a kindly farmer took the "runaway jacks" in and fed them.

Hoping their luck would hold, they resolved to ask if they could sleep in someone's barn. At another farmhouse, they knocked. The door flew open: "Well, what do you boys want?" On asking whether they could sleep in his barn, the boys heard the man respond, "Get going and keep on going." Cussing, he threatened to shoot them and started to make preparations, grabbing his gun. The boys "scrammed." The farmer's door slammed. "I was never so scared in my life," recalled Case 9. They hid in the center of the field. Meanwhile, the farmer went to his neighbor and "told the other fellow that we ducked into his cornfield. I supposed he didn't have no liking for Indians." The farmers sicced their dogs on them, but somehow the dogs could not find the boys in the cornfield. "That was the strangest part of my story. You know, it wouldn't take a dog long to find anything. I thought about this a lot. They had them guns along. I told my chums, 'Now you pray.' I watched that dog. Once in a while he would bark when he got close. He circled once then went on out of hearing. Finally all was still. And you could hear the door slam again." They got out and ran straightaway down to the railroad tracks, which they followed in a big hurry for three or four miles. "Finally, somebody said, 'Well, what do you think about this?' . . . I said, 'That shows us, we have a long time to live, otherwise that dog would have found us. He would have gone straight to us.' I think about that many times since then."

Hitchhiking and walking, the boys reached their home area, coming finally to several frame houses that "just from the way they looked we could tell they was Indian houses." As if in a dream, the homecomer found his town strangely empty, and the few people he did see failed to know him at first. "Well, when we got home, there was nobody there. I went over to my Aunt's and there was nobody there. And so we went over to other places, but there was nobody there. . . . Nobody knew me hardly. I was taller than when I left, I was about 12 when I left." The next day he went to Keshena to the fair grounds, where his family was staying, and "The first fellow I seen was my dad. 'Hello, dad,' I says. Of course, I talked Indian to him, but he didn't know me. 'Is that you?' he says finally. 'Yah, it's me.' Well, he grabbed my hand, and I believe it's the first time my dad ever kissed me. He was crying, he was so happy. Well, that's how I got home. That's the main incident of my life. I will never forget that trip home. I started with nothing, and I got home with nothing."

Case 9's story "slept" for thirty years, as he rarely told it. The man the boy grew into had always intended to send it to *True Story* magazine but had never gotten around to it ("I heard they give prizes. I think of that a lot"), so when an anthropologist appeared to take down his life when he was forty-four, he readily told the story as if it had already been written, and that is exactly how it reads. He concluded matters by telling the man collecting his life history, "Well, I guess that's all, George. There wasn't much to my life. The main incident was the time I run away from that sanatorium. That's the only thing that seems interesting to me."[2] He had never told the story to an outsider before, and perhaps it would never have been told had Spindler not come along. Like this one, myriad life histories in the data bank remained in a curious unresolved state, preserved but not published, archived but not really available. They rested in limbo. That is where I found them half a century later.

George and Louise Spindler, the married pair of anthropologists who collected Case 9's story, were among the most generous of all donors to the data compendium, contributing four Menominee data sets amounting to 741 pages of texts, charts, protocols, and tables, and constituting a unique resource.[3] They attempted to make their data comprehensive by providing a full range of materials, the substance of which included life histories (generated by a new method of Louise Spindler's invention called the "Expressive Autobiographic Interview"), records of peyote hallucinations (called "personal documents"), and multitudinous projective test records (using the standardized forms of the Klopfer Rorschach scoring method). Entering the Spindlers' data nexus, a reader could access the test results, first memories, stories, dreams, and daytime visions of a particular individual—such as Case 9—or look for patterns across the data—regularities among men versus women, children versus adults, veterans versus non-veterans, social elites versus shack dwellers, or regulars in the "P's Tavern drinking crowd" versus the abstemious. "Get the data" was their mantra in the field. "What mattered was getting the data," recalled George Spindler toward the end of his life—and behind it was an ecumenical wish. Aside from the forty-two books they wrote during a long, shared career, the Spindlers hoped future researchers might take their data and draw their own conclusions, for as George Spindler wrote to Kaplan, "We all have a stake in seeing that . . . the primary research data . . . can actually be used."[4]

The Spindlers, perhaps the most vigorous ethnographers of Menominee life in the mid-twentieth century, believed this culture was demonstrably going or gone (incoherent, disorganized, a hodgepodge of new and old customs serving few Indians well) as of 1948, when they arrived to begin fieldwork, but had been still functioning around the time of their predecessor, ethnographer Alonson Buck Skinner, a generation before. Though earlier ethnographers, including Skinner, had visited the Menominee "before their culture was broken," by the time the Spindlers arrived, "Menomini culture, in its total complexity and subtlety, ha[d] been lost."[5] This assessment implied a decisive loss sometime between 1915 and 1948. Yet in fact, the French had arrived at Menominee settlements some three hundred years earlier and began setting up missions in the 1650s, whence they never decamped. French furriers opened up trade in 1667, in response to which the once semi-sedentary Menominee divided into roving bands, each with its own family hunting grounds. Changes followed. Monogamous marriage and the small family became standard over the formerly dominant extended family group. Buying supplies at French trading posts in summer resulted by wintertime in outstanding accounts and an impoverishing cycle of debt. French army men and administrators entered tribal life and intermarried extensively. The English arrived in 1761, initially resisted but eventually somewhat popular due to extensive gift giving, and the Americans arrived in 1815, less popular with their agents, extensive bureaucracy, and parsimony in the matter of gift giving. No part of this adaptive history did the Spindlers deny, suppress, or attempt to hide.[6] Yet still they found a "threshold" the Menominee crossed just before their arrival, on one side of which lay intact Menominee traditions (a Woodlands aboriginal personality type, very stalwart, mordantly intelligent, emotionally inexpressive), and on the other of which lay distorted personalities, self-consciousness, "marginal men" struggling for self-respect in a vestigial culture, and various less or more successful ways of coping with modern dilemmas, from embracing alcoholism to churchgoing. For these reasons they called the first edition of their case study *Dreamers without Power*.[7]

To some extent, Menominee too thought in terms of crossing thresholds, though for them the threshold of loss tended to be just ahead in the future rather than just recently in the past. When asked what she saw twenty years in the future, a stalwart in Group A (the Spindlers' traditional or native-oriented cluster) told Louise Spindler in the early 1950s, as recorded in Kaplan's data archive, "I think everything's gonna be changed. I think there will be even more White man's stuff. Only a few can even talk Menomini [any more]. We were happy that our boy could talk Menomini to an old man. When the little girl has to, she can talk Indian."[8] These accomplishments were the result of concerted parental effort, effort not exerted by many others. Others discussed the story of Spirit Rock, a Menominee emblem, about which it was said that once it disappeared, so too would the tribe. Due to erosion, by 1952–1954 when the bulk of the interviews took place, Spirit Rock was noticeably diminished. When Louise Spindler drove by this rock during a life-history session (the Spindlers often conducted interviews in their car, using a heavy wire recorder), Case 35 confirmed the story: "That's true. The rock is disappearin' because of Indians marryin' whites. I shouldn't have married a white man. My children are on the tribal role. But my older girls are interested in white fellows."[9] The shock of new things had been predicted, as Case 32 recalled her grandfather prognosticating: "I was gonna see wars—my children flyin'—my grandfather said all that on a birch bark. He said I'd live to see my children in airplanes and I'd see people fall over and die. He wrote me this at Rainbow Falls; he wrote it on the [birchbark] house and told about cars without horses. I cried so. He said 'you're gonna see miracles.' And I've seen them."[10] Many who told their life stories to the Spindlers at midcentury agreed that in coming years there would be fewer and fewer Menominee ways observed.

This Menominee "structure of feeling," to borrow Raymond Williams's phrase, the feeling of just crossing or being about to cross a threshold, which the Spindlers shared in a slightly different register, served both parties well in the memory-extending work of the database of dreams. It gave them a mutual if not precisely parallel interest in getting down these life histories, stories shored against time's passage, stories that both contained what was remembered and stood for what was lost. Or as one man put it when shown the Rorschach inkblot of Card VI, "It is like a dead planet. It seems to tell the story of a people once great who have lost . . . like something happened. All that's left is the symbol."[11]

Today we—and by "we" I mean any interested person with an Internet connection—are in a position to assess the unique, forgotten records the Microcard archive holds, or at least to make a start. Taken together, the fifteen thousand or so pages of information about the inner lives of largely unwitting, sometimes unwilling, and usually illiterate subjects are akin to Freud's definition of the uncanny: things that might have remained secret but have come to light.[12] Groups of people, particularly American Indian groups, have their lives, or parts of their lives, contained there. These lives, as we will see, are both ordinary (familiar in their dailiness) and extraordinary (they seem to come from far away, across an expanse of time, or space, or unfamiliarity itself). This you-are-there capacity, creating intimacy while preserving distance, was designed into the life-history method. This is what it was meant to do.

More than two-thirds of the life-history materials in Kaplan's collection come from Native Americans, a large proportion within a globe-spanning collection. The few exceptions include John Honigman's "The Life History of a Pathan (Pakistani) Young Man" and Mohamet I. Kazem's "Autobiographies of Five Egyptian Young Women." The reasons for this apparent disproportion are key to understanding Kaplan's project and the modern American social sciences more generally. It was researchers' own familiar-strange history—American history itself—with which they grappled in taking the life histories of Indians, who were significant for being accessible (just a road trip away, available for summer research and group tours) and also for occasionally providing a glimpse of the non-accessible, the mysterious. For example, when George and Louise Spindler, new on the reservation, gave a ride to a young woman who asked them to take her to the tavern for a drink, as it was illegal at the time for Indians to buy alcohol, she revealed on their return trip that she was the niece of a medicine man, Shumaysen, who had "'some kind of power.'" "Of course our anthropological tongues were hanging out by then," the Spindlers reported. At his tarpaper shack, they met Shumaysen and became long-time friends. In him, secrets of old ways and powerful medicine promised a continuing if tattered mystique, like findings in a ruin. Eventually they gave him the Rorschach test and took down his life history, labeling him Case 1 of male Menominees. Putting American Indian lives into the format of introspective life histories was an act both of understanding and distortion, as historian David Brumble argues in his discussion of one of the most famous American Indian life histories, Sun Chief.[13]

Running through the Spindler Menominee data are people's stories rendered as data and cold-stored for future use. Just as Case 9's story came accompanied by a palpable yearning to recount the pivot point of his life to a wider audience, other Menominee actively sought the help of the anthropologist in recording their accounts (much as forebears had imparted theirs to earlier fieldworkers), and there was an element of collaboration in some of these efforts. George Spindler recalled how "conservative Menominee, who were

trying so hard to maintain a traditional style of life and hold to traditional beliefs, wanted to tell us stories. They wanted to tell us life
stories."[14] In other cases, participants were reluctant or completely in
the dark about the nature of the exercise in which they were participating. What does it mean to collect a life from someone, to give
a life to someone? Is the process a cooperative enterprise, an exchange, a theft, a gift, a memorial?

The technique of the life history arose to address and grapple
with such paradoxical thresholds and contradictory desires. It was a
method that attempted to both capture what tends to disappear
and bring order to its chaotic yield.

A life story as featured in the 1940s-era *True Stories* magazine
(where Case 9 always meant to send his story), or its current
National Public Radio analogs, *This American Life* and *StoryCorps*,
can be powerful testimony, a seemingly self-evident exercise in
public memory sharing. Yet the act of "taking a life history" is not as
simple as someone telling a story to someone else and is, rather, the
result of carefully crafted specialist techniques, scholarly movements, research imperatives, individual and group efforts, and, at
key points, the new possibilities borne by next-generation recording
machines.

The origin story of the life history might easily be located in
the work of Henry Mayhew in his 1861 *London Labour and the
London Poor: The Conditions and Earnings of Those That Will Work,
Cannot Work, and Will Not Work*, an eclectic four-part bible of
street ways and unseen lives (unseen by the middle classes, that is).
Journalist-entrepreneur Mayhew somewhat cheekily announced his
work this way: "*being the first attempt to publish the history of a people,
from the lips of the people themselves . . . in their own 'unvarnished' language*" and went on to provide a compendium of "street biographies." Within the pages of *London Labour,* the reader dipped into
accounts of crossing sweepers, Punch and Judy entertainers, doll's-
eye makers, dredgers, sewer hunters, mudlarks, thieves, prostitutes,
cigar-end finders, costermongers (barrow-boys or -men selling fruit
or vegetables), tea hawkers, dog "finders" and restorers, watercress
girls, and beggars—the background characters of Dickens's novels
now brought forward, "in the dark confusion of their pestiferous

urban ant's-nest," to front stage. In Mayhew's first encounter of volume 3, *Street-Folk*, he seeks out a rat killer who, on being assured Mayhew was not a dog-tax collector, granted him the privilege of meeting his premier bull dog, Punch, who proceeded to "jump around the room with a most unpleasant liberty," Mayhew noted. The man, originally from the countryside, described how, for a sovereign, he would kill rats with his own teeth, competing against his dog: "I'm the man as they say kills rats—that's to say, I kills 'em like a dog. I'm almost ashamed to mention it." In this way, by excluding his own questions from the text, Mayhew created an original literary genre mixing journalism and soliloquy-like answers. The result was a narrative-documentary with encyclopedic range.[15] Thorough as he was, however, Mayhew was no scientific-movement maker or methodologist, both of which would occupy and shape the activities of new generations interested in writing down people's lives.[16]

By the turn of the twentieth century, Chicago succeeded London as the site par excellence of life-history gathering, maybe because of the surging current of fantastic-themed modernization that coursed through the place. The city itself was like a living thing undergoing some kind of electrification process. Out of the systole and diastole of a new place, a new type of person emerged, one the Chicago sociologist Robert Park called "the marginal man." Caught between two cultures, one old and one new (as Park saw it), he experienced a divided self.[17] "It is in the mind of the marginal man that the moral turmoil which new cultural contacts occasion manifests itself in the most obvious forms," Park wrote in 1928, laying out a program for studying such men's life stories. "It is in the mind of the marginal man where the changes and fusions of culture are going on—that we can best study the processes of civilization and of progress."[18] The city worked like a microscope for such minds, whose owners were not only subjects but also scientific opportunities.

The formal beginning of the Chicago-style life-history method lay in *The Polish Peasant in Europe and America*, William I. Thomas and Florian Znaniecki's 1918–1920 work that "invented" this special type of document by naming it a "life record " (sometimes "life-history record") and collecting the documents en masse.[19]

Thomas and Znaniecki then proceeded, with an aplomb born of scholarly originality and a sheer mass of data opportunities, to analyze them over the course of five volumes. Chicago, with 360,000 Polish immigrants, ranked after Warsaw and Lodz as the third largest Polish-settled city in the world, and such figures provided a saga of uprooting and settlement in the early decades of the twentieth century. With a large grant from Hull House benefactress Helen Culver, sociologist Thomas originally intended to write a "source book" of rich empirical detail for each immigrant group—East Europeans, Magyars, Slovaks, Italians, and Irish. He began with the Poles, and while visiting Warsaw in 1913 met a charming intellectual working at the Polish Emigrants' Bureau named Florian Znaniecki, a philosopher by training who had studied under Henri Bergson in Paris but was unable to secure an academic post in the Russian-controlled universities. With alacrity Znaniecki demonstrated his willingness to relocate to the United States by turning up in Chicago with little warning, and the two co-wrote their compendious classic. Volume 1 contained almost two hundred pages of peasant letters, a total of 764 from fifty families. "The Polish peasant, as the present collection shows, writes many and long letters," and the researchers were prepared to pay ten cents per letter, as they announced in Chicago's local Polish newspaper. Volume 3 was itself the autobiography of a relocated peasant, the "Life Record of an Immigrant," Wladek Wisniewski, which, at 312 verbatim pages (though originally twice that size before being edited down), is considered the first systematically collected sociological life history. Given free reign with his pen and a motivating stipend, Wisniewski completed the task in three months. His personal story begins with his arrival in the village schoolhouse at age six and omits the details of early childhood and first memories that were already becoming Freud-inspired staples of self-narration. Studded with Znaniecki's sociologically informative footnotes on topics from pigeon breeding among Polish youth to styles of corporeal punishment, the text demonstrates (according to the two sociologists) how life records are the "*perfect* type of sociological material."[20] More recently critic Judy Long has pointed out the way that Thomas and Znaniecki's overbearing "narratorial voice introduces, surrounds, and even contradicts the subject's account"—a factor that in turn returns to the

problem at the core of the life history: how to capture a life techni-
cally without allowing one's technique to overwhelm it.[21]

In order to solve social problems such as alcoholism and vaga-
bondage, immigrant stress and the strains of modern life, Thomas
and Znaniecki wrote, societies in the "totality of their objective
complexity" must first be known and then compared. They frankly
admitted that their study did not stem from great interest in the
Polish peasant per se, but "the Polish peasant was selected rather
as a convenient object for the exemplification of a standpoint
and method."[22] Appropriately enough, the first pages comprised a
lengthy methodological note. Of course, the use of biographically
oriented data was not new to *The Polish Peasant*—certainly there was
a long history of literate people writing letters, autobiographical
statements, and testimonials—and enthusiasts for collecting life-
history documents gamely acknowledged precursors ranging from
Sigmund Freud and G. Stanley Hall to William James, whose
Varieties of Religious Experience (1902) could be considered the first
great book to "draw from paragraphs and sentences" of personal
testimonials. In addition to saints' and seers' published words, James
accessed a stack of "topical autobiographies" detailing conversion
experiences of unknown people that psychologist Edwin Diller
Starbuck had collected around the turn of the century via questionnaires. These lent to James's work insights "not in the haunts of
special erudition, but along the beaten highway."[23]

Znaniecki and Thomas's 1918–1920 work was different because it marked the first step in building a self-conscious method,
one that ultimately would become a meta-methodology—that is, it
would be methodologically aware of its own methodological
stance. The reliance on personal documents marked *The Polish
Peasant* as "a new departure," observes Martin Bulmer. "The subsequent use in sociological research of personal documents such as
life histories, letters, diaries, and other first-person material, may in
large measure be traced back to *The Polish Peasant*." As confirma-
tion of this departure, the Social Science Research Council (SSRC)
convened a conference in July 1938 and commissioned follow-up
studies (1939, 1942, and 1945) to discuss the significance of *The
Polish Peasant*. These methodological assessments became so successful and widely read that they eventually upstaged the volume

itself. As Liz Stanley shows, Thomas and Znaniecki's work came to constitute a "lost disciplinary past," an erstwhile classic whose traces came to rest "in palimpsest forms in once seismic theoretical and methodological debates and fierce controversies between competing schools and 'isms.'"[24] The disappearance of *The Polish Peasant* save in brief genuflectional citations, its reputation subsequently upstaged by the methodological stringencies of the "big science" moment in the social sciences, was a paradoxical outcome, for, as mentioned, Znaniecki and Thomas always intended their work to be read as a methodological call to arms.

Meanwhile, between 1918, when Thomas abruptly left Chicago, and the early 1930s, landmark Chicago-style life histories explored Chicago's social netherlands, including Nels Anderson's *The Hobo*, Harvey Warren Zorbaugh's *The Gold Coast and the Slum*, and Paul Cressey's *The Taxi-Dance Hall*.[25] Young Chicago sociologists set out for "the field" (the city), where they undertook informal interviews, close observations, and the intensive collecting of personal documents. They embraced new methods, though no one was very explicit about them—unlike *The Polish Peasant* with its eighty-six pages of methodological notes and unlike a phalanx of researchers to come. If the Chicago School of Sociology had at its core an empirical turn as seen in these street's-eye classics, then Thomas and Znaniecki, though latterly neglected, can in good part be credited for it. As Thomas recalled, he began to explore the city in the early part of the century, after a professor requested he find out "a bit of information about the saloons," the professor having never visited one nor tasted beer. Thomas's friend Robert Park, of "marginal man" renown, was an even more direct progenitor of the up-close method, not afraid to urge his students (as the writer Richard Wright recalled) to "trust their feelings for a situation or an event . . . [and to] stress the role of insight, and to warn against slavish devotion to figures, charts, graphs, and sterile scientific techniques."[26] Chicago School life histories aspired to be a non-sterile science.

In the following decade, the sociological life history evolved as the "own story" emerged, giving a reader word for word the life history of someone unequipped—by lack of education or low social status—to write his own, as exemplified by *The Jack-Roller: A Delinquent Boy's Own Story* (1935). A best seller in its day and considered a masterpiece today, it told the gripping story of Stanley, who grew up the neglected child of Polish immigrants in Chicago's "Back of the Yards" neighborhood—"yards" here referring to railroad yards, packing-plant yards, and livestock slaughterhouses. One of fifteen children in a family riven by tuberculosis and poverty (his mother died when Stanley was very young), he reached confirmed "delinquent" status at an early age. When he measured only 3 feet 3½ inches and weighed in at fifty-six pounds, he was already thieving, hanging out on the streets, failing to go home, and notorious for sexually experimenting with two girls. A confirmed jack-roller—jack-rolling was a hobby that consisted of scoping out drunk, old, or drunk and old men and beating them up for their money—he gained temporary salvation and a new path when he met the sociologist Clifford Shaw, who undertook to help him write his life history as an "own story" document. Shaw, after studying briefly with Park and urban sociologist Ernest Burgess, joined Chicago's Institute for Juvenile Research and used it as an urban center for studying the ills and crises of children labeled delinquent. Stanley lived there for around six years, during which he mostly kept out of crime. (The institute's basement likely still holds thousands of pages of unpublished life histories, sociologist Howard Becker speculated recently. Precisely such uncared-for sociological records were what Kaplan and his committee hoped but eventually failed to preserve in their data bank.)[27] His reform turned out unfortunately to be incomplete, however, and he was in and out of institutions, flop houses, and marriages for the rest of a long and tumultuous life—a life that became by the 1960s a classic "case" in the surging field of criminology. In 1979, emerging from a poker game, he was jack-rolled himself.

What made these count as full-fledged, scientifically valid "life histories" rather than simply people telling or writing their stories was the radical choice of subject—these were voices that were unknown, unheard, and generally unattended to—and also the assumption that these lives held huge significance, not so much in their individual particulars as in the how-the-other-half-lives cultural insights they delivered. This was true of any number of "other halves," not only the urban poor. One could access whole worlds otherwise unavailable to the middle-class social scientist or curious chronicler. There was a proper (if evolving) method for doing this.

As it emerged in its Chicago setting through the 1930s, the life history above all was a "method of data collecting." It was a targeted attempt to capture the uncapturable, the onrushing feel, the texture and rhythm of a life as it is being lived from another person's particular point of view, and in a systematic way. To access this point of view, experiential reality itself would soon become the holy grail of the human sciences. Yet such access was an intractably difficult assignment, the "impossible possible," to adapt a poetic phrase from Wallace Stevens. Could one observe and experience another's person's life as if it were one's own? Could one tap into someone's dreams and know what they meant to him or her? Modernist writers also tried to do the same, and their literary stream-of-consciousness technique as developed in the 1920s was, if nothing else, a concentrated effort to do what the social scientists also aimed to do: gain unmediated access to the inside feel of life being lived by a seemingly unremarkable subject. Along these lines, Progressive-era writing programs urged students to embrace the "find one's voice" paradigm of self-expression and, as literary critic Mark McGurl argues, these were programs of self-making designed to produce "original research and original persons at one and the same time."[28] Such programs increasingly embraced a faith in self-expression through writing, fostered by institutional support. You are Mrs. Dalloway for a day—or almost.

"Many, many" life histories piled up in collections by social-scientific "young hopefuls" in the 1930s, 1940s, and early 1950s.[29] A psychological self-consciousness could be observed not only in the social-scientific frameworks applied, but also in the way subjects structured their narratives. This structure also depended on the degree to which the teller came from a preexisting tradition of first-person narrative; there was such a tradition, for example, among Hopi, and there was not, for example, among Navajo. Dream scholars make a similar point about the unequal salience culture to culture of remembering and telling one's dreams. After 1945, narratives and case studies began to bore directly into the psyche, as seen in contributions such as the Hallowell student Viktor Barnouw's 1949

"The Phantasy World of a Chippewa Woman" or database-of-dreams contributor Edward Bruner's 1957 "Life History of a Fort Berthold Indian Psychotic." A young researcher, J. S. Slotkin, encountered a Menominee peyote worshipper whom he believed to be psychotic and took down his life story, to which George Spindler contributed a Rorschach reading for the Microcard archive.[30] Whereas the phlegmatic British school of social anthropology remained unmoved by the prospect of such documents, as it legendarily preferred kinship trees, American and to some degree French anthropological circles targeted these beguiling sources of insight. (More recently, under the banner names "egodocuments" and "life writings," scholarly centers in the Netherlands, Belgium, and Eastern Europe, some with roots in the immediate postwar years, have flourished—from the Estonian Literary Museum to the Dutch Diary Archive.)[31] With their emphasis on "internalization and motivation," these collections of lives constituted a "Copernican revolution" in data, as Kaplan's friend and data contributor Melford Spiro put it, valuing what had been unvalued.[32]

Methodological consolidation accelerated as researchers, hungry for state-of-the-art work, revived interest in sociologist John Dollard's landmark 1935 *Criteria for the Life History: With Analysis of Six Notable Documents*. Dollard had trained in Chicago, then moved to Yale for the Rockefeller-sponsored seminar in culture and personality, training among others Max Weinreich, the Yiddish-Polish scholar who would in turn spur the collection of some nine hundred life histories from young Jewish residents of shtetls, East European cities, and eventually the Warsaw and Vilna ghettos. Dollard's work in the American South, combined with his Chicago orientation, his psychoanalytic training in Berlin, and his experience in the Yale seminar experiment, stimulated him to attempt a systematic guidebook. *Criteria*, as Jennifer Platt points out, came after the heyday of the sociological life history, although, as Lewis Langness argues, it spurred a new surge of enthusiasm for the technique that would crest at the end of the war with Kluckhohn's contribution.[33] Significant for systematizing, the first sentences of *Criteria* bespoke an almost tangible methodological angst: "Every scientific worker has felt the anxiety associated with an insecure grasp on his object of research. Many of us, too, know the feeling of fumbling while trying to find the right way to think about a given problem. Very often the worker will proceed for his whole scientific life with a gnawing sense, appearing perhaps only at intervals, that he has never gripped his material in a satisfying way."[34] The insecure grasp, the fear of fumbling, the onset of gnawing fears, and the specter of losing valuable materials constitute the emotional tone of the book, which is only just compensated by the book's aim to provide a way to "grip" material in the "right way," satisfyingly, and ultimately to build "a beautiful and satisfying 'science'" right now. The life history, properly treated as usable data, properly remade into an "adequate life history" (rather than a "much-suspected tool") could do this. It was not a matter of treating biographical data like beads to be strung on a string or like a data "dump"—for the material "does not speak for itself"—but the way, when properly presented, the material captured a person within an active environment, being molded and modified, being socialized, and becoming a fully cultural entity. The person via a life history "must be accounted for in his full, immediate, personal reality." In this way, his story will show how cultures shape people and how people transmit cultures. Culture and personality, still in its infancy, still "primitive," promised, with the life history properly pursued and with new masses of data gathered and expertly shaped, to become a full-fledged science—a science defined as "organized realism." This was Dollard's vision, and it inspired many. In fact, he was scheduled to speak about the life history at Weinreich's Yiddish institute in September 1939, the day Hitler's forces advanced into Poland.

By the end of World War II, the life history emerged, or reemerged, as a special weapon in the changing fields of American sociology, anthropology, and psychology (and also criminology, history, social work, and medical practice). Now it was not the life history per se that stood out but its combinatory capabilities, for it could be mixed, switched, and hybridized with other methods such as psychometric testing, direct interviewing, and participant observation. This was a time of methodological hyper-consciousness. Thomas and Znaniecki eschewed face-to-face observation and special techniques of interviewing (Thomas felt they produced a "body of error"), but Park, Burgess, and others urged students to go out and talk to people but did not specify how.[35] Now in the 1940s the research interview and participant observation emerged strongly (along with the survey), and scholars belabored technical questions. Interviewing techniques, and methodological disquisitions thereon, proliferated in these same years. The non-directive interview (1942–1945), the Hawthorne psychodynamic interview (1928–1934), and the focused interview (1943–1956) were only a few that gained followers.[36] They produced a text where none had existed before.

And then, too, in any history of the life-historical interview properly belongs Louise Spindler's invention, the "expressive autobiographical interview" (which she shortened to the EAI), as captured in the data sets she donated to Kaplan's collection. In them, in-depth interview interactions combined with chronological autobiography, a hybrid of techniques Spindler explored in the late 1940s and early 1950s with Menominee women, lending to her data a "person-centered view."[37] An EAI took four to five hours of interviewing to complete. Louise Spindler did sixteen of them among sixty-seven women studied in all, and she tried to cover the different subgroups from native-oriented women who lived around Zoar to peyote-cult members to middle-class Menominee housewives in neat frame houses. Pre-slated questions interposed and sometimes interrupted the telling of the subject's story, some concerning magic and witchcraft, others about the subject's first memory.[38] Louise Spindler's EAI was a pioneering effort because interviews as generators of firsthand data for research purposes were "largely untested in anthropology in 1948." In turn, the EAI differed from interview techniques used by predecessors Cora DuBois, Ruth Schonle Covan, Margaret Mead, and Ruth Landes (all of whom had smaller samples, less well distributed). Note the concentration of female social scientists involved in technical innovation of life-history methods.[39]

Looking more closely, however, one discovers in Kaplan's archival data bank earlier experimentalists—that is, fieldworkers who were self-conscious about their use of interview and life-historical methods. One was Elizabeth Colson, a UC Berkeley graduate student who interviewed three Pomo Indian women in northern California in the summers of 1939, 1940, and 1941 as part of

Berkeley's Social Science Field Laboratory, which trained students in the intensive gathering of personal documents from "non-European cultures."[40] Young but not easily daunted, Colson spent her first summer alternating between dismay over the external facts of the lives some Pomo Indians seemed to be living—families picking through trash for leftover food was not her idea of a native American lifestyle—and determination to record the voices of native women, from whom "lengthy personal accounts . . . are all too few."[41] Sometimes Colson tried straight-out life-history elicitation ("spontaneous, unquestioned, life history"), but this method did not work in every case. Some subjects were shy, some developed migraines, some got bored. Working with sixty-seven-year-old Mrs. Sophie Martinez, who at first "protest[ed] she could remember nothing," Colson found that a dialogue elicited the story more fluently and with more intimacy than simply asking for a recitation. Interviewing Mrs. Martinez while she was in the fields stripping hop blossoms from vines worked best, Colson found, as the rhythms of the labor seemed to "le[ave] her less guarded." Colson reflected that interposed questions such as "How did you feel about this?" and "What did you say?" might successfully spur a reluctant or forgetful subject, but the attempt to obtain material in chronological order by use of the question "What happened next?" almost always failed, and chronological memory, especially for childhood, was tenuous. Later, Colson rearranged the data to fit a standard chronological sequence, even when this rearrangement sacrificed the informant's sequence of thought. Such "drastic revisions necessary to humor the historical bias of our own culture" gave Colson pause, she wrote, but keeping the document in its "raw data" form would have made it unusable for any but "the most hardened and enthusiastic pursuer of the personal document." This battle between fealty and functionality was one most life-history takers had to consider. The "recorder" (Colson herself) worked in shorthand as the women spoke, and being "faced with a notebook and flying pencil" did not seem to disturb them, though the occasional sharing of a cigarette did cause a break in the note taking. All three women spoke English (as their second language), but Colson edited for grammar while attempting to preserve the turns of speech giving flavor to the original. No words were changed or inserted. Perhaps her exactitude concerning methods stemmed from the fact that Clyde Kluckhohn, in addition to her Berkeley advisers, supervised her research.

The life-history interview was plastic and changing, and this mutability became ever clearer by the 1940s and 1950s, when field-workers such as Colson and Louise Spindler began applying much more methodological self-awareness. Some studies traded techniques, some influenced others, and some gave birth to further techniques. Almost all drew authority, to some degree, from the popularity and mystique of the psychoanalytic interview. Whereas some social scientists during the 1930s Golden Age of Chicago-style sociology had eschewed interviews as too lab-coat-ish—redolent of clipboards, clipped speech, and pre-set questions—the give and take of the interview, when adapted and melded with the borrowed ambiance and authority of the psychoanalyst's domain, eventually became an asset. (During the first heyday of the life history, instruction on how-to had been sparse. The chosen style at the time was the "verbatim interview," which was not verbatim in the sense implied today, but rather a recollected "report of the interview.")[42] By the end of World War II researchers had pushed beyond the standard survey interview to favor "non-directive" styles in which the scientist acted as a mirror. Dovetailing with each other, the research interview and the life document both saw vigorous innovation in Euro-American scholarship during these years.[43]

Around this time, some Chicago-trained sociologists of the "second wave" began recording life histories using wire recorders (e.g., Howard Becker with "Janet Frame," the girl heroin addict who told her story in *The Fantastic Lodge*), and some anthropologists, such as the Spindlers, also used (very heavy) equipment. Although Kluckhohn as late as 1945 urged researchers to study speedwriting, by the mid-1950s the "recorder," as both Dorothy Eggan and Elizabeth Colson had referred to themselves in their stenographic-ethnographic roles during the previous decade, now meant a physical device. Further technological innovation in recording devices, as well as the goal of "giving a voice to those who had been silenced," came with the oral history movement. Oral history regularized procedures involved in "taking a life." Manuals abounded: how to record oral histories, including the "live history interview," became highly technical matters. The recorded interview took center stage, its mechanics essential to the practice of oral history (unlike the life history, in which methodological exactitude was often overpassed).[44] A certain (limited and variable) self-reflexivity about such devices arose, though, as Platt pointed out: "Research on the consequences for practice of changing techniques and technologies for the recording of free answers is strikingly absent."[45]

By the 1960s and 1970s, it became de rigeur to use a portable recording machine in the field, wherever that happened to be. Certainly the use of machines shifted the dynamics of life-history giving and taking, as suggested by an anecdote from Andrew Strathern, who recorded and translated the life history of Ongka, a Big Man of Papua New Guinea, in those years. At first he prompted Ongka with questions but soon left him alone with the tape recorder: "I would return to the room from time to time to change the cassette over, and would find him fully engrossed, gesturing and smiling into the microphone."[46] Whatever new relationships might arise with new recording machines, however, this change is not entirely relevant to the history of the experimental archive—aside from the Spindlers' contributions—for the bulk of its constituent data sets was taken down in handwriting on charts, lined notebooks, or standardized forms; subsequently typed up; and finally micropublished.

Along the way, a "different style" of postwar social science emerged, one that valued systems and structures over the interactionism, non-reductionism, active pragmatism, and case-based method of the prewar years.[47] A symptomatic 1940 review article downgraded Thomas and Znaniecki's classic work as "slightly sophomoric after all," marred by use of "inept and worthless tools." In this assessment one gets a sense of the precise flavor of the new approach of the moment, which favored special and super-powered tools. The research situation itself became experimental.[48] Similar transformations, with scientific "rigor" a rallying cry, took place in the other fields that were in the process of coalescing as the American behavioral sciences. "Anthropological [and other] techniques of data collection and analysis bec[a]me explicit," the Spindlers pointed out.[49] So the life history increasingly joined the broader yet (paradoxically) more scientific ranks of the human sciences as a whole. Life histories became one of many multiform "tools" to be employed in the field of research, which was where they stood when Kaplan's project sought to gather them up.

Throughout those shifts, the life history evolved and added a particular valence. Such evolution is how the Spindlers' experimentally designed approach, seemingly at odds with their woodsy lifestyle, makes sense.

The 2,731 Menominee Indians who lived in the woods and small towns of a northeast Wisconsin reservation forty miles from Green Bay represented, to those who studied them, transition more than tradition. What sort of data did they afford, and what sort of place was the Menominee reservation from which the data came? For one thing, it was a spot where the husband-and-wife anthropological team of George and Louise Spindler undertook, beginning in the late 1940s, to gather personal data on a scale that could best be described as experimental. Although today the two receive reverential citation as key founders of educational anthropology and occasional denigration as central figures in the projective test movement—and in fact, as one of their students recalled, the Spindlers often received criticism from colleagues who saw them as "hooked on the Rorschach" and who intimated in the early years through the 1960s and 1970s (their prime Rorschach years) that their embrace of this projective technique could adversely affect their careers—the Spindlers might best be described, especially in their immediate post–World War II work, as meta-methodological experimentalists.[50] They saw the Menominee on their reservation as an opportunity to try out combining anthropological methods in a new way, borrowing techniques from across the human and behavioral sciences, recombining classic approaches with brand-new apparatuses, and reconstituting what it meant to do anthropology in the process. Naming their approach "experimental design," they devoted themselves to making "anthropological techniques of data collection and analysis ... [more] explicit." By the mid-1960s they were editing a series for Holt, Rinehart and Winston, "Studies in Anthropological Method," including one on the life history.[51] They also edited a series of anthropological case studies, including one by Hallowell on the Ojibwe. For them, it was not enough to do, without reflecting on the doing. Still graduate students when they arrived with their five-year-old daughter in June 1948 to camp by the side of Menominee's Moose Lake, the pair noted they had only one sparse instruction from their adviser about how to carry out fieldwork: Observe carefully and think of yourselves as newspaper reporters. Wait for patterns to emerge. This was not bad advice, they felt, but they spent their careers trying to revise it all the same. Through their work and the work they encouraged others to carry out, "processes of gathering, ordering, and interpreting data" should be skillfully employed but also systematically written about, shared, and subjected to improvement.[52]

Between the Spindlers' first and second summers in Wisconsin with the Menominee, George Spindler turned down a teaching assistantship at Harvard to spend an intensive year at UCLA in 1948–1949 studying Rorschach administration in Bruno Klopfer's legendary projective-techniques seminar. The following year, they returned to the Klopfer group bearing their sampling of Menominee males' Rorschach data "in hand" to begin analysis. Further discussion ensued when Spindler's 1955 report, notable for over five hundred statistical tests of differentiation and association in the Menominee Rorschach data, performed by hand by the Spindlers, came out. These qualities of evident "scientific rigor" and statistical analysis were what most impressed participants, though the Klopferians criticized the Spindlers' neglect of certain doctrinaire points. Along with the normally credited pioneers of the projective test, the Spindlers were some of its most vigorous early cross-cultural users. For many, the giving of psychological tests tended to diminish rapport and increase grumbling, for not only were psychometrics often tedious, but also the conclusions drawn, if and when they eventually circulated among test subjects, often came across as tendentious and offensive. In contrast, the Spindlers played on Menominee amusement at the slightly absurd spectacle they presented: "The people thought it was fun to look at the blots, called George Spindler 'Doc Psyche,' and cooperated willingly." Only one potential subject, or perhaps four—the number depends on which source one is consulting—turned them down. Robert Edgerton hailed their Menominee studies as "a landmark in psychological anthropology," noting, "the successful use of the Rorschach to enhance rapport is by no means commonplace in anthropological research."[53] With most of the 129 subjects, they began by giving the Rorschach, using it as a tool to get to know people, take a quick look around their houses, and do a bit of gossiping, only proceeding to take life histories after some time (up to three to four years) of acquaintance, or what was often by that time friendship.

Hardly resembling anything primitive, Menominee economic life by the time the Spindlers arrived was dominated by a "thoroughly modern sawmill and logging industry," whose workers cut 25 million cubic feet of lumber each year and which supplied a net yearly income of over $1 million. The Smithsonian holds stereographs dated between 1900 and 1910 of the then 55,000-feet-a-day sawmill on the edge of Menominee River. Half a century later, despite shrinking reserves of lumber, the mill was going strong. Foresters, lumberjacks, truck drivers, stackers, saw operators, planers, electricians, mechanics, engineers, warehouse workers, accountants, clerks, stenographers, salesmen, and typists came from the Menominee tribe. Other Menominee occupations ranged from the traditional, such as trapping and hunting, to the up to date, such as running a gas station, gathering ferns for florists, or selling "curios" to tourists—all of the non-logging activity adding up to a paltry $80,000 in annual tribal income at the time researchers collected their data sets.

For Menominee, there were forms of worship that were of Native American origin but were not strictly Menominee. The Peyote Cult arrived in 1914 from the Winnebago and before that had undergone a long, meandering spread from Mexico. The peyote worship tent had thirteen poles, one representing Christ and the other twelve his disciples; peyote worshippers chanted a simple phrase for hours, such as "All I know is Jesus save you" or "Hey, no, no, no, no." There was also the Dream Dance (Nimihitwin), introduced around 1880 from Potawatami and Chippewa sources and based on a female messiah dispatched from the West by the "great gentle spirit" Ksamanido with the message to stop drinking, not strike back if struck, and speak the truth; and the Medicine Lodge (Metäwin), which had Algonkian roots but was not strictly speaking Menominee, and had been active since at least the turn of the

century.⁵⁴ A 1925 photographic postcard of the Medicine Lodge an-
nounces, "Greetings from Shawano, Wis," offering tourist onlookers
and post recipients a tantalizing glimpse inside. All three—the
Peyote Cult, Dream Dance, and Medicine Lodge—were adoptive
belief systems that either incorporated Christianity in some way or
rebelled against the incursions of "Whiteman" ways of worship—or
both. Nonetheless, the Spindlers presented the Menominee who
participated in these forms of worship as traditional groups where
the "bulk of the old values" and rituals such as menstrual taboos,
fasting, and dream-seeking were still "vitally functioning."⁵⁵

By the mid-twentieth century, some Menominee lived distinctly middle-class, tea-drinking, bridge-playing lives—those in
the sawmill's upper management, especially—even as others still
hunted, fished, and gathered according to the seasons' patterns. A
1929 newspaper profile in the *Milwaukee Journal* of nineteen-year-
old "Menominee Princess" Alice Oshkosh described this "unusually
pretty girl" as fiercely dedicated to the welfare of her people while
also "in many respects ... a regular American girl" who enjoyed
dancing with white friends at St. Joseph's Academy in Green Bay,
where she attended secondary school.⁵⁶ Waves of religious influence over centuries led to further divergences among Menominee.
Most professed at least a nominal Catholicism, the French having
converted almost all members by a kind of mixed coercion and
coaxing, in all of which Protestant missionaries never gained a
toehold.

According to one account, the last overt resister to Catholicism
took his stand when the man being interviewed, the "Gray
Eminence" of Menominee men, numbered Case 20 by the Spindlers
in their 1957 Microcard archive data set, was a child. Now old, he
recalled to Spindler the last stand of his forebear, Neopit Oshkosh,
around the turn of the century. One day, the local priest, Father
Blaze, was ridiculing the Menominee religion, calling it a form of
devil worship and rudely imitating the " 'Wha-ho-ho-ho!' " of their
dances. Oshkosh became angry and condemned the Christian religion's way of installing itself: the priest collected alms and told the
people they would be admitted to heaven. He acted as if he owned
their souls. He was mistaken. "But it is only the length of your foot
you own here. You have no ownership of the Menomini Indian res-

ervation. You are selling your religion," the old man recalled over-hearing Oshkosh spit back. That was a lifetime before, however, and now he saw (so he told Spindler) that the priest was right and there was no place for the Menominee religion he once knew. Even his own son failed to understand, for his son, Case 20 reflected with sadness, knew nothing. "When I think of his future. . . . He is alone in the world with no understanding of life . . . of the old Menomini way. He is trying to make a poor copy of the white man's civiliza-tion. He'll never know anything himself, even if I was to teach him. He never contradicts me. That is about all I can say about my son." And so he ended his story within a story: "That is all. That is the end of this I wanted to tell you. Maha'w!" The anthropologist politely thanked him for the "interesting story"—and immediately requested to hear more about medicines and secret cures.[57]

Case 20 was a man who had seen one ending, the ending of his own way of life, come to pass and was witnessing its final extirpa-tion—so he felt. He told his story to the anthropologist, George Spindler, who appeared as if on cue, just as the earlier anthropolo-gist, Skinner, had made himself available as scribe for his father. Case 20 was deliberate about using Spindler as his scribe: "It is as Skinner said. Now you come talk to me, like Skinner did my fa-ther." He wanted to make a record of what he knew and what he had lived through because this process would soon reach an end: "But sometime a man will come to my son and he will know noth-ing. . . . Maha'w!" Even the anthropologists, the man fears, will have nothing to write down soon. The last round of Menominee secrets would die with him and his generation. In this he was wrong, for in the future anthropologists would turn to study the process of dispossession and acculturation itself, the "liquidation" or dispersion of the Indian "self" (along with its occasional resur-gence) in all its circumstances. Or perhaps, borrowing Theodor Adorno's words on the subject of childhood, they would build a new kind of anthropology around the "irretrievability of what, once lost, congeals into an allegory of its own demise."[58]

In return, Spindler's reply seems inadequate to his interviewee's emotional circumstances: "That is sad," he averred. "But I am glad I have come to you. . . . Do you wish to tell me how you came to get your medicine?" It was late afternoon, and his subject was getting tired. To a man who had witnessed a loss on this scale—a "cultural devastation," as philosopher Jonathan Lear put it, for even the sacred songs, he says, are now sung in Winnebago as the Menominee words, including those of the Shawano and Thunder cults, have vanished—his interviewer responds like Prufrock's salon ladies, turning to another subject, interested in another meaning, refusing his offering.[59] The strength of the human document, in the form of the life stories and interviews provided in the Spindlers' Microcard archive, is that all of this exchange with its questions, hesitations, things left unsaid, back-stepping, negotiation, and hopeful appeal, is included, unedited, even now—as it was not in the published anthropological work.

A single life or the life of a family—through the influence of earlier life-history researchers such as Dollard, Allport, and Kaplan's teacher Kluckhohn—was about much more than the totality of its details. Such records spearheaded "an endeavor to discern through the lives of individuals or families the broader contours of the social and cultural landscape." Thus a life history was not an autobiography at all, not from the point of view of social science, at least. It may have seemed similar in form and was likewise shaped by the course of a life, but it was guided by other priorities. As Chicago-trained sociologist Howard Becker put it, a life history, in contrast with the "more imaginative and humanistic forms" of autobiography, was a different beast, a beast devoted to its master—the data collector or sociologically minded inquisitor. "More down to earth," "less concerned with artistic values," and more devoted to "a faithful rendering of the subject's experience and interpretation of the world he lives in," the life history's goal was to give one the chance to see the world literally as its subject did, with the help of whatever technological add-ons or aids that might require: "To understand why someone behaves as he does you must understand how it looked to him, what he thought he had to contend with, what alternatives he saw open to him; you can only understand the effects of opportunity structures, delinquent subcultures, social norms, and other commonly invoked explanations of behavior by seeing them from the actor's point of view."[60] Unlike projective tests, life histories allowed the teller the right to make meaning of his own life, to interpret her own life. At the same time it allowed the listener to make a science based on "organized reality" (Dollard), to "keep[] the game honest for us" (Becker). It asked, "How close could you get to another person's point of view?"

The life history was a procedural technology, and with it a new scope of data could be contained, recorded, kept. When the database of dreams came together in its final form, it combined these and other technologies on many levels.

CHAPTER NINE

New Encyclopedias Will Arise

As the result of a congeries of techniques, tools, and technologies stretching over decades, the Kaplan archive began to work. If you walked into the library of UCLA or the University of Wisconsin in 1958 (say) and looked in the card catalog for "Kaplan, Bert" or "Primary Records," you would find a listing that resulted in your being given a neat set of $3'' \times 5''$ glossy white cards. Well, neat or not so neat: if you asked for the whole collection that year, it would fill a good part of a drawer. As the Kaplan-issued brochure hastened to assure potential users, it was a "standard library card file drawer." (As another brochure mentioned, any Microcard collection "can be conveniently stored in file cabinets that accommodate 3×5 inch file cards" so that it can be "filed and retrieved as easily as a standard library index card.") A message was in this way communicated. Here was something new: stacks of mini-pages with unseeable words condensed like textual thumbprints on cards, yet still they fit into existing drawers. Exactly which drawers—those of the card catalog—suggested an ambiguity Kaplan's group had not yet faced publicly.[1] Such problems as indexing and privacy would only later become urgent. Meanwhile, the fact that the Microcard's inventor, Fremont Rider, had trained as a student of the legendary American librarian Melvil Dewey, became Dewey's secretary, wrote Dewey's biography, married Dewey's niece, and cultivated a long-standing interest in the finer points of library standards—expending much early career effort on topics such as "A

Tentative System of Subject Headings for the Literature of Military Science"—made his choice of standard-size library-friendly units unsurprising.[2]

Collections one and two, already out in 1958, would soon be followed by three and four. One might look something up by the master table of contents; from there, smaller subsets could be called up on their own. Note that the translucent, eyestrain-inducing microfiche sheets with which most library researchers of a certain era are all too familiar had not yet been invented, although this format would later eclipse the Microcard. Meanwhile, a welter of earlier micro-storage efforts had fallen out of favor—the Photoscope, Film-O-Graph, and Fiske-O-Scope among them. As described above, this made the Microcard the up-and-coming text-storage technology during the time of Kaplan's experiment.

The 131 cards in the first run, each one holding approximately forty 8" × 10" pages of data, occupied scarcely more drawer space than a clementine: "The entire set for the first volume will occupy only 1 ½ inches," proclaimed Kaplan's brochure announcing the availability of this new resource, and, if one reads it carefully, even today, one can glean that there was something truly revolutionary, in the committee's view, about this debut collection: "Now, for the first time, hitherto unpublished source material, Rorschachs, life histories, dreams, collected by anthropologists and psychologists in the field appear readily obtainable in compact and permanent form."[3] While it is true that by 1958 many other things were available on Microcard, Kaplan's collection was still an unprecedented happening. Normally Microcard delivered medieval manuscripts, old English literature, technical reports, blueprints, or billing receipts. There were signs of an increasingly experimental spirit as well. The same year, the Council on Library Resources announced in a press release that the first scientific journal to convert itself exclusively to Microcard format, *Wildlife Disease*, would begin publication in January 1959, experimenting with optimal layout as well as portable readers.[4] What was new and different in Kaplan's enterprise?

These materials were unprecedentedly intimate. They came direct from the psyche's newly named terrain that was just being charted: inner space. They also came from people unused to providing such things—American Indians and representatives of "cultures other than our own," as Kaplan sometimes put it—people, one might say, who were themselves being charted, in some cases for the first time. For American Indians, this charting occurred in an ongoing process Western historian Thomas Biolsi calls "internal pacification." In a study of Sioux history between the 1880s and 1940s, Biolsi shows how investigations of Sioux life delved increasingly into psychological domains. Such research did not go deeper and deeper into hidden unconscious areas so much as further and further into the psyche, which was treated as a kind of territory to be mapped. But the mapping also helped change the territory. Not evenly or regularly, but painstakingly, a transformation of the Sioux "self" was under way, as Biolsi describes it, and the process of being measured, counted, quantified, and (eventually) tested served to aid and abet the subjective changes taking place. In the early 1940s, the Sioux Project with its staff of nineteen fieldworkers, ten test analysts, a supervisor, a field assistant, an editor, and an advisory committee almost twenty strong amassed reams of data—of which a portion, Royal Hassrick's projective protocols, made its way in 1961 into the Kaplan data archive.[5]

Selected to "encompass[] representative samples from widely varied areas," Kaplan's data were intended to be not only specific, but also generalizable. In addition, a prefatory account from each contributor accompanied his or her data, explaining "the nature of the study being conducted at the time the material was collected, the locale and characteristics of the cultural group concerned, the methods used and conditions under which the samples were taken."[6] Each set also came tagged with a geo-cultural code such as AB43, FX10, or NN1915 (Okayama, San, or Tuscorora respectively). These tags derived from the work of Yale anthropologist George Peter Murdock, who established for anthropology "a uniform system for the classification of societies and cultures comparable to the systems used in the biological sciences for the classification of organisms."[7] The codes told the researcher, at a glance, where in the world and who in the world had generated the data found therein.

The data were representative and they were samples, then, in addition to being highly specific and rather intimate.

Also in Kaplan's archive were to be found accompanying materials that expanded on the dreamer's identity and told you (among other things) his or her life goals, sexual complexes, scores on Rorschach tests, and storytelling urges. In the case of certain particularly meticulous researchers, Rorschachs and other tests came collated within the data set with dreams and life histories so that a full picture could be resurrected. (In most cases, correspondences, when possible, had to be traced by means of the pseudonyms or codes.) Perhaps it is true that seeking to render dreams as actual too-solid things is a doomed project—a critical pile-on has followed, or rather predicted, the argument of one of Jorge Luis Borges's late lectures: "The study of dreams is particularly difficult, for we cannot examine dreams directly, we can only speak of the memory of dreams," Borges urged in 1977, toward the end of his life, speaking at the Teatro Coliseo in Buenos Aires. "And it is possible that the memory of dreams does not correspond exactly to the dreams themselves. . . . If we think of the dream as a work of fiction—and I think it is—it may be that we continue to spin tales when we wake and later when we recount them." Yet of such methodological impediments mid-twentieth-century methodologists were well aware. They worked daily among them. Dream specialists in the culture-and-personality movement were careful to study the way dreams worked in each culture and not to assume uniformity. Erica Bourgignon collected and wrote about Haitian dreams, arguing that they entailed two levels of perception: "the dream is already partially interpreted when it is told," observed Bourgignon of Haitian dreamers being questioned.[8] The dream and its telling were two spun tales. Even if one believed in a dream that perfectly preexisted its telling, the researcher might never be able to access it. Nonetheless, there had to be a record amenable to study and, for modern researchers, manipulation. "Strictly speaking we should call our project a study of what people say they dream about. This has ever been the case and probably always will be, for no one has discovered a means of transcribing a dream while it is being dreamed," remarked Calvin Hall almost wistfully.[9]

Collecting such human documents became a problem of scientific technique and "how to" because Kaplan and like-minded researchers saw dreams in a technical manner, phenomena that could be understood as functions within a system. The key premise, for Kaplan and his cohort, was that dreams were powerful examples of "common projective experiences" and thus yielded insight into other "culturally defined projective systems," with which they merged.[10] Dreams were a way for the researcher to access directly the meaning of how people lived and even the processes by which meaning (shared, cultural, or personal) was actually made. When pooled with other subjective materials and understood by means of insights from across the human sciences, dreams were levers in reaching a "common goal—the global understanding of man." Coordination of effort was key: proceedings from an international convocation of oneirologists called in 1966 for "urgen[t] . . . interdisciplinary cooperation" so that results and problems could be shared across neurology, anthropology, psychology, and sociology.[11] It was a high point of international-scale utopianism, and such calls for cooperation were common.

But just what was a "global understanding of man" urgently pursued? To say that it hardly needed explanation among certain types of researchers circa 1958, and that it very much does today, is to hint at the eclipse of a once taken-for-granted paradigm about the nature of human nature. In the process of looking for answers (discovering through hard work what was man), it postulated pre-accepted premises along two lines: mankind's universality and mankind's particularity. The first (universality) was an inheritance of Enlightenment thinking, the second (particularity) of Romantic historicism, and these two together marked a path by the end of which eighteenth-century ideals and nineteenth-century historicism had become twentieth-century techno-utopian pragmatics. On the one hand, dreams and other subjective materials were clearly unique to each tribe and place—only the Haitians valued Haitian dreams in a particular Haitian way. Yet on the other, Haitian, Ifalukan, Ilongot, and Hopi dreams spoke to a common condition. No matter how different, there was always a potential sameness, as Nils Gilman has described for allied mid-twentieth-century modernizers.[12] Although this notion recalled older "psychic unity of mankind" debates among anthropologists, it was also the result of a post–World War II movement to encourage democratic personalities that would exist beyond petty nationalism and the blood-and-soil-based exclusions of fascism.

Let's pause for a moment to visit a contemporary event similar to but far better known than Kaplan's archive. The group photographic exhibition *The Family of Man* opened its doors in the early spring of 1955 at the Modern Art Museum in New York and eventually, over the next decade, hosted some 7.5 million visitors via portable shows that toured regularly. Although it was an experimental exhibition rather than an experimental encyclopedia, its values were similar in important respects. Play between the same seeming irreconcilables, the universal and the particular, was also on display there. Out of a repository of 2 million submitted photographs, Edward Steichen and his team chose some 500 images by 273 photographers from around the globe. Displaying panoplies of intimate black-and-white images that captured love, war, happiness, pestilence, "dream," birth, death, and the stations in between; offering biblical and Navajo texts scrawled on its walls; and purveying an all-around proto-immersive design, the show was meant (for the artists and the designers, as well as the financial supporters and the public visitors) to "show Man to Man across the World." And like the global Microcard archive, it contained contradictions: explicitly anti-racist and anti-totalitarian, the exhibit promoted freedom but also functioned by stimulating and mining subjectivity. The viewer was a participant, encouraged to be free but not too free, to wander through the exhibit and yet emerge with a basic, shared message. The very symbolic and semiotic act of walking through the 3D environment—each viewer navigating his or her way through its image-maze of hanging, miniature, wall-sized, surrounding, and at times intrusively jutting photographs—taught the viewer to manage alternatives for how to find meaning in a still new postwar world.[13]

Kaplan's database also offered a compendium of alternatives for making one's way in and through its data maze.

Sit down, then, and take a look. What do you find? It is something close to raw data, yet it is carefully arranged and gathered, almost curated. On one level, it was a fairly simple research tool as it stood. You could choose what you would like from the catalog and look it up. "These cards are, therefore, easily catalogued, and individual contributions may be readily selected and extracted for reference," announced Kaplan's accompanying missive.[14] Calling up Dorothy Eggan's collection of Hopi dreams, for example, produces a set of cards not searchable by word or specific topic, although there is a guide at the front to page numbers and types of dreamers (man, woman, or child)—just as in a full-sized publication. To navigate the tiny pages, move row by row, adjusting the machine. To find a particular cache of dreams—say, from children eight to sixteen years old—shift the Readex lens to the relevant card and skim the pages. Or if you have a pocket reader, you can do this in a more leisurely and perhaps even ambulatory manner.

Kaplan was taken with the Readex hand reader, despite its frustrating shortcomings, which were the subject of new research. In 1960 a "Review of Progress" noted new grants to improve the hand reader for micro-opaques "either through reduction in cost of the existing devices or through the improvement of its optical characteristics. It was found possible to convert this reader for use with ambient light instead of with lighting equipment and thus to reduce the cost material."[15] Held close to the body, promising portability, and functioning like eye or limb extensions, such devices, even when not quite doing what they were supposed to do, retained their appeal. Battery operated or plugged into 110-volt current, they came with a transformer-plug, connecting cord, and protective plastic cover. Their billing as "pocket-size" and "slightly larger than a package of king-size cigarettes" emphasized how easily grabbed they were.[16] Offering such access was a dogged refrain for American scientists interested in cultural and personal materials. They sometimes evoked a data set's ease of use by phrases naming body parts such as "at hand" (Kaplan) and "near the elbow" (Rider). Data-pioneering predecessor George Peter Murdock spoke of having a world of information "in readily accessible form" at the fingertips, dispensing with the need for time-wasting "legwork" and trips to the stacks.[17] The hand trumped the legs. In this spirit, Kaplan's data sets both perpetuated and partially fulfilled a promise.

If you sat in a library looking at someone's dreams, what were you seeing? It was something close to raw data arranged in a network—a "pooling of data," as Kaplan put it in an early vision, linked by new technology. The network extended across the United States and tentatively into Europe. Thirty full sets printed by the Microcard Foundation as a first run sold out immediately, within three weeks of the initial announcement—and it "astonished and delighted" Kaplan, who noted that another ten orders were on hand and would have to wait for a second printing. A letter from

Raymond Firth at the London School of Economics suggested the psychology department under Hilde Himmelweit would "certainly be interested." In this lateral handoff he confirmed the oft-made observation that British anthropologists were allergic to adding psychological approaches and kept their disciplinary domains neatly apart; but for a more comprehensive set of social anthropology data, Firth speculated, there "might be more demand in this country." Meanwhile, Chief Acquisitions Librarian Rupert C. Woodward of Louisiana State University in Baton Rouge wrote to inquire into purchasing a set. The University of Colorado's anthropologists considered the records most valuable and had a fleet of Microcard readers and a large microfilm and Microcard collection already, while conservative scholars at places such as Dartmouth seemed ready to be won over: "Although like many of my colleagues I have somewhat mixed feelings regarding the culture and personality field (due to the somewhat extravagant claims of some of its proponents)," wrote one such scholar, "I feel quite sure that your cross-cultural project represents an important step in the right direction." At its height, by my estimate, Kaplan's "revolutionary" archive sat in about a hundred research libraries worldwide, as well as some military or governmental organizations and health institutions; at an early stage a handwritten note in Kaplan's paper listed fifty-eight universities, soon succeeded by a count of seventy-five, served by Readex machines and sometimes pocket readers. But we don't know exactly the peak user count because the records, as I will explain, have had a tendency to melt away. This melting away itself is part of the "problem of data" of which Kaplan's project was an early articulator.[18]

In its heyday, the Microcard archive worked and people used it. Sol Tax, then chair of the department of anthropology at the University of Chicago, saw "wonderful progress" being made and expected the numbers of participants "[would] go much higher." Optimism ran so high that soon, within two decades, as Kaplan predicted, augmented systems like his would make "sophisticated . . . search and retrieval" possible. Researchers even attempted gamely to employ the experimental pocket reader. Staff clinical psychologist Evan L. Wolfe of the U.S. Naval Hospital in Oakland wrote Kaplan for information on the pocket Microcard reader so he could consult the cards on the go. Kaplan had to admit he "was perhaps over-optimistic in implying that the pocket reader is an adequate substitute for the library reader," and though he found it "quite satisfactory" himself, others had "some difficulty in using it."[19]

Historian Megan Prelinger's recent catalog of mid-century space-race personnel ads, *Another Science Fiction*, reveals an emerging visual language that was anything but drab as corporate firms such as Boeing and Raytheon attempted to capture the science-fiction-steeped imaginations of the best engineers and lure them to work in their laboratories. Their ads communicated "a sense of fantasy and possibility around the process of technological emergence that was erupting."[20] It was Cold War baroque. There was something of this in the Kaplan data collection, where old-fashioned materials documenting old and (purportedly) disappearing ways of life captured by complex tests came cached in super-high-tech technological micro-reduction machines. In it one saw the combination of "a practical tool" (as the Microcard Foundation's brochure called its technology) and a fantastic liftoff of futurism. The humdrum details of Zuni men's boarding school memories or arguments with their wives, the remnants of long since disclosed secret ceremonies, along with their participants' test scores and psycho-sociological diagnoses, were like cloud seeds, little patterned bits that could add up to design materials with which to work.

The archive's successful run stretching from the mid-1950s to the early 1960s could hardly have been guessed from its ragtag beginnings just after the war.

The Memex was a fantasy machine that was never built and thus never actually existed. Yet history has proceeded as if it had and did.

In 1945, computer engineer and influential Washingtonian Vannevar Bush, writing in the *Atlantic Monthly*, put forth his vision of a "Memory Extender" or "Memory Index." He did not clarify which he meant, but he stipulated that it be made out of an oak desk.[21] The desk was the key—for Bush's epochal vision of how to join human and machine memory systems did not take the form of a robot. It was a kind of thinking furniture.

Memex was a hypothetical assemblage of parts that worked to-gether within a single office desk, making use of both the desktop and the interior. The parts included a voice-recognizing typewriter, a set of electromechanical "keys and switches" and push-button controls, projectors, and two translucent viewing screens made of frosted glass. At its heart was its own built-in library: a set of super-high-resolution microfilm reels that stored records, correspon-dence, books, pictures, and technical papers. On the desktop surface sat the two screens on which the operator, by pressing keys to his right, could project the microfilm data Memex held. "It would look like a printed page, [but] better illuminated and easier to read than the present printed pages," predicted Bush, and it would include the user's "longhand notes . . . also."[22] Moving between screens, calling up one image and dismissing another, the researcher could juxta-pose and link any two records. It was "a device that would supple-ment thought directly rather than at a distance," predicted Bush.[23] It collapsed distance. It was an intimate machine.

The miniaturized library that resided within Memex's desk was meant to grow bigger—almost unlimitedly so. By running a photo-graph, document, or experiment on the desktop, the operator ex-posed the information to "a tiny camera the size of a walnut" he had affixed to his forehead. This strapped-on mini-camera with a universal lens could photograph anything its wearer looked at and would then create a micro-image—that is, microphotographically shrunk down many times smaller than original size. As of the 1925 International Congress of Photography, when a grainless microfilm debuted capable of storing the entire text of the Bible fifty times over on one square inch of film, the medium's capacity for storage seemed poised to revolutionize scientific inquiry and knowledge making. Such spectacular achievements were not, of course, rou-tine (and, furthermore, they raised the question of who needed fifty Bibles end to end), yet microfilm nonetheless seemed to be on its way to fulfilling its promise of potentially storing the totality of the world's information. This was a promise of which Bush was well aware, having worked in the decades before World War II to develop a primitive "search engine" to trawl microfilm records—in the form of Bush's successively unsuccessful Comparator and Rapid Selector information machines. In fact, it was while working on these failed machines in the mid-1930s that Bush first envisioned Memex, though he did not publish his account until the end of the war. But Bush was not only a narrow-gauge inventor, nor a mere technological visionary. Bush saw the engineer as an inheritor of an American pioneer spirit, and in the 1930s he wrote of the need for engineers to cultivate a wider vision attuned to social and political questions of modern life even as they forged into new territory; the engineer could thus be a new kind of frontier hero, creating "trails in the technological advance."[24] One could argue that these metaphorical trails realized themselves more literally in Memex's data trails.

All the while working microfilm developed in capacity, and soon the medium would be able to shrink data by a linear factor of one hundred and still be able to project them to their original size. "The *Encyclopaedia Britannica* could be reduced to the volume of a matchbox," Bush predicted.[25] Thus according to Memex's dreamed-up design, an almost unlimited amount of data could be added with ease because it was stored in miniature. Space was not an issue. Such an achievement was already in the realm of the possible.

A graphite illustration of Memex's expansion-through-miniaturization process appeared some months later when *Life* magazine republished Bush's article from the *Atlantic Monthly*. It presented an ever so slightly balding, bespectacled personage wearing a leather headband with an additional band running atop his head to hold the camera in place. The subtitle, "A Top U.S. Scientist Foresees a Possible Future World in Which Man-Made Machines Will Start to Think," conjured up fear and hope, and the term "possible future world" with its overtones equally cautious ("possible") and bold ("future world") was alive to the contradictions and accidents already becoming commonplace in the nascent information age. The gaze of the man in the picture, his eyes focused downward and slightly cross-eyed, his spectacles cross-hatched like targets, gave off the kind of blank engagement that may result from the labor of envisioning worlds to come.

Current commentators heap Memex with credit as the precursor of everything from the Internet to the World Wide Web and tack on personal computers, Wikipedia, and hypermedia to the list. Tech blogs name it "the Internet's first visualization" (or expressions along these lines), and scholars of Internet history treat it to "constant acclaim." Meantime, this never-built machine latterly secures Bush's place as the "father of contemporary information science" alongside his many other more visibly successful roles—overseeing the Manhattan Project, for one, and in addition managing six thousand research scientists' contributions to the war effort.[26] Whether as an "image of potentiality" for the information revolution or a retro-futuristic blueprint for desktop computing, Bush's Memex and what it symbolized—the arrival of the cyborg into that most un-sci-fi of objects, the work desk—has thrived. It has been taken as an augury for just about every new thing: "If there was a more eerily prescient piece of prose, fiction or otherwise, written in the first half of the twentieth century," observed William Gibson of Bush's Memex article, "I don't know it."[27]

On the other hand, all this eerie prescience may be overdone, a Pavlovian itch to locate a proper visionary from whom the current data-centric moment seems to spring, such a visionary needing always to be credited with prophetic powers that range right up to the exact present instant but not beyond. As historian of computing Colin Burke argues, Memex, despite elevating Bush to seminal status, was "an ambiguous and not too original concept," in part due to the fact that Bush was isolated from state-of-the-art librarians and *avant la lettre* information scientists. According to Burke, he was a "giant in engineering," but his "role in the emergence of associational indexing and computerized information retrieval has been greatly exaggerated."[28] Some of his most important cataloging machines, after all, either failed or did not exist in physical form.

Bush never did build his Memex, but many others, including Kaplan, did. They built up and built on the spirit of the unbuilt microfilm machine, with its desk-centered construction, its fingertip promises, its experimental approach. In bits and pieces, by means of awkward constructions and forced hybrid hacks, but with a strange doggedness, they pursued the task of putting within arm's reach a whole array of deeply personal knowledge, all based on the then most promising technology available for intensive data storage and retrieval: microfilm—specifically, the Microcard. If Memex was, according to the prominent mid-century librarian Verner Clapp, a "private memory device," then Kaplan's machine was a device actually made of private memories. Its design was to pool large amounts of data, where the data in question were dreams and dream-like phenomena.[29]

In this sense, Kaplan and his group were not just using innovative storage techniques but were innovating at different levels. They bootstrapped a whole array of high- and low-tech elements— sociological methods, fieldwork designs, testing protocols, dream-rendering strategies—and assembled them in new ways. Theirs was a memory extender extended, dog-eared perhaps, but made into a real, material thing.

And yet it was not only that. Recall that the other speculative meaning of Memex, aside from memory extender, was "memory index." The *Oxford English Dictionary* gives the earliest meaning of "index" as the finger that points (ca. 1390), and the index refers essentially to pointing, or things that point (such as the needle of a compass, the hand of a clock, or the alidade of a surveying instrument), for the following two hundred years. Then, in the sixteenth and seventeenth centuries it came to mean, among other things, an alphabetical or otherwise ordered list of contents, a table of contents, or a commercial reference list of "the names of Men, Wares, and Ships." Indices were for tracking, listing, tallying—and often entailed tedious work to compile. Soon enough, historian Thomas Macaulay could speak of "starving pamphleteers and index-makers." Finally, several computer-associated meanings of "index" seem to emerge just after Memex and the Microcard archive, including this one: the "Glossary of Terms for Automatic Data Processing" in 1962 described an index as "a sequence or array of items with keys, used to identify or locate records."[30] Thus a "memory index," if that is how Bush conceived Memex, pointed to a lacuna in indexing capabilities, something not yet possible (but just about to be possible), for an index, aside from index cards, generally had a set order. Memex was a dream of movement.

Unlike the sorts of objects that seem to sit, merely material and merely passive, Memex was a thing with purpose, a thing that combined the efforts, tools, plans, and dreams of a range of people (including inventors, adopters, contributors, and institutions). This confluence of effort and technique created, in effect, a new kind of assemblage, something that philosophers of technology call "distributed agency." Political theorist Jane Bennett recently described assemblages this way: "a material cluster of charged parts that have ... affiliated, remaining in sufficient proximity and coordination to function as a (flowing) system." She is describing a modern electrical power grid, which she understands to be a system with distributed agency, but this insight can be adapted to this mid-century neo-reference device: think of knowledge, machines, dreams, gathered together, electrified.[31] It was the creation of a group, not a single inventor, the extension of a movement rather than an individual scientific will.

In a nutshell, Memex demonstrates how imaginary machines become real. Before "big data" existed as a phrase conveying equal parts hope and a leviathan-like threat, Kaplan's group and other obscure researchers began to amass data on a new scale and to fill their archives with types of never-before-collected materials. Their goal was systematically to link these to each other through their pooling of data or series of accessible data sets.[32] Although Bush had not specified what exactly Memex would hold aside from the fact that it would serve scientific inquiry, Kaplan's Microcard cohort targeted new sorts of data. This was one of its original steps.

When in 1945 Bush declared, "Wholly new encyclopedias will ... appear," he was referring to the generative quality of the "trails" researchers could trace through Memex's stores. But it was not only trails through information that generated new encyclopedias. It was trails through the previously unconnected sets of things the very workings of the technology brought together. Not to be too ardent, but it was, finally, trails through America that it offered, especially with its preponderance of materials from American Indians. So once again Bush had been prescient. Kaplan and other social scientists following him, latterly unknown, were those new encyclopedists.

The Microcard archive was not a great machine, exactly—not in the sense people now hail Memex, not in the technological determinist spirit by which precursor inventions and inventors receive periodic anointing today—and yet the process by which it came to exist was the more extraordinary for that fact. A team of specialists from different academic areas and technological subspecialties built this unique data-pooling device before there was even a well-articulated need for such a thing. The device then performed a function not yet fully articulated or seen as even desirable. William Gibson uses a phrase to characterize extreme forms of the seemingly limitless capacity for repurposing: "The street finds its own uses for things."[33] It entails the hacking of purpose. But it is another thing to build devices without quite knowing beforehand what their capacities must be—working collectively and driven by vying parameters, freewheeling conversations, tentative or decisive experiments, dueling imperatives, and the weighing of values and needs. For as we've seen, Kaplan's committee wrestled over months and years with the parameters of what their device would be and do and hold. Would it contain soft or hard data? Would it use microformats or filing cabinets as storage? Would it incorporate test protocols or life-history materials? Would it build on existing infrastructure or create its own? How would it network the data? Comprehensiveness of content vied with universality of function as goals. The privacy of individuals sparred with the ecumenical desirability of sharing data. These debates guided development, and the most valued values were always shifting.

They came to adopt a non-rule-driven, non-canonical, improvisatory style of working without a real plan in unknown territory. This style might be called "epistemological pluralism," a term Internet theorist Sherry Turkle and computer engineer Seymour Papert invented to describe an alternative way of interacting with the digital world and its demands.[34] At the cusp of the digital era with analog machines such as Kaplan's, epistemological pluralism happened too. In the realm of proto-big-data, tinkerers created oddball contraptions using existing techniques, tools, and human labor, innovating with what was at hand, for purposes not always completely clear.

Half a century ago, Claude Lévi-Strauss in *The Savage Mind* introduced the concept of bricolage as a way of describing such a process of ad hoc, haphazard building. Lévi-Strauss saw mythmaking as the preeminent area of bricolage and opposed this to the deliberate actions of the engineer. However, a broadening of the concept in the work of Jacques Derrida painted all knowledge as the fruit of bricolage: "If one calls bricolage the necessity of borrowing one's concept from the text of a heritage which is more or less coherent or ruined, it must be said that every discourse is bricolage." In the information sciences, in particular, the concept proved fruitfully descriptive, as when Sherry Turkle in *Life on the Screen* advocated the "bricoleur-style" of programming as valuable in contrast to the more top-down, abstract, rule-driven "planner" approach. When researchers, each with his or her own immense set of observational data, try to talk about their scientific projects across fields, they must establish metadata and rely on certain forms of improvisatory exchanges, the authors argue. This is increasingly common.[35] Influential concepts such as "trading zones" (Galison), expert witnessing (Shapin and Schaeffer), and immutable mobiles (Latour) are ways of describing these comings-together that were not at all planned.

A contrasting case is the history of the card catalog, at least as historian Markus Krajewski paints it. In *Paper Machines* he argues that the long history of this device, extending back to the early modern period, reveals the formation of an emerging type that developed through channels and streams. There were "elements of risk" in the new uses to which it was successively put, but the catalog persisted and evolved. As a possible progenitor to a looming inheritor (the computer), the card catalog was a type, a paper machine, that Krajewski argues was, in effect, a Turing machine. One can tell the card catalog's history through its different iterations, which amount to a series of linked failures, perhaps, but are still a distinct, alternative lineage: one of experiments in data processing and mobile card slips. There was a preexisting "what" that carved out a space for this universal machine. From this example, Krajewski argues, we see that the production of innovation never happens ex nihilo but "always includes a recombination of disparate or similar elements. In short, the production of innovations is always based on the fortified recombination of the existing."[36]

Here are two recent views of how modern technology develops, then: the unplanned (as in bricolage) and the proleptic (a term I've borrowed to mean the apparent anachronism of something that is dreamed of before it actually exists or, as *Merriam's* defines it, "the representation or assumption of a future act or development as if presently existing or accomplished"). In the case of the curious dream archive, there were elements of both at work. On the one hand, there was no "what." Distinctively, Kaplan's unlikely machine arose at the intersection of anthropology and psychology and was sui generis, a thing for which there was no type. Yet it did pioneer certain functions—in particular the fast mining of ever more personal information—that appear to be part of a distinct lineage. The Microcard-based collection seems proleptic: it embodies a projected future development. On the other hand, it was so obviously a product of improvisation, scrambling, and accidental encounters, making use of top-down structures when possible but never entirely guided by them, that "epistemological pluralism" sounds like an understatement.

In this sense—that of targeted haphazardness—Kaplan and his collaborators were pioneers, Memex inheritors. They used already operative systems, just-created tweaks, oddball ideas, and clever bootstrapping to create something new, and in the process they nudged an idea about the relationship among subjectivity, memory, and data-storage machines, or a significant part of that idea, into becoming real. An analog device that combined and redistributed the fruits of experiments, tests, techniques, wayward conversations, and governmental policies, it was a machine for holding the most human kinds of data, data that could be and often were known as human documents.[37] Via its links, Memex mimicked the brain's associational organization. On the other hand, Kaplan's psycho-anthropological data storehouse did not achieve this kind of personalized functioning. What it did was to take the artifacts of the brain's workings themselves—those subjective materials, traces of waking life, dreams and daydreams—and make them accessible from remote technological outposts.

Between Memex's hypothetical machine and Kaplan's actual data clearinghouse one sees a step from an imagined technology to a real one, a transformation that in part answers the question, How do ideas take on physical form? This transformation was part of what Memex's creator hailed as a coming "mental revolution" that would be much more far-reaching than the industrial revolution, for "this time steps in the thought processes are becoming mechanized."[38] Step by step, Kaplan's "database" grabbed hold of certain kinds of thoughts—"subjective materials" direct from the inner life—and mechanically stored them in an altered form. Nothing quite like this had been done before.[39]

Questions about the privacy of subjects arose, returning to debates about circulating secrets back in southwest sojourns at Zuni and Hopi. Among data collectors who contributed to the Microcard "database," there were worries: "A number of workers have been troubled by the possibility that their subjects might in some way be embarrassed or hurt by the publication of their life histories, dreams or projective test data," Kaplan wrote. "This is a general problem in the publication of personality materials and it is even more important in the Microcard series in which original records are being presented with a minimum of alteration." Kaplan's solution to this "general problem," which became even more problematic in his own enterprise: in such cases, workers must remove all identifying names from the records and substitute some sort of code or pseudonym. Life history, of course, made this remedy relatively ineffective. Anyone who knew the Hopi village or Fort Berthold society, where the subject lived, would be able to crack the code (unless many elements were withheld or rearranged). In response, Kaplan suggested limited distribution as a limited solution. The documents would not be available in any old local library or Indian Bureau but in research institutions and high-level policy circles. (In fact, Harvard experimented with censorship in its related Five Cultures project, instructing the Ramah library and the BIA not to carry its publications on the area and all researchers to use Scotch tape on outgoing correspondence to provide a barrier, however flimsy, to prying eyes at the local post office.) Kaplan's attitude was more forthright than the Scotch-tapers' in acknowledging front and center the "delicate issues" that arose when "basic data" circulated, and he called on users of the Microcard resource to exercise judgment and care in not revealing sensitive information—an injunction this author likewise has attempted to heed.[40]

A reading of Judith and Herbert Williams's 1962 Microcard data set, containing some two thousand pages of dreams and personal information, hauntingly invokes some of these concerns. The set begins with a young mother from a small village called Hadchite, in Lebanon, settled by a mix of Maronites (Christian Lebanese) and Muslims. Her own mother had died of an acute illness while the young woman was a baby on her breast, and she was raised the cherished and somewhat spoiled daughter of a well-regarded priest. As a grown woman, she has many family members living in a Maronite settlement in Michigan and is relatively well off according to village standards. She is interviewed a couple of times, beginning in early January 1950, but only during the third session does she begin (according to the researchers) to speak freely instead of haltingly. She responds to a request that she describe her childhood with the response that she has already told everything and is now eager to speak about her marriage. Nothing will deter her, even targeted questions about other subjects. It is as if a stream of words has broken loose, write the Williams: she becomes suddenly animated and begins to gesture as her words are taken down in dictation, only interrupted by the woman's inquiry, recorded in the anthropologists' parentheses: "(Here the subject inquired whether the material would be kept confidential.)"[41] No doubt she was assured on this point, though the anthropologist does not say so in the text, because she continued speaking, and in fact her name was changed in the document to "Rachael."

From the reader's position today, overlooking such intimate revelations, there are layers of resonance: the level of detail is so thick over Rachael's seventy-four pages of "dream autobiography" that no doubt she would be easily identifiable to anyone who knew the village, perhaps even today, for the records are only around sixty-five years old. And the fact that they have been available for fifty of those years in microfilm rooms of research libraries all over the world is striking. But beyond that is the fact that I can read it now—on a pdf on my personal computer, macrophotographed, formatted, and digitized through OCR software. The fate of these data is not yet known and makes me aware of the potential consequences of my own research. As more and more of what we do is recorded and stored online, observes Catherine Crump, an attorney at the American Civil Liberties Union's Speech, Privacy, and Technology Project and an affiliate of Stanford's Center for Internet and Society, "It's important to focus on the more mundane ways in which, on a personal level, each of us now has what we say and do, where we go, recorded in a relatively permanent and constant way."[42]

Perhaps because of the Kaplan committee's limited and (as seen from a long-term perspective) obviously untenable solution, their creation's "failure to thrive" can be seen as sidestepping an issue that would eventually emerge with great force. Yet it already existed in nascent form. It spoke to a conflict in access, a divide of security, and an unbridgeable gap between privacy and exigency that would become ever more acute.

The beauty of the "double experiment" described in this book was to take materials collected under extravagant and sometimes hallucinatory theoretical imperatives—such as A. I. Hallowell's Rorschach romance, or Jules Henry's Freudianism, or Cora DuBois's experimentalism—and prepare them for other uses not yet known. This does not mean they became (like a remade virgin) somehow raw or pure data once more, renovated after being falsely interpreted. I am romanticizing neither the data themselves nor the possibility of "data themselves"—that is, interpretation-free information.[43] As mentioned, the data were full of (and forged through) privacy abridgements, ethical lapses, and complex human relations. Yet one can see, also, the curious creations that inhere in the archived words, as when a Salteaux woman pointed with an orangewood stick at painted inkblots to describe what she saw, or an anthropologist shared a cigarette to encourage the sparse stanzas of his informant. They do look like discarded treasures, from a certain point of view, an observation made too by historian Alan Rosen, who examined the abandoned TAT records of European Jewish "displaced persons" in 1946, some of the first to be interviewed after they were liberated from concentration camps. If the projective test did not do what the researcher wanted, Rosen remarks, still it is "meaningful today for the poetry it evoked."[44] This kind of paired and sometimes group project was a different form of literary-cum-scientific creation. It bespoke an alternative tributary, data packaged and envisioned not as raw or pure but as usable units prepared for future access. They describe an already existing possible future.

CHAPTER TEN

Brief Golden Age

B Y the late 1950s, money for the database-of-dreams project was drying up, and desultory meetings dragged on in less and less distinguished hotels. Vague talk of "some uncertainty in the situation" cropped up in correspondence.[1] The data bank, however, continued to expand, and major Microcard-borne publications appeared in 1957, 1962, and 1963. A flowering of new kinds of data was taking place, but no one seemed to heed it, at least not those in positions to continue its funding. The project was being bureaucratically buried.

Not that there was a lack of interest. Some of the most prominent names in social science supported the Microcard data bank and recognized its visionary qualities. Hans Wallach of Swarthmore was "all for the project" and "happy that someone was thinking of such things," as observational material tended to lose its value when reported and "original protocols are what one wants to see."[2] Even Kaplan's language was ahead of its time, and this fact some recognized. The well-known psychologist Solomon Asch affirmed Kaplan's foresight in targeting data as a problem: "You are quite right to talk about the challenge of data."[3] In 1957 there were very few people in the world who could easily employ the phrase "challenge of data" in this way. The word "data" had not really entered common parlance at the time, and it certainly did not yet bear the connotation of the problems that now perennially shadow it. (A list of data problems includes, for example, National Security Agency

[NSA] global surveillance; personal privacy in the midst of ever more intrusive social media; and the quandaries associated with the ongoing digitizing of books, newspapers, university teaching, health-care services, and other entities—otherwise known as "ephemeralization" or sometimes just "disruption." Even the much-used "big data" saw its first deliberate deployment as a computing term in 1997, when two NASA scientists pointed to the "problem of *big data*.")[4] Sixty years ago Kaplan and his cohort trafficked daily in data challenges and data opportunities. Meanwhile, the phrase "data bank" had its first usage in 1966, not incidentally around the same time a welter of phrases sprang up to describe what one could do with massive amounts of data, often with the aid of computers— "data processing" (1954), "data transfer" (1959), "data handling" (1964), and "data capture" (1966), for example.

Taking off from Kaplan's original vision, senior psychologist Solomon Asch (Kaplan's one-time teacher) imagined him leading a roving experimental group of data experts who could lend a hand at the planning stage of any project: "In social psychology especially the validity of studies can be checked. I would not restrict this to the qualitative side. Your group could have a servicing agency and provide advice as to how to arrange raw data of the future so that they could be most useful. Your group should be present at the planning stages of research projects. You should experiment in this. Our own data: I'd be glad to have it done. . . . These things of ours should be arranged before they get too cold. Many of my interviews are on tape."[5] Here was a vision even grander than the grand role the Primary Records Committee at times entertained for itself. The "raw data of the future" required a master plan and people willing to experiment with the best ways to keep, store, and share them. "These things of ours," in short, needed an infrastructure.

Today, this problem—the "curating of data"—is only beginning to inspire a sufficiently urgent response. In a recent *Science* issue on data, over 80 percent of scientists surveyed (representing the full range of sciences) said they lacked sufficient funding to preserve their laboratory or research group's data. "There are many tales of early archaeologists burning wood from the ruins to make coffee," remarked one environmental scientist responding to the survey on data preservation. "If we fail to curate the environmental archives we collect from nature at public expense, we essentially repeat those mistakes."[6] From this point of view, it is not surprising that the project of saving the "raw data of the future"—precisely Kaplan's aim circa 1957—failed to inspire funding agencies to continue their support. Certain individuals Kaplan met during a year-long road trip—actually three intermittent trips he undertook in order to canvas the "state of data" according to bigwigs in the sociological, psychological, and behavioral sciences—expressed personal concern for the security of their respective fields' data. Roger Russel of Stanford and Edwin Boring of Harvard discussed with Kaplan the need for a central archive on the history of psychology to be housed somewhere safe, and regional experts, distressed at the "deterioration of valuable records on the Pacific Islands"—as well as other places—saw the urgency of massive micropublishing to form archive centers to preserve them.[7] Others admitted to having fantasized about building or at least accessing such a "central registry" as Kaplan hoped to construct. Yet few concerned themselves with systemwide scientific data preservation.

An undercurrent of feedback to Kaplan's project suggests an answer to why the money people neglected this important activity, then as now. Some responded to Kaplan's queries by remarking that his undertaking seemed worthwhile but it was not where the fun lay; as psychiatrist Maria Rickers-Ovsiankina of the University of Connecticut observed, undoubtedly it was important to preserve data sets, but, really, "Who wants to do all this work?" She further suggested that it was only those with an "anal tendency" who would be drawn to the boring task of rescuing endangered data sets. Certainly the step from "original data" to published study involved a loss of "life," but regrettable as such loss might be, it was dull as toast to be the one assigned to redressing it by preserving others' sheaths of original data.[8] This was a common view, yet not one that either troubled Kaplan or seemed to penetrate his consciousness. In an ecumenical and future-focused way, buoyed by his "endangered data" sensibility, Kaplan was undeterred. Yet the project would founder in part because of his inability to express to others exactly why it was urgent. By 1964 Kaplan would effectively shutter it.

What went wrong? Some sensed Kaplan had not promoted his project with sufficient savvy and that otherwise it might have succeeded and even revolutionized the fields it touched—for as events have surely shown, buying future options on databasing was a speculator's goldmine. Today, one might argue with a degree of retrospective flourish, we are living in an "age of Kaplan" in the social sciences, even if much of Kaplan's specific work toward this goal has been forgotten. Had he only caught the promotional bug, the thinking goes. Had he only knocked people over the head with his invention. In fact, he was too diffident in his entrepreneurial expressions, if not in the project itself. During the La Salle Hotel heyday of the NRC committee, George Spindler—ethnographer and master data gatherer to the Menominee of Wisconsin—complained in a gentle-but-firm letter to Kaplan that he could put more effort into simply making the data bank's existence known. Why do so much and yet fail to promote the project? If he wanted to go from having at most 10 percent to making 80 percent of potential users aware of the Microcard set's existence, he would have to (effectively) shout from rooftops via departmental notices, oral announcements, a good deal of hand-shaking, and a general call at the annual meeting of anthropologists. "We all have a stake in seeing that this attempt to make primary research data available is publicized enough so that . . . the materials can actually be used," Spindler wrote, assuring Kaplan he did not mean to criticize.[9]

To some extent, this failure to push was linked to a rather lovable character trait in Bert Kaplan: he was not particularly self-seeking or self-promotional. If he picked someone up at the airport for a job interview in his department, he might very well greet the person carrying a volume of Hegel or speaking of nothing but tennis (a topic that was to become of pedagogical interest to him). He was always full of revelations but rarely a determined publicist for his own projects. To some extent it was always on-to-the-next with him. "Nobody ever knew quite what to make of Bert. And he didn't write a lot," the man he greeted with Hegel recalled many years later.[10]

Yet Kaplan in his modest way felt he was making efforts, and he printed up 150 copies of a brochure to promote the Microcard data bank: "I am very pleased with the way the brochure looks. If it cannot do a good sales job, nothing can," he wrote to his senior colleague Hallowell. Living as he was at the dawn of a great age of persuasion, of the ad man and the PR campaigner, Kaplan evidenced a naïve view of what a promotional campaign could be asked to do, if nothing else. He professed to be "astonished and delighted" that the thirty sets of cards the Microcard Foundation had printed were already sold out and ten additional orders were on hand to await the second printing.[11] As of 1957, spirits were still running high despite money-flow questions. He continued to rely on the project essentially to "sell" itself, and such reliance—along with some technological developments—doomed it to an early if not complete obsolescence.

Looking right into people's heads, as if one had opened up a window to the self or applied an X-ray to the psyche, was another way Kaplan carried out cutting-edge research interests of the day. Others dreamed of or partly enacted such projects. Robert K. Merton, the most prominent American sociologist of his day, fantasized in a 1956 footnote to an otherwise workmanlike manual for giving group interviews, *The Focused Interview*, about the ultimate social science tool—the "introspectometer." Buried in the how-to instructions for use of the interview technique, a careful reader could find a single, seemingly offhand mention of this intriguing "hypothetical machine," at once admittedly imaginary yet possibly real. The introspectometer, according to Merton, would act like a movie camera, capturing the data of an actor's life while he or she engaged in it: "A technological contrivance—an introspectometer, so to say—. . . would record, in accurate and intimate detail, all that the individual perceives as he takes part in social interaction or is exposed to various situations. . . . It would provide, in other words, a motion picture of the individual's stream of experience as he is engaged in the situation." The machine, however, must work in secret, and the subject must "not be aware that the apparatus was at work" in order to secure the most undisturbed data flow.[12] It would be like a 360-degree camera filming the inner happenings of its targets. Full information could be tracked in real time. The closest existing technique to the vaunted machine was the literary method of stream-of-consciousness, Merton observed, citing James Joyce's *Ulysses*.

Even though the introspectometer naturally could never justifiably exist—if it did, it could become a "collective nightmare" as easily as a scientific boon—Merton suggested, in a seeming contradiction, that the beginnings of such an instrument "have of course been made" in several nascent forms, including the Lazarsfeld-Stanton Program Analyzer and his own focused interview.[13] The imagined capacities of the introspectometer are reminiscent of the actual workings of projective tests, at least as framed by their users.[14]

Cold War–era look-inside-your-head fantasies flourished even as "cooling" sets of data lay neglected, and, meanwhile, Kaplan began looking for a new angle. First he sought a successor to carry on and maybe renovate the project. (He found one, but the passing of the baton seems to have been for naught.)[15] He was now a "hot" professorial property, and in 1964 Rice University tendered him an offer to move from the University of Kansas to its psychology department, where he had free rein to found a new cross-disciplinary institute for the study of behavior. "Rice out of the blue it seemed offered him a job as head of the Department of Psychology," Kaplan's wife recalled. Despite the fact that her husband "did a lot of research on Rice and on Houston and discovered, among other things, that the British embassy people got hardship pay for working in Houston," the couple with their young family found it not to be a hardship at all. It was nice living in a Houston dominated by oil barons' wives and astronauts, she adds: "We were invited to all kinds of dinners of that sort."[16] Living in proximity to the Manned Spacecraft Center, NASA's hub for human space flight, itself established in 1961, seemed to top off Kaplan's space-age approach and his personal trajectory within the behavioral sciences. The university itself was the creation of a wealthy department store founder, its libraries full of books whose pages had not even been cut, and Kaplan had a standing offer to order any book he wanted any time, a munificent detail that set the tone for his new project.

At Rice, Kaplan took direct aim at public life, for he felt that the cross-cultural research on the "psychology of peoples" he and others engaged in could be useful in practical affairs. "Behavioral scientists have in general experienced a frustrating difficulty in establishing direct links between their work and pressing national and international problems," and he proposed that his institute at Rice change that—with further financial help from the Ford

Foundation.[17] Around this time he abandoned his vision to save a whole world of psychological data sets and began focusing on a new project researching Navajo mental health, as he had gotten along so well with the Navajo. He pursued the Navajo research over several summers (starting in 1961), filling notebooks with detailed observations of Navajo mental illness and cultural pathology, when he ran aground. A *Newsweek* columnist, Saul Friedman, came at the behest of the renowned sociologist David Riesman (who was a big supporter of Kaplan's ideas) to visit "the Rez" and wrote about Kaplan's work there—with life-changing and unforeseen results.

Journalist Friedman was a talented wordsmith who, three years hence, would win the Pulitzer Prize for his coverage of the 1967 Detroit riots. Perhaps a bit too talented with words, he managed to sum up some of the painful paradoxes both of reservation life for the Navajo and of Kaplan's attempts to capture that life in social-scientific terms. His *Newsweek* article, titled "The Sick Navahos," led by recounting two incidents of "bizarre" Indian behavior. In one, a Navajo man named Frank, after an evening of "drinking cheap Tokay" at a tribal dance, stopped his car off U.S. Route 66 in New Mexico and beat his wife to death. In the second, another Navajo, also after a night of drinking, charged off a cliff still riding his horse. Such acts, according to Friedman, citing Kaplan's research, qualified for the Navajo category of *tsí-ni-did-aa*, or "wild and reckless" behavior. And although these incidents might appear on the surface to be standard cases of Indians proving unable to hold their liquor, the author ascribed to "painstaking anthropological research" insights beyond cliché: to the journalist, in fact, Kaplan's work "revealed the Navajos as members of a sick society."[18] Along with his dedicated team of research associates, Kaplan, as described by Friedman, spent four summers covering several hundred homicides and about one hundred suicides attributable to *tsí-ni-did-aa* on the 18-million-acre reservation. His methodological tool kit, Friedman mentioned, included a strategy to "lounge in bars frequented by the Navajos and draw them into conversation, especially the wine-drinking men from 20 to 50 who are most prone to violence." (This detail is likely accurate, for Kaplan himself had written earlier of using a beer-centered technique to ply Rorschach answers from Spanish Americans in bars near Gallup.)

From such encounters Kaplan concluded that the Navajo style of drinking and violence was unique and not to be found in Zuni, Taos, or Ute groups; nor did he believe it was attributable to a genetically inherited sensitivity to alcohol. It arose, Kaplan was quoted speculating in Friedman's article, from "an ultrasuperstitious way of life." In a taboo-laden society such as Navajo it was impossible to avoid violating taboos, which were like landmines in a war zone, and the constant fear this ubiquity occasioned became a way of life—and also a set of pathologies, a "whole casebook of aberrant behavior."

Other "bizarre disorders" appeared in the article as well, as Kaplan's research topics: there was "Ghost Sickness," caused by witches casting a spell via a part of the victim's body, and "Moth Craziness," based on the "belief that a moth has gotten behind the eyes of the victim and is pressing him toward a flame." Symptoms were similar to epilepsy. In addition, schizophrenia was common, such that one-half of the adult population, by Kaplan's estimate, suffered from at least one of four disorders: *tsí-ni-did-aa*, Ghost Sickness, Moth Craziness, and schizophrenia. The first three were what would be called today "culture-bound syndromes," and the last one, schizophrenia, Kaplan also saw as the product of cultural interactions and ego patterns. The article's title, "The Sick Navahos," was, then, a pretty accurate representation of its contents.

This dismaying portrayal of pervasive Navajo mental sickness—based on Kaplan's work—did not serve his project well, to say the least, and Friedman's article "essentially got Bert banned from continuing his research," according to an anthropological colleague, Richard Randolph.[19] In a letter to the editor, Kaplan defended both the Navajo and his work from the unfair *Newsweek* portrait. "The Navahos are not a 'sick' people," he wrote. "They are a flourishing and a vigorous group who have been astonishingly successful at maintaining an exciting and satisfying way of life." Much as his work was directed toward explaining what mental illness was like in this group, such pathologies should "not overshadow the balance, wisdom, and beauty that can be seen everywhere among the Navahos." He also claimed to have been misquoted in the article and asserted he had never said that Navaho mental disorder "arises largely from an ultrasuperstitious way of life."[20] (*Newsweek*'s editorial staff meanwhile compounded the problem by headlining Kaplan's self-defense with the insouciant less-than-contrite "Redman Defended.") Protestations and clarifications notwithstanding, plenty of damage had been done, at least from Kaplan's point of view. He was distressed at the thought of having caused pain to the Navajo he knew and worked with. Moreover, his "boycotting" from Navajo meant he was not allowed back on the reservation by common tribal agreement, despite the fact that his project had just received a generous renewal grant from the U.S. Public Health Service. This debacle was professional pain of a new sort, as Randolph pointed out: "Losing access to the Navajo was also losing membership in the group of social scientists who worked in the Southwest. These people all knew each other and formed a pretty cohesive group." It would be hard to continue his work at the new Rice institute with no more "ethno" to his own ethno-psychiatry. But a rescue route accompanied the debacle: a call to join the University of California at Santa Cruz came just a month or so after the article appeared. Bert was to be "lead man" in psychology there. As Randolph recalled, "He never talked to me about it, but I believe the Navajo disaster was critical." It caused him to change course. Not only did Kaplan leave Rice after only a year, abandon his chairmanship of the institute there, terminate his cross-cultural research, drop his Navajo manuscript on the verge of its publication, and accept a new job, but he also never (to his colleagues' knowledge) published a single monograph and scarcely an article after the summer of 1964— certainly not on American Indians.[21] It is not that he became unproductive, but his scholarly life, and his life with his family as a whole, took a completely new direction.

Yet before moving on to the cliffsides and caverns of California, we can see this episode not as Kaplan's final break from the database but as an extension of the problems of data and interpretation it posed. Pause for a moment and consider the "Sick Navahos" argument not only as a blunder, a turning point, and an example of misreportage, but also a telling document.

Let me be clear: Kaplan was horrified at *Newsweek*'s depiction of his work, as we have seen. Yet it was actually not so far off base when one stripped away its tone of incendiary journalese. Examine the single article to emerge from Kaplan's multi-year project. Readied for publication just before the Friedman debacle, it contained Kaplan's and co-author Dale Johnson's argument, in notably anodyne scholarly language, that "there is an important and perhaps primary sense in which psychopathology and deviance have a positive or functional significance in the social system."[22] Navajo society, during the three years had Kaplan studied it thus far, ran by and through pathologies of heavy drinking, assaultive patterns including wife beating, "crazy violence," ghost sickness, and untoward aggression. Through intensive data collection from police files, hospital records, ethnographic interviews, and observations in bars and at AA meetings, Kaplan and his team of junior researchers aimed to compare systematically the dynamics of Navajo mental health with those of Apache, Zuni, Sioux, and Ojibwe, where they were also hard at work collecting: "A very large amount of material has accumulated," he wrote in his successful grant application, "and it is very clear that the work will extend over the next two years." It did not.

A few elements made Kaplan's approach stand out from run-of-the-mill culture-and-personality studies along the lines of *Warriors without Weapons* and *Dreamers without Power.*[23] Not least of these was the fact that Kaplan intended to make "the actual data of the study available in our publication" as much as possible.[24] By this innovation he offered a way for others to draw their own conclusions. His group project—whose contributors, incidentally, included the authors of both *Warriors without Weapons* and *Dreamers without Power*—was an attempt to allow the unexpurgated voices of subjects to act as "raw data of the future" and to have a place, on their own terms, in the social scientific literature.

In addition, Kaplan was one of the least vociferous of his cohort on the "sick Indians" point. The elegiac tone, and the assumption that the loss of the American Indian's essence, self, or soul was some sort of fundamental transition or "threshold" being crossed, was not so much voiced in mid-century social science as understood before research even began. (Perhaps, some would argue, it was a part of modern thinking itself. Perhaps it was a way for "modernity" to conceal the "very unevenness that marked its own moment of formation and which constituted the very history it was determined to overcome," as one critic put it recently.)[25] More to the point, Kaplan himself had always confronted the overgeneralizing tendencies in culture-and-personality research since his dissertation days. He favored the charting of idiosyncrasies, the hearing of unheard voices, and the cataloging of sometimes disturbing accounts of madness or alternative forms of consciousness. He was not given to ponderous generalities but, borrowing the subtitle of an edited volume he published the same year as the *Newsweek* article, sought access to "First Person Accounts of What It Was Like."[26] It is a possible irony that he lost more than most social researchers for a view he did not particularly hold and which he countered by the highly unusual step of offering his data up to alternative analyses.

Still, the diplomatic nightmare ensued for Kaplan, and, his legitimate protestations aside for a moment, when the *Newsweek* article appeared, the Navajo for the first time had a chance to read about what he was doing. It was not common practice in the 1940s, 1950s, and 1960s—nor in earlier decades—to inform one's informants of ongoing anthropological findings. In fact Harvard's Five Cultures project, many of whose researchers were close colleagues of Kaplan's, worked out an anti-freedom-of-information drive in the southwest. Project heads worried about "security problems" in publishing scientific results. Questions arose about who should have access to Harvard's results and who, for one's own good and for the good of the project, should not—and therefore the argument went that Harvard would be conducting "a useful experiment in security for the anthropological profession at large" by controlling the spread of published research.[27] As a result the 1951 book *Navaho Veterans*, the first monograph to come out of the team project, was "not being distributed to local people or libraries in the Ramah or Gallup area" so as to protect rapport with Navajo and to keep facts from being known that might disrupt the lives of informants or "contribute[] to the suffering."[28] A copy of the book at the BIA office in Gallup was not to be made available to Indian residents. Security in the local mail was a more immediate problem: recall that anyone with mail going through the Ramah post office had to use Scotch tape on top of the gum seal as an extra precaution—an internal departmental memo stipulated—so as to save residents from their own curiosity, for too much curiosity about what exactly Harvard was up to was sure to be a damper to the rapport that was necessary for the project to continue.[29]

This apparently cynical bargain returns us to a conflict at the heart of modern social science, to which the problem of data was intimately related. Anthropological work went smoothly when it was limited in circulation to elites and experts. One could make trades "in the field," have good or bad relations with subjects, and return to the university relatively secure in the knowledge that one's work would not be immediately available, if ever, among the people who had generated it. Books made their way back to their origins sometimes, but it might take years or decades, not every subject who contributed was able to read them, and they were not always widely available. Kaplan's case highlighted how ideas perfectly acceptable to say or publish among scientific colleagues suddenly looked very different when printed in words that circulated in national high-profile magazines. The fact that the Navajo habitually read *Newsweek*, that some even had weekly subscriptions, or that (as the contents of the database of dreams revealed) a Zuni man dreamed of pictures seen in *Life* magazine of bombed-out European cities while a Marshall Islands girl glimpsed Japanese battleships in Rorschach inkblots meant, among other things, that the challenge of data was already arriving full force across the world. Data qua data—the rendering of testimonies, materials, and relationships as circulating nodes in a network—were less fungible when they traveled.

Kaplan's case took place against the backdrop of Harvard's preceding "security problems" from 1951 to 1957 on the same reservation (as well as others in Ramah). Sensitivities were already running high. Even if Friedman's description of his research was, according to Kaplan, wildly inaccurate, the Navajo continued to take offense after Kaplan's correction was published. Perhaps rightly so, though Kaplan was hardly the only offender.

And perhaps, after all, all people increasingly participated in this "difficulty." For social scientists, American Indians were convenient vehicles for observing changes through which all were going, sometimes faster and sometimes more slowly, with more or less enthusiasm. In the mid-twentieth century through the early 1960s, Native Americans and all kinds of non-Western groups became foci for observing the fleeting, disappearing quotient of reality in an especially dramatic way. Powerful emotions attached to things and people seen as just now disappearing, and these converged as a "structure of feeling"—Raymond Williams's term for the phenomenon—that social scientists employed as a research tactic.[30] "What will happen in ten years?" was one of anthropologist Louise Spindler's standard questions in her Expressive Autobiographical Interviews, as preserved in the database of dreams.[31] Case 11, a Menominee native speaker who first learned English at a Catholic school staffed by German and Polish nuns (hence her peculiar accent), answered the question as follows: "I think it's gonna change quite a bit. I don't know just how, but my dad used to talk to us. At dat time dere wasn't even cars and my father told us stories about some kind of engine runnin' through the woods and it all came true. I don't think dere will be anything left of the Indian way." Case 53 was a middle-class-leaning woman of thirty with three children and a successful husband who understood a little Menominee but mostly spoke English, kept a clean house, and didn't drink. "Do you dream very much?" Spindler queried. "I only dream about grocery lists and cooking," she replied. Yet as Spindler categorized her based on her Rorschach results, this woman nonetheless had a "modal personality"—that is, a traditional Algonkian psychogram type that—or so the anthropologist believed—could in light of her modern ways only be attributed to "acculturational convergence." When sometime around 1952 Case 53 held Card III in her hands—the same card to which Hermann Göring just a half decade earlier had reacted so strongly by imagining doctors dividing up someone's innards—she saw "a couple of colored people" and, in the second round, some "birds out of one of those spooky comics."

The new generation of post–World War II anthropologists, psychologists, and sociologists—especially those contributors to the "database of dreams" who targeted dreams and dream-like materials—wanted to study not just what was "lost" about cultures, or being lost, but the mystery of change and persistence itself. Engaging in technological advances in recording, the Spindlers lugged their seventy-pound wire recorder into the field, and though they camped in tents for months at a time, they found it easiest to gather life histories while driving around in their car. Their goal was to be able to generalize about the changes they were seeing. At the same time, they habitually collected an excessive amount of data, far more than they, even over forty-five years as professional anthropologists, could make sense of. As the totemic figure and Kaplan mentor A. I. Hallowell remarked in an interview toward the end of his life,

> I made a complete list of all the plants and all the animals and all the insects and so on for books . . . and I checked on every one of the things. And I had big sheets, and . . ., I had not only the native name but then I had about a dozen columns showing how it was used. And I still have these, but this got to be . . . a really complicated thing. . . . I collected a lot of information on native medicine and some case material. . . . The problem now is, and I guess this has been true of other people in anthropology, . . . what do you do with all of your material? Nobody really wants to publish all these details, you know. Therefore, in everything I've written I've always selected.[32]

Excess was not new. Excessive collections had been a feature of anthropological endeavors at least since the mid-nineteenth century, as Stephanie Gänger, a scholar of Peruvian and Chilean antiquarianism points out. As early as the 1840s among European and American pre-Columbianists, according to Gänger, one sees expressions of the faith that if one collects all the stuff of cultural life, scientists in the future will know what to do with it.[33]

These scholars were in search of experience as experienced in its totality and "allness." Despite their specific findings, today it is compelling that the database-of-dreams approach did not overlay or underlie any privileged interpretative layer. Reading through the data, one finds "native-oriented" Menominee women who saw Card IV of the Rorschach as a Bikini-style atomic bomb and, in other cards, church-going middle-class women who glimpsed traditional Indian maps, which they had seen only in museums. Among professionals, new tools developed to extract the direct stream-of-consciousness experience/expressions of undergoing the changes discussed above, from men who were not psychologically inclined and women who, as almost all of Louise Spindler's 449-page data set attests, were "for the most part, reluctant to talk extensively about their personal lives."[34] Not just shards in the form of second-hand recollections were to be passed around, or generalized portraits, or sentimental dorm-room soliloquies. There was now the sense that researchers could collect and perhaps preserve whole on-rushing realities, the dynamics of choice and of self-definition among the emerging new selves and styles of self-consciousness. Kaplan and others may have entertained far-fetched and soon out-dated "sick Indians" hypotheses, but their data allowed other possibilities. If nothing else, the database of dreams is testimony, from Sioux Indians and South Seas islanders in their own contexts: this is what it was like to struggle with modern contradictions and complexities, which could seem to resemble the eclipse of one world and the dawn of another. Traces of unexamined "native" lives and traces of a widespread obsession in twentieth-century life with not fading away, these testify to the perishability of memory and the desire to keep in ether or otherwise cold-store modern life itself.

Finally, the *Newsweek* incident revealed not only an epistemological problem within anthropology, but also a further "problem of data" that occurred when data piled up in excessive amounts. Such piling-up was more and more the case. Researchers habitually collected too much (due to new and better recording devices, among other things), resulting in more than they could use.[35] That Kaplan thought to create a container for this overflow was his contribution. But the excess, overlap, and ever more free circulation patterns in the system of knowledge exchange would soon bubble up in many new places and create crises, after which the social sciences, science, and the Western world's relationship to its non-literate or have-not other half would never be the same.

By then, Kaplan had moved on.

Santa Cruz was Kaplan's next post, and in taking it, he cut almost all links to his previous research. In 1964, he began the experiment that was the new university and also co-founded the "History of Consciousness" program there.

Santa Cruz in 1965 was a place where the roses insisted on growing as big as a person's head and Harvard expatriates often could not resist settling. But there was just a marine mammal station and hardly any university there, although dreams for founding one went as far back as the 1930s. In 1963 the University of California system decided to go ahead on the site of a former ranch looking out over the sea. The school named its first college, Cowell, for the ranching family and appointed the college's first provost, Page Smith, luring him from UCLA with the promise of full privileges in choosing to build the new college as iconoclastically as he liked. And he did like. In particular, he liked Kaplan.

Hardly had any more detailed decisions been made when Kaplan arrived in Santa Cruz, at the behest of Smith, who invited him on the basis of what might best be called the inscrutable quality of *Menschlichkeit*—something like "good-guy-ness"—for in practice Smith's procedure was assiduously to hire scholars with whom he felt a strong humanistic bond; some years later, it would be called a "good vibe." (There was also a strong recommendation from David Riesman, the eminent sociologist and public intellectual who "saw something in Bert" and urged Smith to take him on.) Smith wanted people who in his mind opposed the dominant academic tropism that leaned toward "hard-nosed" empiricists. Smith described himself as a "soft nose"—despite "the size of the proboscis."[36] As it turned out, Kaplan became an important new faculty member in the foundling college but not an "empire builder"; others had a much stronger power drive than Bert, Smith recalled, and Bert was a "private person, much more preoccupied with teaching."[37] He eventually rose to legend status for his teaching, and even today, walking the Santa Cruz market, his family is sometimes accosted by former students who say Bert changed their lives.

Still, Kaplan may have been a controversial choice. There was tension between Provost Smith and Chancellor Dean McHenry, who remarked that he worried that the Cowell College hires via Smith were "lopsided" in the direction of softness and an overly philosophical bent. They also set up a precedent, arguably one that continues to be followed today:

> I think it's fair to say that Page's interest ... was in what is usually called in most fields the "soft" approaches to various subjects: instead of a philosopher who is a logician and interested in analysis, very hard-boiled analysis, he ran to philosophers who were interested in rather—well, the soft

approach in philosophy of religion and ethics and things
that were less tangible and less rigorous according to modern philosophers. In psychology was where we differed perhaps the most. I wasn't arguing against having a soft
psychologist. I was arguing against getting something so
completely . . . getting all the people from out of the mainstream of psychology.[38]

McHenry worried about too many Kaplan-style figures who were
not interested in a strict laboratory approach based on running rats
through mazes: "Page Smith said to me one time when we were flying between Dartmouth and New York, 'I just can't imagine Cowell
College with a rat psychologist.' (Laughter.) I wanted an animal-
behavior man quite early, a psycho-biologist, a bio-psychologist,
and he resisted it very strongly." (At one point, when Kaplan and
his allies were feeling expansive about the possibilities for building
a comprehensive data center, they had speculated about *saving* the
results of "rat's maze learning," along with gibbon, baboon, chim-
panzee, and ape data from different field stations, not neglecting
the potentially alarming "Columbia University's colony on an island off Puerto Rico which has been overrun by monkeys." But this
larger data center, of course, never came to pass.)[39]

Smith won on Kaplan, McHenry admitted, but he did manage
to hire, in another college, a psychologist "who had worked with
cats a little bit, and incidentally with fraternity men." Of Kaplan,
McHenry said, "But the main psychologist . . ., of course, the senior
one, is Kaplan. And Bert Kaplan has done very important work,
cross-cultural and that sort of thing. But he's kind of a philosopher
of psychology. And . . . is essentially soft." In the English department, too, McHenry lamented there were "nothing but blue-
birds."[40] Opinion at Stanford University at the Ford Foundation's
well-funded Center for Advanced Study in the Behavioral Sciences
(CASBS) likewise seems to have been that Kaplan was not quite the
man the center wanted. Fellows there pursued such "hard" topics as
matrix algebra, factor analysis, and attribute statistics with Paul
Lazarsfeld (refresher courses offered in 1958–1959) and engaged as
well in long discussions about the future of the behavioral sciences.
In a letter, Clyde Kluckhohn reported to Dorothy Eggan (both of

them slated to spend time at the center, he as a fellow, she as the wife of one) that the attempt to find Kaplan a place had run aground: "Since I wrote you last from Albuquerque I have heard," he confided, "that the psychological 'referees' on B. Kaplan from the Stanford Center have not backed him strongly enough to make an invitation at all likely. And I doubt that we 'outsiders' can do much to remedy this picture."[41] By the early 1960s, Kluckhohn and Eggan had both finished their stints at CASBS—each would suffer a premature death shortly afterward—and Kaplan was embarking in a new direction, no longer in line to be one of the next big thinkers in the postwar behavioral sciences.

In this way Kaplan metamorphosed as a soft-singing, philosophically inclined bluebird of psychology who alighted in the gardens and beaches of Santa Cruz, its library inhabiting a redwood grove, its colleges nestled above the sea, and its hills housing Mima mounds, mysterious structures possibly built by archaic giant moles. Here Kaplan spent the rest of his career, and, innovative as it was, renowned as his teaching soon became, and sparse as his publications record eventually stood, the database-of-dreams experiment lay mostly forgotten even by Kaplan himself.

The history of technologies, especially experiments in the large-scale storage and use of data, is often envisioned as a grand march toward a future so inevitable it seems to be capable of surging backward to meet the onward rush of progress. The push and pull of present and future clears a path. This is the historical vision we casual data users of the world and "dwellers in total information" now assume.[42] Revolutionary technologies demand revolutionaries, "pioneers of data," plucked from obscure Palo Alto garages, perhaps, but of a certain stature. Wondering How did this come to be? often yields just-so stories of this or that inventor tinkering away in this or that place until the Aha! clicked or the imagination's chemistry took off. But morasses of contributing ideas, machines, social forms, and personalities can turn out to be far less direct, far less linear. Assemblages of humans, machines, ideas, accidents, and institutions arise that have great and unanticipated effects; they build on earlier inventions and jerry-rig them into new ones; they involve quite a bit of fumbling in the dark. Many true pioneers of data gained stature for other reasons and remain unheralded for their pioneering efforts. Many did not know what they eventually contributed. (Kaplan is an example.)

The British historian of computers Jon Agar recently researched the arrival of stored-program electronic computers in different sciences during the late 1950s and early 1960s, examining the commonly made claim that they resulted in sudden new capacities and "extended seeing."[43] What he found is that "extended seeing" preceded the actual advent of computers in most cases. People were already doing by hand through mechanical techniques what the electronic computer would soon be able to do much more quickly. When scientists performed the first 3D modeling of vitamin B12 in the mid-1950s, it seemed to some observers that computers had brought this about out of the blue. "This was made possible," J. D. Bernal wrote, "only by the extensive use of computing machines."[44] In fact, Agar shows, existing manual and mechanical methods were already moving ahead without computing machines—just more slowly. Similarly, the Botanical Society of the British Isles began a comprehensive mapping of British flora post–World War II by the use of an "essentially drill-like" approach to data gathering—teams of plant watchers fanning out carrying seven regional cards, each with nine hundred species. All surveyor teams were made up of amateurs who became increasingly skilled when they persisted over the years. Such routines preceded the arrival of computers and in the process "nature was squeezed to fit" the collecting cards. Mountains were flattened and subspecies limited to six due to the capacities of the cards and the inability to represent the full range of nature with mechanical methods. Yet when computerization did arrive, these routines and mechanics already in place prepared the way. Earlier practices strained at their limits but continued to be workable; at this point computerization of certain elements (not all and not all at once) propelled research.

The development of Kaplan's analog archive was not a case of the shock of something wholly new followed by an absolute rupture in scientific possibilities, but instead a case of borrowing, fusing, and stumbling toward progress, as with punch cards, which "combined human, machine, and method."[45] The history of analog and hybrid computing, as well as accounts of the more general crossing of the "analog-digital divide," testifies to the fact that such "data journeys" (to borrow philosopher of science Sabina Leonelli's term from a slightly different context) do not tend to run smoothly.[46] During this pivotal transitional period from the mid-1950s through the early 1960s, Kaplan too envisioned an information-science approach to the self before there were electronic databases readily available for such an approach. By the mid-1960s, as we will see, a few others would jump on this data-storehousing bandwagon, advocating the creation of massive analog archives to hold social science information, but Kaplan was a good ten years ahead even of the avant-garde in his and related fields. He was a full fifty years ahead of the rest.

It took some time, but Kaplan did prove visionary. Rescuing trash-bound treasures is now a leading edge of research in the social sciences, with the revival of lost archives or resuscitation of lost data more and more common. These data offer themselves as de facto laboratories, allowing all sorts of previously unavailable topics to be probed scientifically. "In 1969, the sociologist Morris Zelditch asked, rhetorically, 'Can you really study an army in a laboratory?' Nearly half a century later, the answer appears to be yes," wrote prominent Harvard social analyst Nicholas Christakis recently. You can do this through data, which offer the possibility to test hypotheses as in an experiment. Christakis himself is famous for re-analyzing the "Framingham Heart study" data set to draw original conclusions about epidemics of obesity and social influence. (The Framingham study had conserved information from 1948 about the residents of the Massachusetts town, but Christakis asked new questions of the data, which they proceeded to answer.) Take a walk, or, better yet, click your way to the Harvard "Dataverse" or the ISPCR at the University of Michigan, or the UK Data Archive. There are others emerging all the time, of course—for example, a fledgling project called Qualitative Data-Processing Software for Large-Scale International Cross Comparisons (QUALIPSO). Inspired by comparative linguistics and comparative anthropology, Sorbonne researcher Anne-Sophie Godfroy hopes to "provide reliable epistemological frameworks and effective tools to collect, to browse, and to analyse large amounts of data, heterogeneous data," by searching across varied formats and multilinguistic materials.[47] Another example is a new archive of medical charlatans in early modern Italy, assembled by historian David Gentilcore. Worlds of data sit waiting where digitized records from longitudinal and massive studies are stored. As Jo Guldi and David Armitage argue (to some controversy) in their recent *The History Manifesto*, such stockpiles may constitute the future of historians' work: in the "era of digitised knowledge banks, the basic tools for analysing social change around us are everywhere."[48] Digital humanities trade in hopes along these lines.

Another related direction entails the building up of Leviathan-scale stores of data from cell phones, Netflix accounts, online activity, or any behavior that can be tracked. The new field of computational social science relies on immense stores of data that arise like dunes in the Sahara due to shifting winds or like trash piles in Rio, where whole cities exist to live off them—stores that dwarf Kaplan's 1950s effort and its fifteen-some thousand pages. The Library of Congress in 2010 purchased the entire Twitter archive, which now serves as a resource for researchers; started in March 2006, it is stacking up at a rate of 20 million messages per day. The whole collection numbers in the ever-growing billions of items. The possibilities of what researchers, scholars, and artists can do with it are also growing. A recent study, for example, traced 74 million "diffusion Events" (re-tweets or RTs) among 1.6 million Twitter users over a two-month interval in 2009 to study the cascade patterns of influence.[49] Such activity, as many boosters do not hesitate to point out, is the future of social science, or at least one version of it. Or, as a possibly overexcited Library of Congress blogger put it, "It boggles my mind to think what we might be able to learn about ourselves and the world around us from this wealth of data. And I'm certain we'll learn things that none of us now can even possibly conceive."[50]

Such a success story did not unfold at the time of Kaplan's experiment. The Primary Records Committee lost its money. No sooner did its pilot machine come into existence than it was effectively suspended. It received some citations, yes, but few scholars used it in the way Kaplan intended. Theirs was a "brief Golden Age."[51] Like other analog machines, it was not a failed technology, but neither did it live up to the once vaulting hopes that attended it.[52] Few used it at all, a fact that in itself explains how it fell short of Kaplan's vision. Instead, it rested at the Library of Congress and several university libraries in a form that became increasingly less accessible as the Readex technology became obsolete, the Microcard lapsed, ideas about anthropology altered, and the world itself changed.

For decades after its platform viability collapsed, the Microcard archive remained available but generally unaccessed. Elizabeth Colson's contribution to the first series of *Primary Records in Culture and Personality*, in 1956, was her "Autobiographies of Three Pomo Women," which made use of data she gathered from 1939 to 1941. It sat largely unmolested on Microcards until anthropology professor Robert Heizer of UC Berkeley in the early 1970s "found a reference to the Microcard publication in his assiduous search for materials relating to the history of Native Americans of California." Colson then agreed, at his urging, to prepare it for re-publication in the form of a soft-cover mimeograph "to make it more available to those interested in the history of California and the experiences of its people." The new version's preface declared with palpable relief that now "the autobiographies finally appear in an easily accessible format," owing to the work of Berkeley's Archeological Research Facility in rescuing it. It had itself become an archeological find.[53]

Some years later, historian David Brumble stumbled across the rare life-history materials held in the Kaplan data collection. He was then a graduate student at the University of Pittsburgh, where he worked with John Roberts, a former colleague and co–graduate student of Kaplan's at Harvard's Department of Social Relations. "Pitt had a complete set of the Microcards," Brumble recalled recently. "Probably it was Roberts who put me onto them—but my sense at the time was that no one had ever looked at them before I did." The cards by then were capable of yielding the sort of scholarly thrill that attaches to the rare and neglected source to which one is uniquely privy. Meant to rescue data from obscurity, the cards had themselves become obscure, even to the most thoroughgoing scholars. "And in all the years I worked on American Indian autobiographies, I don't remember ever coming across a reference to them. It's possible, of course, that after thirty years, I'm forgetting something—but it is certainly my sense that all that work was done and then dropped down a scholarly well. Of course the unconventional form of publication didn't help."[54] There were references, as mentioned, but what is notable is Brumble's dominating sense of being the first to discover or rediscover the resource, along with the aura of its having been "dropped down a scholarly well."

Kaplan at the heyday of the microform absorbed the dreams it spawned of perfect data retrieval and universal stores of information push-button style. Although the 1930s had seen innovations such as Kodak's Recordak for storing especially dull and repetitive records, storage and retrieval was not just about bank checks and blueprints. At root was the fantasy of a book of all things—even those things least like things, such as dreams. Utopians and dreamers sought, via microfilm's capacity to miniaturize, "the Universal Book." Shrinking the size of tributary information was key, as one pioneer remarked in 1898: "A . . . radical assumption would consider that all knowledge, all information could be so condensed that it could be contained in a limited number of works placed on a desk, therefore within hand's reach, and indexed in such a way as to ensure maximum consultability. In this case the world described in the entirety of books would really be within everyone's grasp." This desk of all books would then be ("very approximately") "an annex to the brain, a substratum even of memory, an external mechanism and instrument of the mind but so close to it, so apt to its use that it would truly be a sort of appended organ, an exodermic appendage."[55]

Like Paul Otlet's Mundaneum, Albert Kahn's Archive de la Planète, Félix-Louis Regnault's ethnographic film archive, Alan Lomax's Global Jukebox of folk recordings, and H. G. Wells's World Brain—all earlier enterprises—the database-of-dreams project too became a sort of ruin. But this ruin was unique in certain ways due to the technology it employed. As it turned out, Kaplan had picked the particular style of microform doomed to fail most quickly. Declared officially dead by 1965, two years after the last Kaplan volume appeared, the Microcard proved—to venture into anachronism—the Betamax of the data-storage world.[56] It lost out to microfiche and microfilm despite a more legible and higher-quality format. Doomed by the so-called "microfiche revolution," the Microcard slipped into a sort of oblivion of stasis. "Rider's ideas dropped like lead weights in bottomless wells of the library community."[57] Despite its delightful glossiness and its paper-based materiality, it did not survive the greater standardization capabilities microfiche offered, and it remained in collections but no longer a viable form. Even today this aura of failure persists. Microfiche is always easy to find at an up-to-date university research library (although it too has fallen out of fashion), but Microcards are another thing entirely. At the Library of Congress, the series was occasionally not available or had to be rummaged for at some length. Such an outcome was more favorable than that at Harvard, where a large portion of Kaplan's Microcard sets is still missing. Moreover, Readex machines dedicated to reading opaque Microcards were being jettisoned in North America and Europe and sometimes— with an air of pathos perhaps invisible to most passersby—found in or next to dumpsters.

In the same manner as all videotape technology (VCR and Betamax) ultimately went out of date, so too has microform (Microcard and microfiche), and for approximately the same reason. The dates of the Microcard's heyday, 1950–1963, coincide with the most active span of Kaplan's data experiment, 1954–1963 (considering the narrowest of possible dates). Yet the movement as a whole was continuous with the growth of computing and digital computer processors. Ever-smaller dual-core processors rely, at root, on the same photographic processes as did the lineage stretching from microphotography to "modern" microforms. Common to all types of micro-documentation, and responsible for its failure to fulfill the nearly boundless expectations pinned on it, was the inability to perfect the "swift and precise retrieval of items"[58]—although the Kodak Minicard System of the early 1960s, never commercially available, did manage to select and print microfilmed documents. By the 1960s, however, the cutting edge of information storage and retrieval moved away from microformatting to the computer and eventually to DBMSs.

Meanwhile, Kaplan's committee met for the last time in 1959 in an undisclosed location, chosen, as the minutes that day noted, "to save money," doubtless a step down from the La Salle Hotel. Volumes 3 and 4 appeared in 1962 and 1963. The Microcard

proved the wrong technological horse to bet on, but the real problem was that there was no right horse—not quite yet. (By "right
horse," I mean one that would provide search capacity, flexibility,
and the ability to create new and different arrangements with data;
to store "trails" through data; and to perform complex operations
on them.)[59] Within only a few years, digital storage capacities would
exist to capture the scientific data on a scale Kaplan and others had
envisioned. They would also make the data "workable" in a way
Kaplan had never envisioned. During these retrospectively interim
years, many social scientists shared a *Waiting for Godot* feeling; as
another NRC group put it in 1961, "Until electronic data processing and data storage devices rescue scholars from the limitations
of the traditional research tools," they must remain content with
"abstracts, annotated bibliographies, and inventories."[60] Electronic
databases were just around the corner, and their would-be users
languished in a historical pause.

Yet the Kaplan group members had not been content simply
to wait for electronic solutions. They had forged ahead with the
Readex-Microcard-projective-test-fieldwork hybrid solution, by
which the best data-storage and -retrieval techniques met a super-
personal knowledge base.[61] Their MacGyvered alternative—using
available technology before dedicated machines actually existed to do
what was needed—remained in a kind of limbo. It quickly moved to
the Library of Congress, available to all researchers, yet its technology platform, along with some of its theoretical assumptions, slowly
sank into oblivion.

Conclusion

HAT are we to make of this "database of dreams" today, as we stand in the midst of seemingly unprecedented changes in the way we manage our memories, pasts, human relationships, institutions, and almost all other aspects of social life in the ongoing wake of massive, disruptive, churning technological change? What, in short, does this alternative history tell us? Aside from what it says about how other people sought and found and lost meaning, what is its message?

Kaplan's experiment in data collection was not only obscure but also prescient, and its significance, to me, lies in toggling between the two, looking back (at a dead end) and looking ahead (from a pioneering resource). To explore its significance felt akin to thrift store history, finding value in discards others have ceased to value. In the case of the "database of dreams," a castoff revealed itself as a wellspring for "a number of pasts, a hodgepodge of pasts, a spider's web of pasts, a jungle of pasts," all massed together in a collection of collections: a once futuristic data repository.[1] In these old odds and ends of data were elements connected directly to the present, it struck me, yet also disconnected, as if they formed a strange, Greek-chorus commentary on the ever-growing urge to build a total archive.[2] Just so, the lost data repository illuminates forgotten or disrespected elements of our own moment, all that is lost in pursuit of the all, and how even that residue leaves a trace. For example, what are the sounds that are lost when music is compressed to make it readily portable and easy to use? Ryan Maguire, a PhD student in Composition and Computer Technologies at the University of Virginia Center for Computer Music, recently inaugurated a project called "The Ghost in the MP3," of which the first installment is a song made with only the left-out sounds lost when rendering Suzanne Vega's "Tom's Diner" from a large RAW Audio uncompressed file to MP3 format. In the early 1990s, this was the song that famously—famously, at least, to music-format enthusiasts—served as the prototype for listening tests that the Moving Pictures Experts Group (MPEG) ran to develop the MP3 encoding algorithm. As Maguire points out, such listening tests, "designed by and for primarily white, male, western-european audio engineers [*sic*]" and "using the music they liked" (including also a Haydn trumpet concerto and Tracy Chapman's "Fast Car"), then went on to refine the encoder—that is, to determine which sonic information was kept and which was discarded.[3] Yet one can salvage the discarded acoustical residue, as Maguire has, and the resulting ghost song (to this listener at least) sounds like a precise record of exactly what it is to be left behind—one could say the orphan sounds are throwaways that have become testimonies.

In this light, Kaplan's "database of dreams" can be seen as a ghost song. With its numerous data sets making up the Microcard archive as well as its unrealized but dreamed-of whole, its cadences capture the mid-twentieth-century social-scientific enterprise and the more general quest to make a science of humanity, leaving no part unexamined. Kaplan's data collection is made up of a variety of hissings, dull clankings, and residual gray, pink, or brown noise. When extracted, the harvested residuals make a work that at times reads as if the researcher were encountering it from very far away in outer space. Often it recalled Claude Lévi-Strauss's 1962 characterization of the work of anthropology: to bring together a "collection of oddments left over from human endeavors."[4] In this sense, the "database of dreams" is equally a late-modern hymn, an extracted residual, and a great-if-problematic, collective-yet-authorless work of art and science. The technologies on which it was based and that constitute it continue to leap into the foreground yet also remain in the background to facilitate further findings. A key claim about "big data" is that it marks a qualitative (not just quantitative) leap in knowledge—"More data create not just more knowledge, but a qualitatively different kind of knowledge," as Hallam Stevens argued recently in the context of current developments in bioinformatics— and perhaps Kaplan's endeavor can serve as a map of *how and where to look* for these different kinds of knowledge. Like the "oldest message in a bottle," recently found on the shores of the Baltic Sea near Kiel, in northern Germany (its message effaced, the inky characters not quite legible, the bottle actually not that old, only a century or so), it is significant for having washed ashore.[5]

What the data tell us, to put it more succinctly, is that "human materials" are at the root of big data. In bumper sticker form or the tagline to a movie, the moral would be: "Big data is people!" This has both utopian and dystopian implications.

Despite the growing obscurity into which it slid over the decades, the Microcard collection remained accessible (even if it occasionally went missing when I called at certain libraries that posited its presence in their online catalogs, only to be found in physical form a day or two later, and once, at a prominent research center, librarians discovered it had mistakenly been thrown in the dump). It also, despite the perishability of its chosen format, remained legible. At around the time Readex machines were being finally discarded, new super-magnifying devices became available at most well-equipped libraries, and these could "read" any microformat developed in the past hundred years, including a Microcard, by blowing it up to sufficient size. At the Library of Congress Microform Reading Room on June 19, 2008, and August 3, 2009, for example, I consulted the Kaplan data via an aged Readex Microprint machine, the Opaque Viewer Model 7, which could neither search nor scan, nor even reveal the data particularly well. When I returned in the summer of 2010, the excellent ST200X Series Digital Film Viewer sat in place of the Readex—"an all-new, all-digital solution" to reading all micro-formats—and I could now make my own digital versions of the data archives by copying them, combining these images as pdfs, and running them through text-recognition software. Whereas traditional machines such as the Readex "sh[o]ne the light of a hot halogen light bulb through a series of interchangeable reduction lenses, Fresnel condenser lenses,

large mirrors, and then ... reflect[ed] it onto the back of a translucent viewing screen panel," resulting often in a dim and unevenly
lit image, the digital machines clean up micro-images nicely, "remove dirt, brighten dingy backgrounds, and remove scratches,
greatly improving both the viewing experience and the scanning
quality," after which much can be done with the files.[6]

   This turn of events emphasizes a point central to the story told
here: it is only in the second decade of the twenty-first century, with
the ever-growing ambit of Google Books digital library (with the
stated goal of encompassing every book ever written, it is possibly
the largest non-numerical database in the world today) that a text-
based, "words-as-data" database has become the leading edge of a
significant amount of data-based inquiry.[7] In the case of Kaplan's
collection, it is only with the widespread availability of scanning and
Optical Character Recognition (OCR) that the collection becomes
searchable and potentially assimilable with other data. (Recall that
Vannevar Bush in the 1930s had attempted to construct machines
capable of searching micro-text, the Rapid Selector among them,
but was never successful.) Unlike the HRAF at Yale, which moved
years ago to digital format by means of its eHRAF extension, a transition made easier by the fact that the constituent data were already
"processed" and accompanied by metadata, the Kaplan Microcard
archive did not make this leap to digital (or any other leap) initially.
Yet it sits, now, in a position of potential usefulness it was not quite
able to hold during its debut years. In this way, it fulfills advice
Kaplan received from Verner Clapp when he interviewed the prominent librarian in 1957: "His advice in the selection and recording of
data was to be as forward looking as possible and to try to keep in
mind the potential usefulness of data rather than be oriented only to
the uses that were clearly envisaged at the present time."[8] Likewise,
my own efforts have merged with Kaplan's and with those of the
other contributors. I have hoped to do justice to their experiment
by describing its unforeseen potentially proliferating uses—one of
which, one could say, is this book.[9] Among those uses is its only-
now-appreciated beauty.

   The message of the experiment lies in the curious dream-like
quality of the *data themselves* (as opposed to the dreams). Today one
can casually find images of Katchina or clown ceremonies just by

Googling them, and a brief search calls up flickering YouTube footage of circling Hopi Katchinas dancing in 1913 for Theodore Roosevelt. Before the Internet's operations put all such images "at one's fingertips" in a seemingly relentless, reverse-avalanche mass upload of history's dusty ephemera, Kaplan's data bank, equipped with the data of John Adair, A. I. Hallowell, and many others, would do something similar. With primitive tools, it would make possible a networked, miniaturized, and putatively permanent storehouse of data that rendered the inaccessible accessible. In libraries across the land by 1956, less than a decade after John Adair had paid a Zuni war veteran (Miguel A.) for his words and Kaplan had accessioned them, one could find and tap into the direct records and transcripts of Miguel A.'s recollected experiences. However, Kaplan's archive worked only briefly, as mentioned, before falling into inertia.

Just as the data briefly circulated and bubbled up, they were inclined, too, to disappear, like the nighttime visions they attempted to pin down. They seemed almost to flicker in and out of existence, like the scenes of a busy street in an old black-and-white film. One moment they were there, and the next they were gone. They reminded one of Thomas Pynchon's Icelandic expert Professor Svegli, who explained that certain problematic scientific artifacts, including maps and specimens, "begin as dreams, pass through a finite life in the world, and resume as dreams again." Here the specimens were themselves the data of dreams, creating a feedback loop of oddness. (This looping effect was already present in the origin of the Rorschach test, which issued from a dream Hermann Rorschach had while still a medical student—a "very special" dream of his own brain cut into slices as if in an autopsy, the slices falling forward, one after another. The resultant series of inkblots, carried around the world, would inspire Dorothy Eggan to consider Hopi Indians' dreams as akin to projective test data.)[10]

My research for this book began eight years ago with an interest in dreams and ended with a realization about data. With the layering-on of rapid technological, institutional, and humanistic changes through which we are all living, as well as the quick if uneven dissolution of shared institutions and infrastructures, the "database of dreams" seemed to reveal itself over the years—as if, finally, through a new lens—as a story about the mystery of our ongoing and ever-changing relationship to data. How do data promise to make the past live once more and the future spring into being? How did the "fantasy of total information," a conceit with a lively history for the last four hundred years or so, spring up anew, with fresh force and an irresistible urgency, in the mid-twentieth century, with resonance all the way to today? How did some of the most ambitious social scientists of the postwar world unerringly target the most "difficult to capture" quarry, the very lives and selves of those no-longer-called-primitive people still of great interest to researchers? The book was supposed to be about them (other people, usually far away, as sources of knowledge and targets of ever more intimate information extraction), but it is equally about us (sources too). It was supposed to be about dreams, but it ended up being about technology. But a better way to put it might be that it is about the relationship between the two (research subjects and objects)—even as the distinction between them was collapsing. And it is about the relationship between technology and subjectivity, even as that distinction may likewise be ebbing away.[11]

As a way of describing the "database of dreams," one could say, for example, that there are 341 Hopi dreams, or here one can find "Modified T.A.T.s of 33 Javanese Men and Women" or the "Life History of a Fort Berthold Indian Psychotic." One can look at the contents, what they hold, for a long time. Yet another way of understanding the database is equally revealing—through its constituent technology. This approach yields a panoramic view of the ruins it was choreographed to abet or counteract (for example, the mass microphotographing of books was embraced to save them from bombing, flood, fire, limited shelf space, and other disasters). Yet to this ongoing catastrophe, the melting away of solid form, paradoxically, it became a foremost contributor. In the search for something immortal and enduring, something lasting beyond the human scale and human lifespan, the data contributions of which added up to make a sum total greater than any humble separate contributions or any silly nightly dreams, a new iteration of "total information" had its brief hour, ending, itself, as a new sort of ruin, a data ruin.

This ruin tells us something, too, about how dreams of total information are also being renewed. The octopus reach of supercomputing, the responsiveness of its algorithmically designed tracking systems, and the pervasiveness of data extraction via trails and click-streams result in a constantly expanding aggregation of data. The goal of National Security Agency surveillance drawn from phalanxes of massive commercial and governmental databases storing ever-greater stockpiles of personal information is staggeringly close to total and bound to become more so. Even the meaning of "total information" is changing as digital data collecting can penetrate night-time realms of REM sleep and autonomic activities such as breathing. Entities private and public are *already* gathering up something resembling the "all" and packing it away in ever-denser digital formats. What these complex systems are hoovering up and archiving is the collected data describing the inner and behavioral inclinations of (potentially) all 7-plus billion residents of this planet—all the jotted e-mails, all the online purchases through the interwebs, all the dumb and clever tweets, and all that is "deleted" as well. The "database of dreams," from this perspective, takes on an aspect that Kaplan and his confederates who began their ambitious though somewhat melancholy project did not predict. In a hybrid Orwellian-Huxleyan world, where *1984*'s big-brother-style surveillance combines with *Brave New World*'s advanced emotional engineering, all augmented with scarcely imagined tools, it is quite possible instantaneously to predict (with increasing accuracy) human behavior acts and even states of mind such as voting, buying, preferring, falling in love, the tendency to gamble to exhaustion, to addict oneself to video games, or to fall into states of despondency and infatuation. The world we live in is one whose taproot is the unintended consequence of the benign dream of Kaplan. His strange machine augured, even if it did not imagine, this unfolding world-scale experiment in ever more personal data collection, an experiment in which we all are living and all de facto participating.

Notes

Introduction

1. Edward Shils, "Social Inquiry and the Autonomy of the Private Sphere," in *The Calling of Sociology and Other Essays on the Sociology of Learning: Selected Papers of Edward Shils* (Chicago: University of Chicago Press, 1980), 3:422.

2. "Microcards: A practical tool for industry, government and education"; four-color brochure published by the Microcard Corporation, West Salem, Wisconsin (n.d.); in Bert Kaplan Papers, Santa Cruz, CA. [Hereafter: Kaplan Papers.]

3. "A vast resource": letter from executive secretary of the Division of Anthropology and Psychology of the National Research Council to Helen Brownson of the National Science Foundation, April 11, 1956; Committee on Primary Records, Division of Anthropology and Psychology Records Group, National Academy of Sciences Archives, Washington, DC. [Hereafter: A&P, NAS Archives.]

4. "Herculean efforts": phrase used by Theodore "Ted" Schwartz; interview quoted in Peter Black, "Psychological Anthropology and Its Discontents: Science and Rhetoric and Postwar Micronesia," in *American Anthropology in Micronesia*, ed. Robert Kiste and Mac Marshall (Manoa: University of Hawaii Press, 1999), 236.

5. Jean Macfarlane of the Institute for Child Welfare, UC Berkeley, interview with Bert Kaplan, in "Report of the Executive Secretary of the Committee on Primary Records" (Not for Distribution or Reproduction), April 1, 1957; Committee on Primary Records, A&P, NAS Archives.

6. Bert Kaplan, typed unpublished memoir, 5; Kaplan Papers. Kaplan quotations in the next paragraph are also from this document.

7. Defining PTSD is best described as a process rather than an outcome. From battle fatigue and shell shock to traumatic neurosis and PTSD, the diagnosis continues to morph as, for example, the recently published *DSM V* emphasizes behavioral symptoms rather than an individual's immediate fear response. Within the military it is an injury, not a disorder. American Psychiatric Association fact sheet, "Posttraumatic Stress Disorder," http://www.dsm5.org/Documents/PTSD%20Fact%20Sheet.pdf. Scholarship on PTSD's history is vast, but to begin, see P. R. McHugh and G. Triesman, "PTSD: A Problematic Diagnostic Category," *Journal of Anxiety Disorders* 21, no. 2 (2007): 211–222.

8. Hermia Kaplan interview with author, July 29, 2011, Santa Cruz, CA.

9. "Perishable format" is my coinage as far as I know, but the same idea, if more narrowly focused, is conveyed by the term "digital obsolescence" or "file format obsolescence." Such terms describe an element of risk built into large digital-format collections. David Pearson and Colin Webb, "Defining File Format Obsolescence: A Risky Journey," *International Journal of Digital Curation*, 3, no. 1 (2008): 89–106. The problem of the perishable format can be observed in analog data collection devices as well.

10. Recent historical work makes use of the large data sets of medical and psychiatric records (beyond simply dreams) that existed at hospitals such as Bellevue and Mass General at the turn of the twentieth century; see, for example, Aude Fauvel, "Cerveaux fous et sexes faibles (Grande-Bretagne, 1860–1900)," *Clio* 1, no. 37 (2013): 38–61, and Akihito Suzuki, "Modernism and the Experience of Mental Illness in Tokyo 1925–1945," presentation at Tokyo Conference on Philosophy of Psychiatry, University of Tokyo, September 21, 2013.

11. All quotations in this paragraph are from Jorge Luís Borges, "Tlön, Uqbar, Orbius Tertius," in *Everything and Nothing* (New York: New Directions, 2010), 18, 16. (A slight alteration from "license" to "incense" has been made.)

12. On Basile Luyet, latency, and the history of the term, see Joanna Radin, "Latent Life: Concepts and Practices of Human Tissue Preservation in the International Biological Program," *Social Studies of Science* 43, no. 4 (2013): 487–488. On atomic trash, "reborn equipment and refurbished laboratories," see Jahnavi Phalkey, "Atomic Trash: Cold War and Global Histories of Reconstituted Nuclear

Physics Laboratories," presentation at Bild Wissen Gestaltung: An Interdisciplinary Laboratory, Berlin, July 1, 2014. On zombie satellites, see http://www.nytimes.com/2014/06/15/science/space/calling-back-a-zombie-ship-from-the-graveyard-of-space.html?smid=re-share&_r=0]. On scholarly anxieties over format, see Rodrick Ewens e-mail to Oceanic Anthropology Discussion Group, ASAONET Listserv, December 21, 2013, 12:53 p.m.

13. Daniel Solove, *The Digital Person: Technology and Privacy in the Information Age* (New York: New York University Press, 2004), 1–2. For cautionary work on the implications of digital personhood, see Evgeny Morozov, *To Save Everything, Click Here* (New York: Public Affairs, 2013), and Jaron Lanier, *You Are Not a Gadget* (New York: Vintage, 2010).

14. The Kaplan Microcard archive receives no mention in either of two encyclopedia entries covering anthropological databases and data archives respectively: Melvin Ember, "Databases, Core: Anthropology and Human Relations Area Files (HRAF)," *International Encyclopedia of the Social and Behavioral Sciences* (Oxford: Elsevier, 2001), and Richard C. Rockwell, "Data Archives: international," ibid., 3225–3230, esp. 3226–3227. (Note: the second edition of this encyclopedia is just being released as this book goes to press.) Aside from its omission of the Primary Records collection, the latter article contains pertinent reflections on data archiving, privacy, technological change, and "an ethic of sharing" (p. 3228). An exception to the historiographical omission of Kaplan's Microcard series is David Brumble, *American Indian Autobiography* (Lincoln: University of Nebraska Press, 2008) (see chapter 8).

15. See William Clark, *Academic Charisma and the Origins of the Research University* (Chicago: University of Chicago Press, 2006), on "little tools of knowledge," rationalizing devices by which the modern academy was transformed: "an armory of little tools—catalogues, charts, tables (of paper), reports, questionnaires, dossiers, and so on. Such things comprise the modern, mundane, bureaucratic repertoire of paperwork and much of the power of the modern academic comes from such trifles" (p. 6; also p. 19). Robert Park quoted in James Bennett, *Oral History and Delinquency: The Rhetoric of Criminality* (Chicago: University of Chicago Press, 1988), 155. Lawrence Stone, "Prosopography," *Daedalus* 100, no. 1 (1971): 49. Prosopographies often involve databases as mid- or end-products of research, but in this case it is a portrait of actors who interacted with a data bank and the constituents thereof. See John Bradley and Harold Short, "Texts into Databases: The Evolving Field of New-Style Prosopography," *Literary and Linguistic Computing* 20 (2005):

3–24. On the HRAF, see Rebecca Lemov, "Filing the Total Human: Anthropological Archives from 1928 to 1963," in *Social Knowledge in the Making*, ed. Michèle Lamont, Charles Camic, and Neil Gross (Chicago: University of Chicago Press, 2012).

16. See, for example, Martin Hand, *Ubiquitous Photography* (Cambridge, UK: Polity Press, 2012), introduction.

17. On the history of the emergence of "relational" databases and Database Management Systems (DBMSs) in the 1970s and 1980s, see David Gugerli, "The World as Database: On the Relation of Software Development, Query Methods, and Interpretative Independence," *Information and Culture: A Journal of History* 47, no. 3 (2012): 288–311, and Thomas Haigh, "How Data Got Its Base: Information Storage Software in the 1950s and 1960s," *IEEE Annals of the History of Computing*, 31, no. 4 (2009): 6–25.

18. On the history of data, see Dan Rosenberg, "Data before the Fact," in *Raw Data Is an Oxymoron*, ed. Lisa Gitelman (Cambridge, MA: MIT Press, 2013). Databases form a more recent part of the history of data, the usage of which extends to the seventeenth century (data were "the givens" of an argument). While working for General Electric in the early 1960s, Charles Bachmann developed the first DBMS, called Integrated Data Store (IDS), which shipped in 1964. Developed to be a generic manufacturing control system, its broader potential took some time to emerge. Not until the late 1970s would relational databases become available, via Oracle, with object-oriented, object-relational databases invented in the 1990s. See Gugerli, "The World as Database," and Charles Bachmann oral history, ACM Oral History interviews, Interview No. 2, September 25–26, 2004, http://dl.acm.org/citation.cfm?id=1141882. Ted Codd invented the relational database: E. L. Codd, "A Relational Model of Data for Large Shared Data Banks," *Communications of the ACM* 13, no. 6 (1970): 377–387. Gugerli in "The World as Database" retells this story in an intriguing way.

19. Louise Spindler, "Sixty-one Rorschachs and Fifteen Expressive Autobiographic Interviews of Menomini Women," *Microcard Publications of Primary Records in Culture and Personality* 2, no. 10 (La Crosse, WI: Microcard, 1957).

20. A tool-oriented approach to the history of the behavioral sciences is found in Joel Isaac, "Tool Shock: Technique and Epistemology in the Postwar Social Sciences," *History of Political Economy* 42, suppl. 1 (2010): 133–164. For work that highlights a turn to methodology in the history of the social sciences at mid-twentieth century, one might begin with Isaac or Sarah Igo, *The Averaged American* (Cambridge, MA:

Harvard University Press, 2010), as well as contributions from David Engerman, Jamie Cohen-Cole, Joy Rohde, and the other authors whose works are collected in *Cold War Social Science*, ed. Mark Solovey and Hamilton Cravens (New York: Palgrave Macmillan, 2012).

Chapter 1. Paperwork of the Inner Self

1. Göring's full Rorschach record appears in "Bats and Dancing Bears: An Interview with Eric Zillmer," *Cabinet* 5, no. 2 (2001): 89. See also José Brunner, "'Oh Those Crazy Cards Again': A History of the Debate on the Nazi Rorschachs, 1946–2001," *Political Psychology* 22, no. 2 (2001): 233.

2. Douglas Kelley, *Twenty-two Cells in Nuremberg: A Psychiatrist Examines the Nazi War Criminals* (New York: Greenberg, 1947), 12; quoted in Brunner, "'Oh Those Crazy Cards Again,'" 241.

3. See overviews in Molly Harrower, "Rorschach Records of the Nazi War Criminals: American Experimental Study after Thirty Years," *Journal of Personality Assessment* 40, no. 4 (1976): 341–351; Eric Zillmer, Robert Archer, and Robert Castino, "Rorschach Records of Nazi War Criminals: A Reanalysis Using Current Scoring and Interpretative Practices," *Journal of Personality Assessment* 53 (1989): 85–99.

4. Gustav Gilbert: *Nuremberg Diary* (Cambridge, MA: Da Capo Press, 1995 [orig. 1947]), and *The Psychology of Dictatorship* (Westport, CT: Greenwood Press, 1979 [orig. 1950]); Douglas Kelley: "Preliminary Studies of the Rorschach Records of the Nazi War Criminals," *Rorschach Research Exchange* 10, no. 2 (1946): 45–48, and *Twenty-two Cells in Nuremberg*.

5. "Latency" refers to a period of suspended animation, dormancy, or delay—as in a network, where latency is the time during which a packet of data has not yet arrived at its destination. "Latent life" refers to biological specimens kept frozen for future use; here I intend "latency" to apply to paperwork that lapses into a similar state, kept for potential future use in a condition of possibility yet obscurity. It could perhaps be called "latent information" or "latent knowledge." On latency, see Joanna Radin, "Latent Life: Concepts and Practices of Human Tissue Preservation in the International Biological Program," *Social Studies of Science* 43, no. 4 (2013): 484–508, esp. 487–488.

6. Language of "securing" the records is from Kelley, "Preliminary Studies of the Rorschach Records," 45. On the fate of the paper copies of the Nazi Rorschach records, see Harrower, "Rorschach Records of the Nazi War Criminals," 342ff.

7. Florence Miale (psychologist) and Michael Selzer (political scientist), in their co-authored *The Nuremberg Mind: The Psychology of the Nazi Leaders* (New York: NY Times Books, 1975), undertook a detailed content analysis of sixteen Rorschachs made available to them. They found the Nazi leaders to be psychologically disordered psychopaths prone to depression, violent acting out, and narcissism, much like their devotees in personality structure—that is, not normal. Yet Barry Ritzler, in "The Nuremberg Mind Revisited: A Quantitative Approach to Nazi Rorschach," *Journal of Personality Assessment* 47 (1978): 344–353, found Miale's and Selzer's interpretations flawed by their preexisting full knowledge of the identities and historical actions of these actors, and a similar critique was in Richard Rubenstein, "Review of the *Nuremberg Mind,*" *Psychology Today*, July 1976, 83–84. For a synopsis of 1970s research, see Zillmer, Archer, and Castino, "Rorschach Records of Nazi War Criminals," 86–87.

8. Harrower's study presented ten "acknowledged authorities in Rorschach interpretation" with eight Nazi records and eight "control" records from a variety of mental patients and normal people. She found that experts working blind could not distinguish group similarities from the Nazi responses. Harrower concluded that productive and well-integrated personalities may become involved in large-scale "horrors." Also, she stressed the finding of many individual differences that outweighed the similarities and concluded, "It could happen here." Harrower, "Rorschach Records of the Nazi War Criminals," 343–347, and Molly Harrower, "Were Hitler's Henchmen Mad?" *Psychology Today*, July 1976, 80.

9. Hans Frank and Hermann Göring quoted in Zillmer, Archer, and Castino, "Rorschach Records of Nazi War Criminals," 96, 97.

10. Joan Ryan, "Mysterious Suicide of Nuremberg Psychiatrist," SF GATE, February 6, 2005; http://www.sfgate.com/bayarea/article/Mysterious-suicide-of-Nuremburg-psychiatrist-2732801.php. See also Jack El Hai, *The Nazi and the Psychiatrist: Hermann Göring, Dr. Douglas M. Kelley, and a Fatal Meeting of Minds at the End of WWII* (New York: Public Affairs, 2013).

11. Hannah Arendt, *Eichmann in Jerusalem: A Report on the Banality of Evil* (New York: Penguin, 1992 [orig. 1963]), 26.

12. For criticism of the "very ill-conceived" work, the "bad example" set, and even the false reporting by Hannah Arendt, "who by the way was not a mental health professional," see Alberto Peralta, "The Adolf Eichmann Case: Contradictions, New Data, and Integration," *Rorschachiana* 23, no. 1 (1999): 79. For a convincing interpretation of

Eichmann's Rorschach as not ordinary at all, see Robert S. McCully, "A Commentary on Adolf Eichmann's Rorschach," *Journal of Personality Assessment* 44, no. 3 (1980): 311–319.

13. Peralta, "The Adolf Eichmann Case," 79.

14. Kerner, a well-known psychologist with wide interests in the paranormal and "mythopoetic" functions of the unconscious, turned to inkblots late in life, when he was becoming blind and had fallen into depression; his pastime was to make inkblots on a sheet of paper, ascribing the fanciful creatures that emerged to "Hades." This project later inspired Rorschach. Justinus Kerner, *Klecksographien: Mit Illustrationen nach den Vorlagen des Vergassers* (Stuttgart: Deutsche Verlags-Anstalt, 1857).

15. Hermann Rorschach, *Psychodiagnostik: Methodik und Ergebnisse eines wahrnehmungsdiagnostischen Experiments* (Bern: Ernest Bucher, 1921). English translation by P. Lemkau and B. Bronenberg (New York: Grune and Stratton, 1942).

16. Henri Ellenberger, "Hermann Rorschach, M.D., 1884–1922," *Bulletin of the Menninger Clinic* 18, no. 5 (1954): 206. "Total failure" refers to the publication as a whole and "slovenly work" to the myriad printer's errors (ibid.). "Intriguing unanticipated variation": Naamah Akavia, *Subjectivity in Motion: Life, Art, and Movement in the Work of Hermann Rorschach* (New York: Routledge, 2013), 10. Hermann Rorschach and Emil Oberholzer, "The Application of Interpretation of Form to Psychoanalysis," *Journal of Nervous and Mental Diseases* 60 (1923): 225–248, 359–379. Rorschach's student Hans Binder did publish about shading, or *Heildunkeldeutung*, in Binder, "Die Heildunkeldeutungen im psychodiagnostischen Experiment von Rorschach," *Schweizer Archiv für Neurologie und Psychiatrie* 30 (1932): 1–67, 233–286.

17. On the Japanese trajectory of the test, see Kenzo Sorai and Keiichi Ohnuki, "The Development of the Rorschach in Japan," *Rorschachiana* 29 (2008): 38–63. For a history of its relatively delayed British adoption, see Justine McCarthy Woods, "The History of the Rorschach in the United Kingdom," *Rorschachiana* 29 (2008): 64–80.

18. Klopfer interview, 1959 (about radio show) and 1971 recollection by a Klopfer student; both quoted in Leonard Handler, "A Rorschach Journey with Bruno Klopfer: Clinical Application and Teaching," *Journal of Personality Assessment* 90, no. 6 (2008): 534, 530.

19. Hertz traces the lineage among clinical psychologists from Rorschach, who taught Oberholzer, who taught David Levy, who taught Samuel Beck, the last two of whom taught Marguerite Hertz, and the last three of whom had already published a few articles by the mid-1930s in the United States. However, Klopfer's arrival changed the lay of the land

because he was an energetic institution builder and because he "attracted students not only from psychology but also from psychiatry, general medicine, education, anthropology, and other social scientific disciplines." Marguerite Hertz, "Rorschachbound: A 50-Year Memoir," *Journal of Personality Assessment* 50 (1986): 398.

20. Leonard Handler quoted in Jerry Wiggins, *Paradigms of Personality Assessment* (New York: Guilford Press, 2003), 40.

21. Bruno Klopfer and Douglas M. Kelley, *The Rorschach Technique: A Manual for a Projective Method of Personality Development* (Yonkers-on-Hudson: World Book, 1942); Bruno Klopfer, *The Rorschach Technique: A Manual for a Projective Method of Personality Diagnosis* (Yonkers-on-Hudson: World Book, 1945); Bruno Klopfer, *Developments in the Rorschach Technique* (Yonkers-on-Hudson: World Book, 1954); and Bruno Klopfer et al., *The Rorschach Technique: An Introductory Manual* (New York: Harcourt, Brace and World, 1962).

22. "Extreme subjectivity": Hertz, "Rorschachbound," 400. The fifth rival circle was David Rapaport's.

23. Dean Skadeland, "Bruno Klopfer: A Rorschach Pioneer," *Journal of Personality Assessment* 50 (1986): 360.

24. Zygmunt Piotrowski, *Perceptanalysis: The Rorschach Method Fundamentally Reworked, Expanded, and Systematized* (New York: Taylor and Francis, 1987), 252. On shading as one of the most perennially controversial areas of Rorschach research, see Sorai and Ohnuki, "The Development of the Rorschach in Japan." Klopfer's first discussion of shading was in Bruno Klopfer, "The Shading Response," *Rorschach Research Exchange* 2 (1938): 76–79. Further discussion is in Klopfer, *Developments in the Rorschach Technique*, 346–347.

25. The vagueness of the estimate is itself of interest; figure from Scott Lilienfeld et al., "What's Wrong with This Picture?" *Scientific American Mind* 16 (2005): 52. The Rorschach continues to grow in popularity in Japan, where distinct Japanese "norms" have been pursued (Sorai and Ohnuki, "The Development of the Rorschach in Japan," 53), and the Rorschach is experiencing a resurgence of interest in the United Kingdom (Woods, "History of the Rorschach in the United Kingdom," 74).

26. Quoted in Annie Murphy Paul, *The Cult of Personality Testing: How Personality Tests Are Leading Us to Miseducate Our Children, Mismanage Our Companies, and Misunderstand Ourselves* (New York: Free Press, 2004), 20. On the Rorschach's publication history, see also Peter Galison, "Image of Self," in *Things That Talk: Object Lessons from Art and Science*, ed. Lorraine Daston (New York: Zone Books, 2004).

27. Carl Jung: *The Visions Seminars: From the Complete Notes of Mary Foote* (Zurich: Spring Publications, 1976), and *Visions: Notes of the Seminars Given in 1930–35* (Princeton, NJ: Princeton University Press, 2007).

28. The first author in the initial publication was listed as Christiana Morgan, but by the third publication—the major and influential 1943 volume, which went on to become the second-highest seller in the history of Harvard University Press—her name had "dropped off" the cover. On the "vexing question" of the authorship of the test, see James William Anderson, "Henry A. Murray and the Creation of the Thematic Apperception Test," in *Evocative Images: The Thematic Apperception Test and the Art of Projection*, ed. Lon Gieser and Morris Stein (Washington, DC: American Psychological Association, 1999). Christiana Morgan and Henry A. Murray, "A Method for Investigating Fantasies: The Thematic Apperception Test," *Archives of Neurology and Psychiatry* 34 (1935): 289–306; Henry A. Murray and Christiana Morgan, *Explorations in Personality* (New York: Oxford University Press, 1938); Henry A. Murray, *Thematic Apperception Test: Manual* (Cambridge, MA: Harvard University Press, 1943).

29. For a bibliography of the preexisting tests, see Gardner Lindzey, *Projective Techniques and Cross-Cultural Research* (New York: Appleton-Century-Crofts, 1961), 35.

30. Interview with Murray in 1974; quoted in Anderson, "Henry A. Murray and the Creation of the Thematic Apperception Test," 37.

31. Murray quoted in Anderson, "Henry A. Murray and the Creation of the Thematic Apperception Test," 25, 37.

32. William Runyon, "Coming to Terms with the Life, Loves and Works of Henry A. Murray," *Contemporary Psychology* 39 (1994): 701–702.

33. Jung, *Visions*, 660. Pictures 6 and 12 described in Morgan and Murray, "A Method for Investigating Fantasies," 296, 297.

34. "What's happening": quoted in Roy Schafer, "How Was This Story Told?" *Journal of Projective Techniques* 22, no. 2 (1958): 181; Murray, *Thematic Apperception Test*, 1; Morgan and Murray, "A Method for Investigating Fantasies," 289. "X-ray picture of . . . inner self" is from Henry A. Murray, *Thematic Apperception Test* (Cambridge, MA: Harvard University Press, 1943), 1.

35. August 23, 1947 (date of test), subject Eddie M. Bert Kaplan, "Rorschachs of Sixty Navaho Adults and Children and Modified TATs, Murray TATs and Sentence Completion Tests of Fourteen Navaho Young Men," *Microcard Publications of Primary Records in Culture and Personality* 1, no. 20 (La Crosse, WI: Microcard, 1956).

36. Roy Schafer, *Projective Testing and Psychoanalysis* (New York: International Universities Press, 1967), 116.
37. Schafer, "How Was This Story Told?" 188 (paragraph breaks added).
38. Yehudi Menuhin, personal communication with Wesley G. Morgan, October 31, 1993; quoted at http://web.utk.edu/~wmorgan/tat/tattxt.htm.
39. John Exner quoted in Galison, "Image of Self," 264.
40. Galison, "Image of Self," 275.
41. "Mechanical objectivity": from Lorraine Daston and Peter Galison, *Objectivity* (New York: Zone Books, 2007), 115–183. On "pencil-and-paper measures": Hertz, "Rorschachbound," 397.
42. Discussion of Freudian projection in Lindzey, *Projective Techniques*, 28ff. Quotations from Freud 1911 on p. 66 and Freud 1913 on pp. 64 and 65.
43. Cf. Max Weber, "Science as a Vocation," in *From Max Weber: Essays in Sociology*, ed. and trans. H. H. Gerth and C. Wright Mills (New York: Oxford University Press, 1946), 129–156 [originally "Wissenschaft als Beruf," speech at Munich University, 1918]; Lorraine Daston, "Objectivity and the Escape from Perspective," *Social Studies of Science* 22 (1992): 607.
44. John Tresch, *The Romantic Machine: Utopian Science and Technology after Napoleon* (Chicago: University of Chicago Press, 2012), 309.
45. Daston and Galison, *Objectivity*, 10.
46. Steven Shapin, "The Sciences of Subjectivity," *Social Studies of Science* 42, no. 2 (2011): 172.
47. Marguerite Hertz, "Objectifying the Subjective," *Rorschachiana* 8 (1963): 25–54.
48. On perspectivity and aperspectival objectivity, see Daston, "Objectivity and the Escape from Perspective." The "view from nowhere" is philosopher Thomas Nagel's expression describing the supposedly positionless position of the objective observer. Nagel, *The View from Nowhere* (New York: Oxford University Press, 1986).
49. Daston, "Objectivity and the Escape from Perspective," 601.
50. Margaret Mead's description of Frank quoted in Ann Hulbert, *Raising America: Experts, Parents, and a Century of Advice about Children* (New York: Random House, 2011), 104. On Frank as "impressario" of others' ideas, see Peter Mandler, *Return from the Natives: How Margaret Mead Won the Second World War and Lost the Cold War* (New Haven: Yale University Press, 2013): 23–24. Lawrence K. Frank, "Projective Methods for the Study of Personality," *Journal of Psychology* 8 (1939):

389–414. Morgan and Murray used "projection tests" and "projection methods" in *Explorations in Personality*, 529, 387.

51. Lindzey, *Projective Techniques*, 36; Rebecca Lemov, "X-Rays of Inner Worlds: The Mid-Twentieth-Century American Projective Test Movement," *Journal of the History of the Behavioral Sciences* 47, no. 3 (2011): 251–278.

52. Frank, "Projective Methods," 402.

53. On the proliferation of variations of the Rorschach, see Hertz, "Rorschachbound," 502.

54. Ben Jonson, *New Inne*, 1631, iii, ii, 177; *OED* online.

55. Http://www.hogrefe.com/program/r-rorschach-test.html.

56. Mónica Guinzbourg de Braude, "From Ambiguity in Chinese Painting to Rorschach's Inkblots," *Rorschachiana* 29 (2008): 26.

57. "*Rorschach Archive* Bibliography: Books Articles, Thesis [*sic*], Dissertations, Congresses, Periodicals," Holdings (pdf) as of May 2013, http://biblio.unibe.ch/rorschach/pdf/RA-Bibliography-2013.pdf.

58. The sparse historical literature addressing the test includes Franziska Baumgarten-Tramer, "Zur Geschichte der Rorschach-tests," *Schweizer Archiv für Neurologie und Psychiatrie* 50 (1942): 1–13; Henri Ellenberger, "The Life and Work of Hermann Rorschach (1884–1922)" [1954], in *Beyond the Unconscious: Essays of Henri F. Ellenberger in the History of Psychiatry*, ed. Mark Micale (Princeton, NJ: Princeton University Press, 1993), 193–236; and more recently Galison, "Image of Self," as well as Akavia, *Subjectivity in Motion*.

59. Letter from Christiana Morgan to Bert Kaplan, March 6 (no year, handwritten); Bert Kaplan Papers, Santa Cruz, CA. [Hereafter: Kaplan Papers.] Although the letter is undated, it likely corresponds to the semester during which Kaplan taught at Harvard (ca. 1949–1950).

60. Akavia, *Subjectivity in Motion*, 2.

61. On the non-standard nature of the standardization process, see Stefan Timmermans and Steven Epstein, "A World of Standards but Not a Standard World: Toward a Sociology of Standards and Standardization," *Annual Review of Sociology* 36 (2010): 69–89. For reasons of space, I have not been able to explore the topic of experimentalism in the projective testing movement, but it is a rich one with regard to Rorschach, Murray, and many other projective enthusiasts.

62. Frederick Crews, "Out, Damned Blot!" *New York Review of Books*, July 15, 2004; http://www.nybooks.com.ezp-prod1.hul.harvard.edu/articles/archives/2004/jul/15/out-damned-blot/.

Chapter 2. The Varieties of Not Belonging

1. Title of this chapter is courtesy of James Wood's lecture, "On Not Going Home," *London Review of Books,* February 20, 2014. Twenty-six-year-old Zuni male described in Bert Kaplan, "Forty-nine Rorschachs, Six Modified T.A.T.s, Twelve Murray T.A.T.s and Seven Sentence Completion Tests of Zuni Young Men," *Microcard Publications of Primary Records in Culture and Personality* 1, no. 24 (La Crosse, WI: Microcard, 1956): 65.

2. S. F. Nadel quoted in Jules Henry, S. F. Nadel, William Caudill, John J. Honigmann, Melford E. Spiro, Donald W. Fiske, George Spindler, and A. Irving Hallowell, "Projective Testing in Ethnography," *American Anthropologist* 57 (1955): 247.

3. Jules Henry contribution to Jules Henry et al., "Projective Testing in Ethnography," 245.

4. Some researchers define "psychometrics" as tests that adhere to a strict, quantitative standard; thus according to some traditions, projective tests were the opposite of psychometric tests (although prominent lineages of both Rorschachs and TATs have indeed attempted to make the tests strictly quantitative). Here, however, I use the term to refer more broadly to tests that measure and otherwise scientifically secure subjective personality data. They aim, in the words of Rorschach expert Marguerite Hertz, to "objectify the subjective" ("Objectifying the Subjective," *Rorschachiana* 8 [1963]: 25–54).

5. Bruno Klopfer and Douglas M. Kelley, *The Rorschach Technique: A Manual for a Projective Method of Personality Diagnosis* (Yonkers-on-Hudson, NY: World Book, 1942), 184. The six pages included (1) Identification, (2) Scoring list, (3) Tabulation sheet, (4) Summary of relationships among categories, (5) Picture sheet, and (6) Scoring symbols. Some Rorschach workers included in their data sets, eventually collected in the Kaplan archive, all six pages for each testee, resulting in very bulky sets; others included only a list of responses and the picture sheets; some, simply a list of responses.

6. Horace W. Miner, "Rorschachs of Arabs from Algiers and from an Oasis: 64 Men, 3 Women," *Microcard Publications of Primary Records in Culture and Personality* 3, no. 3 (1961). Since the greatest proportion of data comes from the Rorschach, I focus primarily on the Rorschach in this chapter; there is, however, a good deal more to be written about the riotous variety of other tests, especially as fieldworkers grouped them into "batteries" of three to eight tests to be co-administered.

7. On these large-scale "big social science" projects, see Rebecca Lemov, "Filing the Total Human: Anthropological Archives from 1928 to 1963," in *Social Knowledge in the Making*, ed. Charles Camic, Neil Gross, and Michèle Lamont (Chicago: University of Chicago Press, 2011), 119–150, and Joy Rohde, *Armed with Expertise: The Militarization of American Social Research during the Cold War* (Ithaca, NY: Cornell University Press, 2013).

8. Marguerite Hertz, "Projective Techniques in Crisis," *Journal of Projective Techniques and Personality Assessment* 34 (1970): 449–467.

9. See James B. Waldram, *Revenge of the Windigo: The Construction of the Mind and Mental Health of North American Aboriginal Peoples* (Toronto: University of Toronto Press, 2004), 65. The testing of Vietnamese peasants and Viet Cong POWs occurred as part of military contract work by Simulmatics Corporation of Cambridge, Mass., funded via the U.S. Defense Department's Advanced Research Projects Agency (ARPA), which in 1966 was the biggest funder of social research in Vietnam, amounting to a total of $4 million. Professors from MIT, Harvard, Columbia, and elsewhere contracted with Simulmatics. Joy Rohde, "The Last Stand of the Psychocultural Cold Warriors: Military Contract Research in Vietnam," *Journal of the History of the Behavioral Sciences* 47, no. 3 (2011): 232–250.

10. Although the analogy between mobile technologies and mobile psychotechnologies should not be pushed too far, some lessons can be learned. Jon Agar in his history of the mobile phone points out that to see the cell phone as simply a product of new developments and inventions is insufficient; he recalls growing up perfectly happy in the 1960s and 1970s with a cranky black state-issued telephone: "To say that it did not happen because the technology was not there misses an important part of the story. Technology only becomes 'there' when it fits the wider world" (*Constant Touch: A Global History of the Mobile Phone* [Cambridge, UK: Icon Books, 2003], 26). This is true in obverse: once the "wider world" of social science and social life no longer supported the Rorschach as an all-purpose measure of self, its technology no longer seemed self-evidently applicable, as we will see.

11. Relevant to the question of traveling technologies is history-of-technology scholarship on modification and flexibility within standardization processes. See Amy Slaton, "As Near as Practicable: Precision, Ambiguity, and Industrial Quality Control," *Technology and Culture* 42 (2001): 51–80; Stefan Timmermans and Marc Berg, "Standardization in Action: Achieving Local Universality through Medical Protocols," *Social Studies of Science* 272 (1997): 273–305; Ken

Alder, "Making Things the Same: Representation, Tolerance and the End of the Ancien Régime in France," *Social Studies of Science* 28 (1998): 499–545. Also relevant is Lorraine Daston and Elizabeth Lunbeck, eds., *Histories of Scientific Observation* (Chicago: University of Chicago Press, 2011).

12. Klopfer and Kelley, *The Rorschach Technique*, 8–9. See Klopfer papers listed in Bruno Klopfer, *Developments in the Rorschach Technique*, vol. 1 (Yonkers-on-Hudson: World Book, 1954). Historical scholarship on the Rorschach includes little scrutiny of cross-cultural use of the test save in actors' accounts. For an overview of this scholarship, see chapter 1 above, esp. n. 58. Outside of Rorschach histories, the most incisive historical criticism of the use of projective tests among American Indians is Waldram, *Revenge of the Windigo*, esp. 44–68.

13. On orphaned samples, see Emma Kowal, "Orphan DNA: Indigenous Samples, Ethical Biovalue and Postcolonial Science," *Social Studies of Science* 43, no. 4 (2013): 577–597.

14. A. I. Hallowell, Transcript of Anne Roe interview with Hallowell, January 2, 1963, Series II (Subject Files); Hallowell Papers, American Philosophical Society, Philadelphia. Permission to publish required. Hallowell quotations in this paragraph are from the Roe interview. Samuel Beck, *Introduction to the Rorschach Method* (New York: American Orthopsychiatric Association, 1937). Beck trained in the Rorschach under David Levy, an American psychologist who had studied in Switzerland in the early 1920s with Rorschach's friend and executor Emil Oberholzer (with whom Beck, in turn, went to study in 1934). Beck's unflaggingly rigorous and statistical approach to the Rorschach was often remarked on, alongside his personal and professional distaste for Klopfer's approach, and his obituary writer, John Exner, described how he "worked tirelessly to develop the Rorschach as a scientific tool that would have some meaningfulness for all of psychology" ("Samuel J. Beck, 1896–1980," *American Psychologist* 36, no. 9 [1981]: 986). Once Hallowell encountered Klopfer in 1938 post-fieldwork, he stopped working closely with Beck.

15. Beck book under arm: Hallowell's recollection in Transcript of Roe interview. Since the role of origin stories looms large in this history, note that Hallowell told a somewhat different first-use story many years after this initial recollection in an autobiographical essay, second draft, titled "On Being an Anthropologist: Some Autobiographical Reflections," Series III, Hallowell Papers (later published as A. I. Hallowell, "On Being an Anthropologist," in *Crossing Cultural Boundaries: The Anthropological Experience*, ed. S. T. Kimball

and J. B. Watson (San Francisco: Chandler Publishing, 1972): 51–62. In this statement, made toward the end of his life, he gave the date of his first Rorschach voyage as 1937; he also transposed some important elements: instead of trying but failing to visit Sam Beck in Chicago after administering the tests, he put the failed visit before. Since priority is a matter of concern here, I would suggest that the later date (1938)—which features in published histories, in Hallowell's data set within the Kaplan archive, and more consistently throughout the literature on Hallowell—is correct; if so, it is interesting that Hallowell's revised memory late in life would place him either "first in the field" among American Rorschach workers or very close to first.

16. Hallowell spoke of Berens as "my interpreter, guide, and virtual collaborator"; quoted in Jennifer S. H. Brown and Susan Elaine Gray, editors' introduction in William Berens (as told to A. Irving Hallowell), *Memories, Myths, and Dreams of an Ojibwe Leader* (Montreal: McGill-Queen's University Press, 2009), 9; also quoted in George Stocking, "A. I. Hallowell's Boasian Evolutionism: Human Ir/Rationality in Cross-Cultural, Evolutionary and Personal Context," in *Significant Others: Interpersonal and Professional Commitments in Anthropology*, ed. R. Handler (Madison: University of Wisconsin Press, 1984), 206.

17. Hallowell diary quoted by Brown and Gray in Berens, *Memories*, 3.

18. Hallowell quoted by Brown and Gray in Berens, *Memories*, 6.

19. Hallowell recollection in "Autobiographical Essay," 1972, noted in first and second drafts of "On Being an Anthropologist," Series III, Hallowell Papers.

20. "Playing with inkblots": from Hallowell, "On Being an Anthropologist," second draft, Hallowell Papers. Although this essay, as mentioned above, is unreliable with regard to dates (likely due to Hallowell's relatively advanced age when he wrote it), the session with Klopfer at the 1938 AAA is confirmed, and his recollection of the audience response is of interest. Other personal information in this account comes from Hallowell, Transcript of Roe interview. Note that Anne Roe interviewed Hallowell and gave him the Rorschach in 1950 as part of a study about scientific creativity; she gave a follow-up interview in 1963, taped and (imperfectly) transcribed.

21. As he noted in his data set's introduction, Hallowell attempted a proper sampling of the population by age and sex, though it was never possible to achieve an "ideal sample" due to exigencies of which children were available in the schools he visited, who wanted to take the test, and other factors. A. Irving Hallowell, "Rorschachs of 151

Berens River Salteaux (Ojibwa) Adults and Children and 115 Lac du Flambeau Adults," *Microcard Publications of Primary Records in Culture and Personality* 2, no. 6 (LaSalle, WI: Microcard Foundation, 1957). [Hereafter: Ojibwe data set.]

22. One outgrowth of Hallowell's comparative Rorschach research with the Ojibwe was his contribution to the "atomism" debates of the 1950s and 1960s, one of the longest-running and most fiercely argued contests in the history of Americanist anthropology. On the origin of the atomism debates, see Joan Lovisek, Tim Holzkam, and Leo Weisberg, "Fatal Errors: Ruth Landes and the Creation of the 'Atomistic' Ojibwa," *Anthropologica* 39 (1997): 133–145.

23. Quoted by Brown and Gray in Berens, *Memories*, 196n14 (emphasis in original).

24. This and the following letter from Berens to Hallowell are quoted by Brown and Gray in Berens, *Memories*, 31.

25. Gordon Berens quoted by Brown and Gray in Berens, *Memories*, 32.

26. Hallowell (and later his student and fellow Microcard archive contributor A. F. C. Wallace) was the model of the "Philadelphia anthropologist"; on Philadelphian anthropology, see Regna Darnell, "The Emergence of Academic Anthropology at the University of Pennsylvania," *Journal of the History of the Behavioral Sciences* 6 (1970): 80–92. "If I went into the field tomorrow": from Jules Henry, S. F. Nadel, William Caudill, John Honigmann, Melford Spiro, Donald Fiske, George Spindler, and A. Irving Hallowell, "Projective Testing in Ethnography: A Symposium," *American Anthropologist* 57, no. 2 (1955): 245. On Hallowell's more subtle and refined approach to culture and personality studies, and on the influence of Edward Sapir, see Robert LeVine, "Anthropological Foundations of Cultural Psychology," in *Handbook of Cultural Psychology*, ed. Shinobu Kitayama and Dov Cohen (New York: Guilford Press, 2010), 40–58, esp. 45–47.

27. M. Bleuler and R. Bleuler, "Rorschach's Ink-Blot Test and Racial Psychology: Mental Peculiarities of Moroccans," *Character and Personality* 4 (1935): 97–114; quotations in this paragraph are from pp. 97–99, except as otherwise noted.

28. Quotation in David Rosenthal, "Introduction to Manfred Bleuler's "The Offspring of Schizophrenics," *Schizophrenia Bulletin* 1 (1974): 91.

29. Quotations in this paragraph are from Bleuler and Bleuler, "Rorschach's Ink-Blot Test," 114ff.

30. Francis L. K. Hsu, "Introduction to Part II: Methods and Techniques," in *Psychological Anthropology: Approaches to Culture and Personality*, ed.

Francis L. K. Hsu (Homewood, IL: Dorsey Press, 1961), 231. Spindler quoted in Henry et al., "Projective Testing in Ethnography," 259.

31. Samuel Beck, "The Frontier Within," *American Journal of Orthopsychiatry* 19, no. 4 (1949): 571. Bert Kaplan, "Cross-Cultural Use of Projective Techniques," in Hsu, *Psychological Anthropology*, 235.

32. Margaret Mead review of Hsu, *Psychological Anthropology*, in *American Anthropologist* 64 (1962): 628; emphasis added.

33. Kaplan, "Cross-Cultural Use of Projective Techniques," 235–255. Critics inveighed against the use of projective tests in the field, and against projective instruments in general, for different but sometimes overlapping reasons. Viktor Barnouw describes the projective testing movement and its critics in "An Interpretation of Wisconsin Ojibwa Culture and Personality: A Review," in *The Making of Psychological Anthropology*, ed. George Spindler (Berkeley: University of California Press, 1980), 64–86.

34. Hallowell, "On Being an Anthropologist," first draft, Hallowell Papers.

35. Alan Rosen, *The Wonder of Their Voices: The 1946 Holocaust Interviews of David Boder* (Oxford: Oxford University Press, 2010), 186–194. For other reasons, Boder abandoned the attempt to give the TAT to a large group of Jewish survivors; after around nine tests, he desisted, continuing only with interviews. In England, by contrast, projective tests took another path of dissemination and reached established heights within the Tavistock clinic environs around this time.

36. Bruno Klopfer, introduction to *Developments in the Rorschach Technique* (Yonkers-on-Hudson: World Book, 1956), 2:8; see also Gertrude Meili-Dworetzki, "The Development of Perception in the Rorschach," 104–171, in the same volume.

37. Klopfer, introduction to *Developments in the Rorschach Technique*, 2: 8. As reflected in the bibliographies in two Klopfer publications from 1946 and 1956.

38. Anthony Wallace's term in his biographical memoir of Hallowell: A. F. C. Wallace, "Alfred Irving Hallowell, 1892–1974" (Washington, DC: National Academy of Sciences, 2000), 200.

39. David Kaiser, *Drawing Things Apart: The Dispersion of Feynman Diagrams in Postwar Physics* (Chicago: University of Chicago Press, 2005), 6; emphasis in original.

40. Ursula Klein: "Paper Tools in Experimental Cultures," *Studies in the History and Philosophy of Science* 32 (2001): 293, and *Experiments, Models, Paper Tools: Cultures of Organic Chemistry in the Nineteenth Century* (Palo

Alto, CA: Stanford University Press, 2003). Klein usefully distinguishes paper tools from laboratory tools: paper tools are intellectual and do not interact with the object under investigation. In this sense, projective tests also bore some characteristics of laboratory tools, for they created material traces stemming from their interaction with subjects. Cf. Bruno Latour on immutable mobiles, emphasizing the increasing imposition of regularity in technoscientific activities: "Visualization and Cognition: Thinking with Eyes and Hands," in *Knowledge and Society: Studies in the Sociology of Culture Past and Present* (Greenwich, CT: JAI, 1986), 1–40.

41. Slaton, "As Near as Practicable," 70. Slaton writes about the technical protocols developed by early-twentieth-century quality-control engineers who were trained scientifically and worked in industrial areas such as concrete construction. Using equipment that was durably portable but also patently scientific in origin and accuracy, young engineers traveled from building site to building site, evolving standards that left room for subjective judgment, mutability, and experiential testing while embracing scientific precision.

42. Hallowell, introduction to Ojibwe data set, 9.

43. Marguerite Hertz, "The Method of Administration of the Rorschach Ink-Blot Test," *Child Development* 7, no. 4 (1936): 237–254; see also Marguerite Hertz, "Rorschach: 20 Years After," *Psychological Bulletin* 39 (1942): 90–129.

44. Hallowell, introduction to Ojibwe data set, 9.

45. Here and in the next several paragraphs quotations are from Hallowell, introduction to Ojibwe data set, 1–21b.

46. For full diacritics, see Hallowell's typed version, introduction to Ojibwe data set, 15n2.

47. Hallowell, introduction to Ojibwe data set, 15.

48. See Beck, *Introduction to the Rorschach Method*, 191.

49. Hallowell, Ojibwe data set, 108.

50. Still displaying "good concentration," to Hallowell's surprise. Hallowell, Ojibwe data set, 127.

51. Hallowell, introduction to Ojibwe data set, 16.

52. Ibid., 16. Cf. Marguerite Hertz, "On the Standardization of the Rorschach Method," *Rorschach Research Exchange* 3 (1939): 124.

53. Hallowell, introduction to Ojibwe data set, 18.

54. Klopfer quoted in ibid., 18.

55. Ibid., 18.

56. Ibid., 42.

57. Many remembered Hallowell as a good dancer, including William Berens's son Gordon Berens, interviewed in 1992: "He was crazy dancing at the Indian dance. He can do it too. . . . Just as good as the

Indians did. Oh, he sure enjoyed that" (Gordon Berens quoted by Brown and Gray in Berens, *Memories*, xxii). "Indian nicknames" recollection is from Hallowell, Ojibwe data set, 19n4. Discussion of Hallowell's other nicknames is in Jennifer S. H. Brown, "Fields of Dreams: Revisiting A. I. Hallowell and the Berens River Ojibwe," in *New Perspectives on the Native North America: Cultures, Histories, and Representations*, ed. Sergei Kan and Pauline Turner Strong, 17–41 (Lincoln: University of Nebraska Press, 2006), 38n2.

58. A. I. Hallowell, "The Rorschach Method as an Aid in the Study of Personalities in Primitive Societies," *Character and Personality* 9 (1941): 239.

59. Jules Henry, introduction to "Rorschachs of Twenty-six Pilagá Children and Adults," *Microcard Publications of Primary Records in Culture and Personality* 1, no. 25 (LaSalle, WI: Microcard, 1956); excerpted by Henry from his "Rorschach Technique in Primitive Cultures," *American Journal of Ortho-Psychiatry* 11 (1941): 230–234.

60. Anthony F. C. Wallace, "Commentary: 'Growing Up Indian': Childhood and the Survival of Nations," *Ethos* 41 (2013): 337. "Ulithians' dyspeptic reaction to the cards": Anthony F. C. Wallace, review of Lessa's *Ulithi*, *American Anthropologist* 57 (1955): 393.

61. George Spindler, "Rorschachs of Sixty-eight Menomini Men at Five Levels of Acculturation," *Microcard Publications of Primary Records in Culture and Personality* 2, no. 11 (La Crosse, WI: Microcard, 1957): 62–66.

62. Description of Rorschach-based research conducted by Theodora Abel and Renata Calabresi, 1951, in conjunction with Oscar and Ruth Lewis's fieldwork in Mexico; described in Gardner Lindzey, *Projective Techniques and Cross Cultural Research* (New York: Appleton-Century-Crofts, 1961), 204.

63. An Alorese seeress named Longfani responded to cards while possessed by a familiar spirit who helped her see things; in Cora DuBois, "Rorschachs of Alorese Men and Women," *Microcard Publications of Primary Records in Culture and Personality* 3, no. 1 (La Crosse, WI: Microcard, 1961): 32; a Manitoban Ojibwe elderly woman reported special entities Indians could see in some of the cards; in Hallowell, Ojibwe data set, 40–41. Wannatcos's protocol in Hallowell's Ojibwe data set, 82–83.

64. Case 3, Hallowell, Ojibwe data set, 41.

65. Case 8, Hallowell, Ojibwe data set, 71.

66. Case 26, July 25, 1949, interview/Rorschach test; reported in George Spindler, "Personal Documents in Menomini Peyotism," *Microcard*

Publications of Primary Records in Culture and Personality 2, no. 16 (La Crosse, WI: Microcard, 1957): 18.

67. On dream dynamics, see Brown, "Fields of Dreams," 21–26; dream from Birch Tree, 24. Deferral of dream offering is Case 19, Hallowell, Ojibwe data set, 108.

68. Letter from Melford Spiro to A. I. Hallowell, September 2, 1947, Hallowell Papers. Bear Hair record, #29, Hallowell Ojibwe data set, n. p.

69. Timmermans and Berg, "Standardization in Action," 275. Looseness can tend to stabilize a technique or protocol rather than undermine it.

70. Kowal, "Orphan DNA," 580. On frozen samples and the transformative role of freezing, see Joanna Radin, "Latent Life: Concepts and Practices of Human Tissue Preservation in the International Biological Program," *Social Studies of Science* 43, no. 4 (2013): 484–508; for a contrasting case of colonial-era sample gathering (biological samples of anthropological subjects), see Warwick Anderson, "Objectivity and Its Discontents," *Social Studies of Science* 43 (2013): 465–483.

71. George Spindler quoted in Peter Black, "Psychological Anthropology and Its Discontents: Science and Rhetoric in Postwar Micronesia," in *American Anthropology in Micronesia: An Assessment,* ed. Mac Marshall and Robert Kiste (Honolulu: University of Hawaii Press, 1999), 235; note that Spindler was speaking specifically of Cora DuBois's and Thomas Gladwin's studies to evoke an era.

72. Kowal, "Orphan DNA," 590.

73. On the "situation" in Cold War social science, see Paul Erickson et al., *How Reason Almost Lost Its Mind: The Strange Career of Cold War Rationality* (Chicago: University of Chicago Press, 2013), ch. 4.

74. Black makes the parallel point that TAT protocols may prove useful for historically understanding the construction of American-Micronesian joint understandings and ways of interacting ("Psychological Anthropology and Its Discontents," 252n14). Others have mined them for their poetry.

Chapter 3. The Storage of the Very, Very Small

1. "The total archive" was explored in a conference by that name, organized by Boris Jardine, Matthew Drage, and Ruth Horry at CRASSH, Cambridge, UK, March 19–20, 2015.

2. "Tiny technologies": I owe this phrase to Hallam Stevens. Nanovisions: Colin Milburn, *Nanovision: Engineering the Future* (Durham, NC: Duke University Press, 2008), 9. At a scale a 5 ¼" floppy disk once held, each

synthesized strand of DNA contains ninety-six bits of storage; at the head of each strand is a nineteen-bit "address" used to order the strands properly. A whole trove could be held in a vat or sprayed on a wall and subsequently resorted into usable data. "Next Generation Digital Information Storage in DNA," *Science* 337, no. 6102 (September 28, 2012): 1628. Also on Church, see Sophia Roosth, *How Life Got Made* (Chicago: University of Chicago Press, forthcoming).

3. "Snuffbox": quoted in Frederic Luther, *Microfilm: A History, 1839–1900* (Annapolis: National Microfilm Association, 1959), 23. Hauled away in a van: Vannevar Bush, "As We May Think," *Atlantic Monthly* 176 (July 1, 1945): 101–108. See Davis Baird, Alfred Nordmann, and Joachim Schummer, eds., *Discovering the Nanoscale* (Amsterdam: Ios Press, 2005), including "The Epistemology of the Very Small." The reckoning of thirty-five pages in the *Encyclopedia Britannica* comes from the foundational document of the nanotech movement in the physical sciences: Richard Feynman's talk, "There's Plenty of Room at the Bottom," given on December 29, 1959, at Caltech; published as "There's Plenty of Room at the Bottom: An Invitation to Enter a New Field of Physics," *Engineering and Science*, February 1960, 21–36. Kaplan, editor's introduction, *Microcard Publications of Primary Records in Culture and Personality* 1, no. 1 (La Crosse, WI: Microcard Foundation, 1956), 9.

4. Dancer autobiographical statement from *Manchester Literary and Philosophical Society* 107 (1964–1965); quoted in "John Benjamin Dancer," in *Encyclopedia of Nineteenth-Century Photography*, ed. John Hannavy (New York: Taylor and Francis, 2008), 377. See also Lawrence Leigh Ardern, *John Benjamin Dancer: Instrument Maker, Optician and the Originator of Microphotography* (London: Library Association, North Western Group, 1960). Tom Wolfe has made remarks along these lines in various interviews—for example, "I don't think the unaided imagination of the writer—and I don't care who the writer is—can come up with what is obtainable through research and reporting" (in George Plimpton, "Tom Wolfe, the Art of Fiction No. 123," *Paris Review* 118 (1991), http://www.theparisreview.org/interviews/2226/the-art-of-fiction-no-123-tom-wolfe.

5. Frederic Luther, "The Earliest Experiments in Microphotography," *Isis* 41 (1950): 277. (Article reprinted in *American Documentation* 2, no. 3 [1951]: 167–170.)

6. Dancer worked in two stages, first making a small negative intermediate (by means of a microscope) and from this producing the final reduction positive.

7. Binney quoted in Luther, "Earliest Experiments in Micro-photography," 278. Michael Hallett, "John Benjamin Dancer (1812–1887): A Perspective," *History of Photography* 10, no. 3 (1986): 237. Hallett explores the Manchester and surrounding knowledge networks of which Dancer was a part.

8. "Extreme interest": Luther, "Earliest Experiments in Micro-photography," 279. On miniaturization of the Bible, see Susan Stewart, *On Longing: Narratives of the Miniature, the Gigantic, the Souvenir, the Collection* (Durham, NC: Duke University Press, 1993), 40. The national anthem is mentioned in Roy Winsby, "The Microphotograph Slides of John B. Dancer and Richard Suter," *Manchester Microscopical Society Newsletter* 26, August 1993.

9. Quoted in *Proceedings of the Literary and Philosophical Society of Manchester* 1, no. 1 (1857–1860), March 22, 1859, 112–113.

10. "Photographic curiosities": "Literary Miscellanies," in H. W. Bidwell, ed., *The Eclectic Magazine of Foreign Literature, Science, and Art* 47 (New York: 5 Beekman St., 1859), 143. Dancer's preeminent role in the invention of the microphotograph was not known to historians until Luther's research appeared in 1950: "Earliest Experiments in Microphotography," and Frederic Luther, "René Dagron and the Siege of Paris," *American Documentation* 1, no. 4 (1950): 196–206. Intervening histories of microphotography's origins do not mention Dancer; see, for example, Paul Jeserich, *Die Mikrophotographie auf Bromsilbergelatine: bei natürlichem und künstlichem lichte unter ganz besonders berücksichtigung des kalklichtes* (Berlin: J. Springer, 1888), 5–7.

11. Quoted in Luther, "Earliest Experiments in Microphotography," 279.

12. Hooke quotations are from Robert Hooke, *Micrographia; or, Some Physiological Descriptions of Minute Bodies Made by Magnifying Glasses, with Observations and Inquiries There Upon* (London: Royal Society, 1665), dedication, preface, and Observations LIII, and LVI. For a description of Hooke as the first microscopy exponent and on Hooke's *Micrographia* as the first book of popular science, see Brian Ford, "The Royal Society and the Microscope," *Notes and Records of the Royal Society of London* 55, no. 1 (2001): 29 and 30. Regarding Hooke's self-fashioning, see Steven Shapin, "The House of Experiment in Seventeenth-Century England," *Isis* 79, no. 3 (1988): 373–404. Hooke's quarters were the primary place of experiment in seventeenth-century England, a place of trial/error and the valorization of curiosity in the method of making experiments.

13. John Mack, *The Art of Small Things* (Cambridge, MA: Harvard University Press, 2008), 2.

14. On the way miniaturization plays and confounds the rift between materiality and meaning, see S. Stewart, *On Longing*, 37–69.

15. "Contained in the space": Winsby, "The Microphotograph Slides of John B. Dancer and Richard Suter." See also Boris Jardine, "A Collection of John Benjamin Dancer Microphotographs," Explore Whipple Collections, Whipple Museum of the History of Science, University of Cambridge, 2006, http://www.hps.cam.ac.uk/whipple/explore/microscopes/microphotographs/; accessed June 19, 2011.

16. Susan Cady, "Machine Tool of Management: A History of Microfilm Technology" (PhD dissertation, Lehigh University, 1994). Such limitations were also true of printers, which had to be repeatedly redesigned and were never fully successful.

17. On Victorian householders' curious attachment to pocket-sized things that represented portability, such as traveling corkscrews and beetle collections, see John Plotz, *Portable Property: Victorian Culture on the Move* (Princeton, NJ: Princeton University Press, 2009), 2: the crux of the matter was a struggle over values, "an emergent rivalry between schemes of value being played out in single objects."

18. The dictionary quote is in Allen Veaner, *Studies in Micropublishing, 1853–1976: Documentary Sources* (Westport, CT: Microform Review, 1976), 88.

19. Luther, "René Dagron and the Siege of Paris," 198.

20. Quoted in ibid.

21. For more on Dagron's method, see ibid., 201.

22. Vernon Tate, "Criteria for Measuring the Effectiveness of Reading Devices," in *Microphotography for Libraries: Papers Presented at the Microphotography Symposium at the 1936 Conference of the American Library Association*, ed. M. L. Raney (Chicago: ALA, 1936), 22.

23. Feynman, "There's Plenty of Room at the Bottom," 24. The whole discussion originated in an old hypothetical problem with a real history: could one print all the books in the world in a small space? Say there are 24 million volumes in the world and one is making silica copies at the same scale (the scale at which the entire *Encyclopedia Britannica* could be printed on the head of a pin—that is, shrunk down twenty-five thousand times); this would now equal the size of thirty-five pages of the *Encyclopedia Britannica*.

24. The two 1950 articles are Luther, "Earliest Experiments in Microphotography" and "René Dagron and the Siege of Paris." Quotation on microfilm in its modern sense is from Luther, "Earliest Experiments in Microphotography," 279.

25. See Joanne Yates, *Control through Communication: The Rise of System in American Management* (Baltimore: Johns Hopkins University Press, 1989); James R. Beniger, *The Control Revolution: Technological and Economic Origins of the Information Society* (Cambridge, MA: Harvard University Press, 1986).

26. Cady, "Machine Tool of Management," 4.

27. "Permitting greater rapidity" is from Luther, "René Dagron, Inventor of Microfilm," *History of Photography* 20, no. 4 (1996): 356; Dagron's "lost art" is mentioned in Robert Binkley, ed., *Manual on Methods of Reproducing Research Materials* (Ann Arbor, MI: Edwards Bros., 1936), 121n1.

28. Note on terminology: "Microphotography" is an encompassing term, but at various times in its history, other terms arose to refer to the whole enterprise, sometimes also called microform (twentieth century) or micrographics (after the 1960s). "Microfilm" referred to strips of film on loops, and as we will see, an additional variety of formats emerged post–World War II, such as microprint, Microcard, and aperture cards (combining IBM cards with microfilm inserts).

29. "Brittleness and abrasion": Raney, *Microphotography for Libraries*, 103. "Microfilm is not the universal panacea": Lester K. Born, "Planning for Microfilm Operations," *ALA Bulletin* 44 (October 1950): 342; also published in *American Documentation* 2, no. 1 (1951): 1–5.

30. Otlet's Mundaneum was a combination of museum, cinema, archive, radio station, and encyclopedia; it used microfilm and index cards to catalog and attempt to file all existing books, periodicals, and journals for purposes of unifying and codifying knowledge. Otlet also experimented, in addition to microfilm as early as 1906, with broadcast media. "Otlet's story is ultimately a story of ideas, of passionately held and unwavering belief in the importance of intellectual life, in the possibility of its transformation by means of new kinds of tools and machines for managing and communicating knowledge, and in the need to achieve at last a new, peaceful, world society" (W. Boyd Rayward, "Knowledge Organisation and a New World Polity: The Rise and Fall and Rise of the Ideas of Paul Otlet," *Transactional Associations* 1–2 [2003]: 5).

31. See Michael Buckland, "What Is a Document?" *Journal of the American Society for Information Science* 48, no. 9 (1997): 804–810. (From Briet, *Qu'est-ce que la documentation?* [Paris: EDIT, 1951]). "From about 1920 [in Europe], 'documentation' was increasingly accepted as a general term to encompass bibliography, scholarly information services *(wissenschaftliche Aufklarung [Auskunft])*, records

management, and archival work" (Buckland, "What Is a Document?" 804). Frits Donker Duyvis, "Die Enstehung des Wortes 'Dokumentation' in Namen des FID," *Revue de Documentation* 26 (1959): 15–16; Carl Björkbom, "History of the Word Documentation within the FID," *Revue de Documentation* 26 (1959): 68–69; Michel Godet, "Documentation, bibliothèques et bibliographie: Essai de définition de leurs caractères et de leurs rapports," *Institut Internationale de Documentation (IID) Communicationes* 5 (1938): 15–18.

32. H. G. Wells and Alan J. Mayne, *World Brain: H. G. Wells on the Future of World Education* (London: Adamantine Press, 1994).

33. First in 1906 and with redoubled energy in 1925, Goldschmidt and Otlet argued for the low-cost film-based microphotography, suggesting that books could be copied on 7.5 × 12.5 cm sheets of film; on each sheet or card would be placed seventy-two micro-images. A title and other identifying information at the top of the card in large legible letters could be read with the unaided eye. Robert Goldschmidt and Paul Otlet, "Le livre microphotique," *Institut International de Bibliographie* 144 (1925): 1–8.

34. Tate's inaugural issue read, "As defined by the International Federation for Documentation (of which the American Documentation Institute is the American affiliate), the term 'documentation' refers to the creation, transmission, collection, classification and use of 'documents'; documents may be broadly defined as recorded knowledge in any format." Vernon D. Tate, "Introducing American Documentation: A Quarterly Review of Ideas," *American Documentation* 1, no. 1 (1950): 3. Microtech compared to printing press: see Lisa Gitelman, *Paper Knowledge: Toward a Media History of Documents* (Durham, NC: Duke University Press, 2014), 73. "The literature on documentation": Michael Buckland, "Emanuel Goldberg, Electronic Document Retrieval, and Vannevar Bush's MEMEX," *Journal of the American Society for Information Science* 43, no. 4 (1992): 290.

35. Jon Agar, "What Difference Did Computers Make?" *Social Studies of Science* 36, no. 6 (2006): 878.

36. These were intended to search microfilm; Bush embarked on the project after his 1931 success in building the Differential Analyzer, a world-famous analog machine for solving differential equations. The subsequent devices were entangled in webs of funding and cross-purposes. Colin Burke, "The Other Memex: The Tangled Career of Vannevar Bush's Information Machine, the Rapid Selector," *Journal of the American Society for Information Science* 43, no. 10 (1992): 648–657. On German microform search machines, see Irvin Stewart,

Organizing Scientific Research for War: The Administrative History of the Office of Scientific Research and Development (Boston: Little Brown, 1948).

37. Philip Power, "Eugene Barnum Power (4 June 1905–6 December 1993)," *Proceedings of the American Philosophical Society* 193, no. 3 (1995): 300.

38. See Power's autobiography in Eugene B. Power and Robert Anderson, *An Edition of One: The Autobiography of Eugene B. Power* (Ann Arbor, MI: University Microfilms, 1990). "This most famous": Thomas A. Bourke, "Scholarly Micropublishing, Preservation, Microfilming, and the National Preservation Effort in the Last Two Decades of the Twentieth Century: History and Prognosis," *Microform Review* 19, no. 1 (1990): 5.

39. The use of a 35mm camera was linked to the use of modified double-sprocketed cellulose acetate motion picture film as the camera film. This gave rise to the library community's "35mm film addiction." The 35mm format is still the most popular and commonly found. Thomas A. Bourke, "Scholarly Micropublishing, Preservation, Microfilming, and the National Preservation Effort in the Last Two Decades of the Twentieth Century: History and Prognosis," *Microform Review* 19, no. 1 (1990): 5.

40. See Binkley, *Manual on Methods*, 115–117.

41. Friedrich Kittler, *Literature, Media, Information Systems: Essays*, ed. John Johnston (Amsterdam: G+B Arts International, 1997), 84.

42. Kathy Peiss, "Cultural Policy in a Time of War: The American Response to Endangered Books in World War II," *Library Trends*, 55, no. 3 (2007): 370–386.

43. "Knighted by Britain," *The Michigan Alumnus* 24 (January 1978): 24. On OSS's role in text preservation via microfilming, see Peiss, "Cultural Policy in a Time of War"; Richards, *Scientific Information in Wartime* (for statistics on damage to the British library system and the supply of periodicals). On documentalists working for the OSS, England's Association of Special Libraries and Information Bureaux (Aslib), and the German Society for Documentation, which played seminal roles in enemy scientific information procurement, see Pamela Richards, "Information Science in Wartime: Pioneer Documentation Activities in World War II," *Journal of the American Society for Information Science* 39, no. 5 (1988): 301.

44. Quotations in this paragraph, except as otherwise noted, are from Peiss, "Cultural Policy in a Time of War," 378ff.

45. Alfred Kantorowicz, "The Burned Books Still Live," *New York Times Magazine*, May 7, 1944, 17.

46. Jonathan Auerbach and Lisa Gitelman, "Microfilm, Containment, and the Cold War," *American Literary History* 19, no. 3 (2007): 753–754.

47. Cady, "Machine Tool of Management," 14. On the microdot, see also Kristie Mackrackis, *Prisoners, Lovers, and Spies: The Story of Invisible Ink from Herodotus to al-Qaeda* (New Haven: Yale University Press, 2015), 198–221.

48. David Noble, *Forces of Production: A Social History of Industrial Automation* (New Brunswick, NJ: Transaction Publishers, 2011), 8on.

49. Martin Jamison, "The Microcard: Fremont Rider's Precomputer Revolution," *Libraries and Culture* 23, no. 1 (1988): 1–17; R. E. Molyneux, "What Did Rider Do? An Inquiry into the Methodology of Fremont Rider's *The Scholar and the Future of the Research Library*," *Libraries and Culture* 29 (1994): 297–325.

50. John Milton, *Paradise Lost*, II.561. Milton described the serpent himself in maze-like terms: "Circular base of rising folds that towered / Fold above fold, a surging maze" (IX.498–499).

51. "99% decrease": Fremont Rider, *The Scholar and the Future of the Research Library: A Problem and Its Solution* (New York: Hadham Press, 1944), 91; "micro pioneers": p. 97. Microprint was the invention of Albert Boni in 1934, and it both inspired and irritated Rider into creating the trademarked Microcard. On the origins of the modern bibliographic system in Konrad Gessner's *Biblioteca Universalis* (1545 and 1548), the first volume of which listed over ten thousand texts and described them by format title, author, place of publication and year, and content summary, see Markus Krajewski, *Paper Machines: About Cards and Catalogs, 1548–1929* (Cambridge, MA: MIT Press, 2011), 10.

52. See Gitelman, *Paper Knowledge*, 53. "The new graphic arts devices are, I believe, capable of working the other way—as implements for a more decentralized and less professionalized culture, a culture of local literature and amateur scholarship." Robert C. Binkley, "New Tools for Men of Letters," in *Selected Papers of Robert C. Binkley*, ed. Marx H. Fisch (Cambridge, MA: Harvard University Press, 1948), 180.

53. "Ingenious proposed solution": Molyneux, "What Did Rider Do?" 298.

54. "Education: Book on a Card?" *Time* 44, no. 10 (September 4, 1944).

55. George H. Davison, "Review of Progress," in *Microtext in the Form of Microcards and Other Micro-Opaques, Transparent Microfiches and Unitised Microfilm or Aperture Cards as Well as Reading and Enlarging Apparatus* (Moorgate, Rotherham, England: United Steel Companies, Research and Development Department, Swinden Laboratories, 1960), 5.

56. Disraeli quoted in S. Stewart, *On Longing*, 38; see also ibid., x.

57. Quoted in Gitelman, *Paper Knowledge*, 53

58. Information scientist Allen Veaner quoted in George Schwegmann, "The Bibliographical Control of Microforms," *Library Trends* 8, no. 3 (1960): 380.

59. According to information scientists such as Veaner, the misunderstanding of "man-machine interface" and the lack of concern with users' needs led to microforms' lack of potential relative to the computer; see Jamison, "The Microcard," 13. On the origin of the database form, see Geoffrey Bowker, *Memory Practices in the Human Sciences* (Cambridge: MIT Press, 2010), 29.

Chapter 4. Data Mining in Zuni

1. Kaplan, introduction, "Forty-nine Rorschachs, Six Modified T.A.T.s, Twelve Murray T.A.T.s and Seven Sentence Completion Tests of Zuni Young Men," *Microcard Publications of Primary Records in Culture and Personality* 1, no. 24 (La Crosse, WI: Microcard, 1956): 2.

2. Kaplan, introduction, "Forty-nine Rorschachs, Six Modified T.A.T.s, Twelve Murray T.A.T.s and Seven Sentence Completion Tests of Zuni Young Men," 2. (All quotations in this paragraph are from p. 2, except as otherwise noted.) It was not uncommon to invoke the European Enlightenment as an imperative for one's studies in the mid-twentieth century, yet the invocation was often made with a brisk assurance that further concrete steps were being taken down the road to knowledge. As Edward Shils put it, describing the interdisciplinary social science projects sponsored by the Committee on Social Thought at the University of Chicago, "Our chosen instrument of enlightenment is systematic research, conducted under the auspices of the best traditions of contemporary social science." Enlightenment was a mandate. Shils quoted in Clifford Geertz, "Disciplines," *After the Fact: Two Countries, Four Decades, One Anthropologist* (Cambridge, MA: Harvard University Press, 1995), 113.

3. Kaplan quotations in this paragraph are from Kaplan, introduction, "Forty-nine Rorschachs, Six Modified T.A.T.s, Twelve Murray

T.A.T.s and Seven Sentence Completion Tests of Zuni Young Men," 2. Memorandum, Rockefeller Foundation's Program and Policy Files, January 21, 1914; RF, R.G. 3, Series 910, Box 2, Folder 10, Rockefeller Archives Center (Tarrytown, NY).

4. The official status in 1947 of Zuni residents was as "wards of the government," a label much resented and finally changed in 1975. A common medieval term, "ward" derived from the care and protection a feudal lord extended to a fief until the fief reached majority. Even earlier, in Old English, it referred to God's guardianship. The legally defined wardship between the Bureau of Indian Affairs and the residents of Zuni Pueblo (in fact, all American Indian reservations in these years) drew on this lineage. In the twentieth century, wardship took the specific form of *administration;* see Richard White, *It's Your Misfortune and None of My Own: A New History of the American West* (Norman: University of Oklahoma Press, 1991), 5.

5. Kaplan, introduction, "Forty-nine Rorschachs, Six Modified T.A.T.s, Twelve Murray T.A.T.s and Seven Sentence Completion Tests of Zuni Young Men," 9–10.

6. Ibid. Other examples circulate of what could be called the monkey-wrenching of social surveys and social scientific projects: in the census of 1885, respondents gave names like "Shit Head" and "Dirty Prick" to census workers to sabotage or undermine their calculations. Thomas Biolsi, "The Birth of the Reservation: Making the Modern Individual among the Lakota," in *American Nation: Encounters in Indian Country 1850–Present,* ed. Frederick Hoxie, James Merrell, and Peter Mancall (New York: Routledge, 2001), 111. On the persistence of research programs justified for "the good of knowledge," despite Native American complaints about purposes and methods, see Kim Tallbear, *Native American DNA* (Minneapolis: University of Minnesota, 2013), 2.

7. Kaplan noted, "Some subjects [were] skeptical about whether I was giving my real reasons for coming to Zuni. This is understandable, aside from the noted Zuni suspicion, since knowledge of a large university and scientific research, was outside the experience of almost all." Kaplan, introduction, "Forty-nine Rorschachs, Six Modified T.A.T.s, Twelve Murray T.A.T.s and Seven Sentence Completion Tests of Zuni Young Men," 11.

8. Bert Kaplan and Paul Wright, introduction, "TATs of Thirteen Hopi Young Men and Women," *Microcard Publications of Primary Records in Culture and Personality* 1, no. 19 (La Crosse, Wisconsin, 1956); tests given to 17–21-year-olds at the Haskell School.

9. Kaplan, introduction, "Forty-nine Rorschachs, Six Modified T.A.T.s, Twelve Murray T.A.T.s and Seven Sentence Completion Tests of Zuni Young Men," 6; emphasis added.

10. Geertz, "Disciplines," 106–107.

11. On Enlightenment science's claims of utility, which often pitted rationality against rationalization, see Lorraine Daston, "Afterword: The Ethos of Enlightenment," in *The Sciences in Enlightened Europe*, ed. William Clark, Jan Golinski, and Simon Schaffer (Chicago: University of Chicago Press, 1999), 495–504.

12. Kaplan, introduction, "Forty-nine Rorschachs, Six Modified T.A.T.s, Twelve Murray T.A.T.s and Seven Sentence Completion Tests of Zuni Young Men," 2.

13. On the projective test movement and its aspirations, see Rebecca Lemov, "X-Rays of Inner Worlds: The Mid-Twentieth-Century American Projective Test Movement," *Journal of the History of the Behavioral Sciences* 47, no. 3 (2011): 251–278. On "tool shock" and the mid-century social sciences as a tool age, see Joel Isaac, "Tool Shock: Technique and Epistemology in the Postwar Social Sciences," *History of Political Economy* 42 (2010): 133–164. On methodological enthusiasm, see Paul Erickson et al., *How Reason Almost Lost Its Mind: The Strange Career of Rationality in the Cold War* (Chicago: University of Chicago Press, 2013). On the Cambridge Torres Strait expedition, see, among a rich historiography, Henrika Kuklick, "Personal Equations: Reflections on the History of Fieldwork, with Special Reference to Sociocultural Anthropology," *Isis* 102, no. 1 (2011): 1–33.

14. Kaplan, introduction, "Forty-nine Rorschachs, Six Modified T.A.T.s, Twelve Murray T.A.T.s and Seven Sentence Completion Tests of Zuni Young Men," 3. Note that the language of taking a "sample" and "sampling" a population via psychological tests evokes a statistical sensibility.

15. Ibid.

16. Ruth Benedict, *Patterns of Culture* (New York: Houghton Mifflin, 1973 [1934]), 223.

17. Bourdieu quotation is approximate, from author's recollection. Critiques of culturalist theories (including Benedict's) characterized them as "billiard ball" visions, by which each culture was seen as a neat and rounded whole allowing no traffic between inside and outside. "By endowing nations, societies, or cultures with the qualities of internally homogeneous and externally distinctive and bounded objects, we create a model of the world as a global pool hall in which entities spin off each other like so many hard and round billiard balls." Eric Wolf, *Europe and the People without History* (Berkeley:

University of California Press, 1982), 6. For another classic critique of the anthropological culture concept, see Johannes Fabian, *Time and the Other: How Anthropology Makes Its Object* (New York: Columbia University Press, 2002).

18. Zuni (Cherokee wife of Zuni man) informant, quoted in Trikoli Nath Pandey, "Anthropologists at Zuni," in *Proceedings of the American Philosophical Society* 116 (1972): 333.

19. Quotation from Benedict and analysis of Benedict's view of Zuni pueblo as the "site for a lapsed puritan's 'city on a hill'" comes from George Stocking, "The Ethnographic Sensibility of the 1920s and the Dualism of the Anthropological Tradition," in *Romantic Motives: Essays on Ethnographic Sensibility*, ed. George Stocking (Madison: University of Wisconsin Press, 1989), 297.

20. Benedict, *Patterns of Culture*, 59.

21. Paul Rabinow, *The Accompaniment: Assembling the Contemporary* (Chicago: University of Chicago Press, 2011), ch. 1, esp. 14–22, 33. In particular Rabinow lauds Benedict's "at-a-distance" analyses of Japanese character as "brilliant" (22).

22. Quotations in this paragraph are from Kaplan, introduction, "Forty-nine Rorschachs, Six Modified T.A.T.s, Twelve Murray T.A.T.s and Seven Sentence Completion Tests of Zuni Young Men," 4–5.

23. "Cultural persistence well after conquest": Sherry L. Smith's book review of McFeely, *Zuni and the American Imagination*, *American Historical Review* 107, no. 1 (2002): 215; "storehouses of knowledge": from Gwyneira Isaac, *Mediating Knowledges: Origins of a Museum for the Zuni People* (Tucson: University of Arizona Press, 2007), 64.

24. Leah Dilworth, *Imagining Indians in the Southwest: Persistent Visions of a Primitive Past* (Washington, DC: Smithsonian Institution Scholarly Press, 1996), 15; Stocking, "The Ethnographic Sensibility of the 1920s and the Dualism of the Anthropological Tradition," 292.

25. The threat of Indian violence, even after it became unlikely, was prominent in encounters with Plains groups, such as the various Sioux bands, and fetishized in spectacles such as Wild West shows and Hollywood movies; see Philip J. Deloria, "Violence," *Indians in Unexpected Places* (Lawrence: University Press of Kansas, 2004), 15–51. On northern and southern patterns of Indian raiding and inter-tribe warfare, the U.S. military wars and subsequent suppression of warfare, and the emergence of the federal reservation system *qua* system, see White, *It's Your Misfortune*, 85–118, "The Federal Government and the Indians." Pueblo Indians had a history of armed rebellion (often successful) going back four hundred years against the Spanish, but by the nineteenth and twentieth centuries they tended

not to adopt violent means of resistance. (Hopi, Ute, Rio Grande Pueblo, and Zuni fighting with Navajo, however, continued.) Between 1887, when the Dawes Severalty Act was implemented to put "allotment" policies into effect, and 1934, Indians overall lost 60 percent of their remaining land (land not allotted to individuals) and 66 percent of their allotted lands (White, *It's Your Misfortune*, 115). Tribal governments were systematically disbanded.

26. "Agricultural hegemony": from Eric Perramond, "Melting the Kachinas: Agricultural Hegemony and Indigenous Incorporation at Zuni Pueblo in the Modern Era," *Journal of Political Economy* 12 (2005): 51–66. For a description of irrigation as a "new technological and social system" considered to embody progressive, scientific, moral, and rational values, see Donald Worster, *Rivers of Empire: Water, Aridity, and the Growth of the American West* (New York: Oxford University Press, 1985), 114. For a confirming historical perspective on the distinctive Sapirian thread of culture and personality (and its persistence in the work of Hallowell and Kluckhohn), see LeVine, "Anthropological Foundations of Cultural Psychology," 45–49.

27. Collins quoted in Perramond, "Melting the Kachinas," 57; "to introduce American technology to the Zuni": quoted in ibid., 55.

28. Frank Hamilton Cushing, "My Adventures in Zuni," *Century Magazine* 25, no. 1, (1882): 191.

29. Quotation in Virginia Morell, "The Zuni Way," Smithsonian mag. com, April 2007, http://www.smithsonianmag.com/people-places/the-zuni-way-150866547/.

30. Letter from Cushing to Baird; quoted in Pandey, "Anthropologists at Zuni," 323. There is still some ambiguity about Cushing's status as a Bow Priest, according to Nigel Holman; see G. Isaac, *Mediating Knowledges*, 69.

31. Quoted in Pandey, "Anthropologists at Zuni," 324.

32. See ibid.

33. Quotation on magic amusement park is from Eliza McFeeley, *Zuni and the American Imagination* (New York: Hill and Wang, 2001), 77; Fred Eggan on Durkheim/Mauss influence from Cushing is quoted in G. Isaac, *Mediating Knowledges*, 74. Eggan was referring to culture, but Durkheim and Mauss spoke more of social facts and society.

34. Quoted in Morell, "The Zuni Way."

35. "When a stranger comes to Zuni, he is often asked: 'Are you an anthropologist?' Several of my informants wanted to know why so many anthropologists keep coming to Zuni." Pandey, "Anthropologists at Zuni," 321n3.

36. Description of Mathilda Cox Stevenson in G. Isaac, *Mediating Knowledges*, 65. For more on Stevenson, see McFeeley, *Zuni and the American Imagination*, and others who emphasize her rigorous totalizing approach, unpopular with the Zuni.

37. Ruth Bunzel, "Zuni Katchinas," "Zuni Creation Myths," "Zuni Ritual Poetry," in *Introduction to Zuni Ceremonialism*, 47th Annual Report, Bureau of American Ethnology, Smithsonian Institution (Washington: Government Printing Office, 1932); Benedict, *Patterns of Culture*; Elsie Clews Parsons, *Pueblo Indian Religion* (Lincoln: University of Nebraska Press, 1996); and Mathilda Cox Stevenson, *The Zuni Indians: Their Mythology, Esoteric Fraternities, and Ceremonies* (Washington, DC: Bureau of American Ethnography, 1904). Dick Tumaka's words as recorded in 1952 are quoted in G. Isaac, *Mediating Knowledges*, 79.

38. John Adair: "A Pueblo G.I.," in *In the Company of Man: Twenty Portraits by Anthropologists*, ed. Joseph B. Casagrande (New York: Harper and Brothers, 1960), 494, and "Life Histories of Six Zuni Young Men," *Microcard Publications of Primary Records in Culture and Personality* 1, no. 23 (La Crosse, WI: Microcard, 1956): 2–107.

39. Adair, "Life Histories of Six Zuni Young Men" (Miguel A.).

40. Also consider the influence of empiricist sociological tradition (discussed more fully in chapter 8 below); see, for example, C. R. Shaw, *Delinquency Areas* (Chicago: University of Chicago Press, 1929), and Elsa S. Longmoor and Erle Young, "Ecological Relationship of Juvenile Delinquency," *American Journal of Sociology* 41 (1936): 598–610. See also George Lundberg "Sociography of Some Community Relations," *American Sociological Review* 2 (1937): 318–335. Sociologists were particularly vigorous in developing the research "situation" approach to do this, thus rendering visible and tangible as many "behavior exponents" as possible (in metricized form).

41. For example, by Laura Thompson, quoted in John Honigmann, "North America," in *Psychological Anthropology: Approaches to Culture and Personality*, ed. Francis L. K. Hsu (Homewood, IL: Dorsey Press, 1961), 109.

42. Dorothy Eggan, "Joab's Comments on His Life," in "Hopi Dreams and a Life History Sketch," Series I, *Microcard Publications of Primary Records in Culture and Personality* 2, no. 16 (La Crosse, WI: Microcard, 1957): 2.

43. Quotations in this paragraph are from Kaplan, introduction, "Forty-nine Rorschachs, Six Modified T.A.T.s, Twelve Murray T.A.T.s and Seven Sentence Completion Tests of Zuni Young Men," 5–8.

44. Adair, "A Pueblo G.I.," 494.
45. Adair, "Life Histories of Six Zuni Young Men," 12; bracketed word is obscured in record. The quotations in this and the next two paragraphs are from the same source.
46. Adair, "A Pueblo G.I.," 502–503.
47. Peter Galison, "Secrecy in Three Acts," *Social Research* 77, no. 3 (2010): 941–974; discussion in this paragraph is from pp. 960–961.
48. Ibid., 942.
49. G. Isaac, *Mediating Knowledges*, 84–85. After Kaplan, a husband-wife research team, Dennis and Barbara Tedlock, would record thirty years' worth of oral tradition and publish sensitive personal information about the families with whom they worked, occasioning a new wave of controversy (G. Isaac, *Mediating Knowledges*, 82–83).
50. Fred Eggan quoted in William Sturtevant, *Handbook of North American Indians: Southwest* (Washington, DC: Smithsonian Institution Scholarly Press, 1979), 224.
51. "Seminal idea": A. I. Hallowell, preface, *Microcard Publications of Primary Records in Culture and Personality* 1, no. 1 (La Salle, WI: Microcard, 1956): 4.

Chapter 5. Possible Future Worlds

1. "Representative sample": Bert Kaplan, "A Newsletter: Microcard Publications of Primary Records in Culture and Personality," March 29, 1956; Committee on Primary Records, A&P, NAS Archives. "Collecting, preserving": untitled document, undated; Committee on Primary Records, A&P, NAS Archives.
2. Joel Isaac, *Working Knowledge: Making the Human Sciences from Parsons to Kuhn* (Cambridge: Harvard University Press, 2012), 33, 29.
3. Jamie Cohen-Cole, *The Open Mind: Cold War Politics and the Sciences of Human Nature* (Chicago: University of Chicago Press, 2014), 2.
4. MACOS social studies curriculum organized by social relations psychologist Jerome Bruner in 1964–1965 with funding from the National Science Foundation and the Ford Foundation; quotation from Jerome Bruner, *Man: A Course of Study*, Occasional Paper No. 3, Social Studies Curriculum Program (Cambridge, MA: Educational Services, 1965), 5; quoted in Peter B. Dow, *Man: A Course of Study: A Continuing Exploration of Man's Humanness* (Washington, DC: Curriculum Development Associates, 1970), 4, http://www.macoson line.org/course/guides/One-Talks%20to%20Teachers.pdf.

5. Quoted in Michael A. Bernstein, "American Economics and the National Security State, 1941–1953," *Radical History Review* 63 (1995): 11.

6. Recent scholarship on the postwar social sciences has explored some of these (putatively) value-free approaches to value-laden areas: on the human "reliability crisis" among maintenance technicians, see Edward Jones-Imhotep, "Maintaining Humans," in *Cold War Social Science*, ed. Mark Solovey and Hamilton Cravens (London: Palgrave Macmillan, 2012); on decision-making processes, see Hunter Heyck, "Producing Reason," Solovey and Cravens, *Cold War Social Science*.

7. See Martin Jay, *The Dialectical Imagination: A History of the Frankfurt School and the Institute of Social Research 1923–1950* (New York: Little Brown, 1973; reissued Berkeley: University of California Press, 1996). Theodor Adorno, Else Frenkel-Brunswik, Daniel Levinson, and Nevitt Sanford, *The Authoritarian Personality: Studies in Prejudice* (New York: Harper and Row, 1950); the project had its roots in studies on rising Nazism undertaken at the Institut für Sozialforschung; it was completed at UC Berkeley, where it was staffed by a Frankfurt School alumnus (Adorno) and three UC researchers; other émigré scholars propelled the founding of the Radio Research lab at Princeton and its later re-situation as the Bureau of Applied Social Research at Columbia, run by Paul Lazarsfeld and staffed by Marcuse, Massing, and Lowenthal (these three full time) and Pollock and Neumann (part time)—at least in the immediate postwar years. In California the "speculative camp" of the Institut was staffed by Horkheimer and Adorno (Jay, *The Dialectical Imagination*, 220). Other European exports pursued the human sciences in tandem or groups. Lévi-Strauss teamed up with Roman Jakobson in New York. On Lévi-Strauss's growing information-science interests, see Bernard Dionysius Geoghegan, "From Information Theory to French Theory: Jakobson, Lévi-Strauss, and the Cybernetic Apparatus," *Critical Inquiry* 38, 2011, 92–126. On Japanese social sciences see Andrew E. Barshay, "An Overview," in Barshay, *The Social Sciences in Modern Japan: The Marxian and Modernist Traditions* (Berkeley: University of California Press, 2004), 36–71.

8. For recent work on the "authoritarian personality," see Martin Roiser and Carla Willig, "The Strange Death of the Authoritarian Personality: 50 Years of Psychological and Political Debate," *History of the Human Sciences* 15, no. 4 (2002): 71–96; Cohen-Cole, *The Open Mind*, 40ff.

9. Milosz responded that despite Americans' ample talents and intelligence, "minds are turning round in a sort of spellbound dance of paralytics." Thomas Merton to Czeslaw Milosz, May 6, 1960, and Milosz to Merton, Fall 1960, in Robert Faggen, ed., *Striving Towards Being: The Letters of Thomas Merton and Czeslaw Milosz* (New York: Farrar, Strauss and Giroux, 1996), 74, 102–103.

10. The four founders were the psychologist Gordon Allport, sociologist Talcott Parsons, psychologist Henry A. Murray, and cultural anthropologist Clyde Kluckhohn, Kaplan's mentor. Also influential in planning the DSR was O. H. Mowrer, trained at the Yale Institute of Human Relations, who was then at Harvard. On the DSR's founding, see Lawrence Nichols, "Social Relations Undone: Disciplinary Divergence and Departmental Politics at Harvard, 1946–1970," *American Sociologist* 29, no. 2 (1998): 83–107, and Joel Isaac, "Theorist at Work: Talcott Parsons and the Carnegie Project on Theory, 1949–1951," *Journal of the History of Ideas* 71 (2010): 295–299ff., discussing the founding of the DSR and LSR. See also Rebecca Lemov, "The Laboratory Imagination: Experiments in Human and Social Engineering, 1929–1956" (PhD dissertation, University of California, Berkeley, 2000), chs. 5–7.

11. "An Announcement Concerning the Department of Social Relations at Harvard University," February 1, 1946, Harvard University Archives, HUF 801.5002.

12. Cf. Ian Hacking, *Representing and Intervening: Introductory Topics in the Philosophy of Natural Science* (Cambridge: Cambridge University Press, 2007), 31ff.; on the instantiation of instrumental rationality, see Paul Rabinow, "Artificiality and Enlightenment: From Sociology to Biosociality," in *Incorporations*, ed. J. Crary and S. K. Winter (New York: Zone, 1992), 234–252; on the increasing simultaneity of representation and intervention, see Lorraine Daston and Peter Galison, *Objectivity* (New York: Zone Books, 2007).

13. Letter from "Norman" at Columbia University's Bureau of Applied Social Research to Professor Parsons, undated but ca. 1950 or 1951, Talcott Parsons Papers, Harvard Archives, UAV 801.2010.

14. Yale psychologist Seymour Sarason quoted in Steve Heims, *The Cybernetic Group* (Cambridge, MA: MIT Press, 1991), 3–4.

15. On Bruner's funding, see Lemov, "The Laboratory Imagination," ch. 5.

16. Bert Kaplan, introduction, "Forty-nine Rorschachs, Six Modified T.A.T.s, Twelve Murray T.A.T.s and Seven Sentence Completion Tests of Zuni Young Men," *Microcard Publications of Primary Records in Culture and Personality* 1, no. 24 (La Crosse, WI: Microcard, 1956): 4.

17. An "X-ray" approach to knowledge, discerning structures through special deep-seeing visual techniques in the 1960s, was noted recently by Lorraine Daston, "Structure," *Historical Studies in the Natural Sciences* 42, no. 5 (2012): 496–499. Daston cites "a cluster of influential books in the humanities and social sciences published circa 1960 [that] raised hopes that the complexities of, say, the plays of Racine or cultural taboos or bargaining might reveal simpler basic structures the way an X-ray revealed skeletons" (p. 499). The X-ray trope of structural revelation appears after World War II with particular vehemence in the unifying social sciences.

18. Letter from Bert Kaplan to Clyde Kluckhohn, November 15, 1954, Correspondence HUB 4490.5, Folder K 1954, Kluckhohn Papers, Harvard University Archives; emphasis in original.

19. Ibid.

20. Undated archival document, "The Committee on Primary Records," Committee on Primary Records, A&P, NAS Archives (draft document with emendations).

21. Bert Kaplan, "Dissemination of Primary Research Data in Psychology," *American Psychologist* 13, no. 2 (1958): 54.

22. Hermia Kaplan quotes in this chapter are from interview with author, July 25, 2011, Santa Cruz, CA. Bert Kaplan, introduction to "Rorschachs of Sixty Navaho Adults and Children and Modified T.A.T.s, Murray T.A.T.s and Sentence Completion Tests of Fourteen Navaho Young Men," *Microcard Publications of Primary Records in Culture and Personality* 1, no. 20 (La Crosse, WI: Microcard, 1956): 5.

23. Charles Dickens, *Bleak House* (New York: Penguin Books, 1994), 65. On secrets, see George Stocking's unpublished manuscript on Kluckhohn, which describes his relations at Harvard with a psychohistorical argument; it was presented at a colloquium at Harvard's Department of the History of Science in spring 2001 and is referenced in George W. Stocking, Jr., *Glimpses into My Own Black Box: An Exercise in Self-Deconstruction* (Madison: University of Wisconsin Press, 2010), 139.

24. Kluckhohn's *To the Foot of the Rainbow* described in http://horsetravel-books.com/N-America.htm; accessed August 9, 2012.

25. Kluckhohn interview included in Harry Shapiro, "Seeing Others as They See Themselves (Review of *Mirror for Man*)," *Saturday Review of Literature*, January 29, 1949, 11. Clyde Kluckhohn, *Mirror for Man: The Relation of Anthropology to Modern Life* (Tucson: University of Arizona Press, 1985 [1949]), 286.

26. Bert Kaplan, "A Study of Rorschach Responses in Four Cultures," *Peabody Museum of American Anthropology Papers* 42, no. 2, vii.

27. Summary of Sapir's objection to Benedict's "reifying" tendencies in William Manson, "Abram Kardiner and the Neo-Freudian Alternative in Culture and Personality," in *Malinowski, Rivers, Benedict and Others: Essays on Culture and Personality*, ed. George Stocking (Madison: University of Wisconsin Press, 1987), 78.

28. Http://vcencyclopedia.vassar.edu/alumni/ruth-benedict.html (on the publication selling over 2 million copies in English and translations into twelve languages).

29. On Kardiner and the "neo-Freudian triumvirate" of Karen Horney, Erich Fromm, and Harry Stack Sullivan and their influence on American anthropology during the interwar years, especially after 1934, see Manson, "Abram Kardiner and the Neo-Freudian Alternative," 89. Cora DuBois, a student of Kardiner's, also adopted the label "modal personality," which became influential.

30. See Anthony F. C. Wallace, introduction to "Rorschachs of 100 Tuscarora Children and Adults," *Microcard Publications of Primary Records in Culture and Personality* 1, no. 14 (La Crosse, WI: Microcard, 1956). Cf. Anthony F. C. Wallace, "The Modal Personality Structure of the Tuscarora Indians, as Revealed by the Rorschach Test" (PhD dissertation, University of Pennsylvania, 1950), 149.

31. Henry Murray, "Analysis of the Personality of Adolph Hitler: With Predictions of His Future Behavior and Suggestions for Dealing with Him Now and after Germany's Surrender" (Washington, DC: OSS, 1943). See Virginia Yans-McLaughlin, "Mead, Bateson, and 'Hitler's Peculiar Psychological Makeup'—Applying Anthropology in the Era of Appeasement," *Newsletter of the History of Anthropology* 13, no. 1 (1986): 3–8, and L. E. Hoffman, "Erikson on Hitler: The Origins of 'Hitler's Imagery and German Youth,'" *Psychohistory Review* 22, no. 1 (1993): 69–86; J. M. Cooper, "Anthropology in the United States during 1939–1945," *Journal de la Société des Américanistes* 36 (1947): 1–14.

32. Kaplan wrote that Kardiner's was "perhaps the most influential conception in the culture and personality field"; Bert Kaplan, "Cross-Cultural Use of Projective Techniques," in *Psychological Anthropology: Approaches to Culture and Personality*, ed. Francis L. K. Hsu (Homewood, IL: Dorsey Press, 1961): 235–236.

33. Clyde Kluckhohn, "A Navaho Personal Document with Brief Paretan Analysis," *Southwestern Journal of Anthropology* 1, no. 2 (1945): 261.

34. Ibid., 262.

35. "Vast new data bases": from Christopher Simpson, "Worldview Warfare," in *The Science of Coercion: Communication Research and*

Psychological Warfare, 1940–1965 (New York, Oxford University Press, 1996), 30.

36. Kluckhohn, "A Navaho Personal Document," 262.

37. Quotations from *Mirror for Man* and *Saturday Review of Literature* (as indicated in this paragraph) are from John Gilkeson, "Clyde Kluckhohn and the New Anthropology: From Culture and Personality to the Scientific Study of Values," *Pacific Studies* 32, nos. 2/3 (2009): 251–253.

38. Concrete sociology was related to different styles of empiricism, including radical empiricism (William James), grassroots empiricism (John Maze), delicate empiricism or *zarte Empirie* (Johann Wolfgang von Goethe), experiments with truth (Mahatma Gandhi), and dust-bowl empiricism (Roger Barker et al.). At least these can be seen as variations on a theme. On concrete sociology at Harvard, see J. Isaac, *Working Knowledge*, 70–80.

39. "A sense of the urgency": Kaplan, "A Study of Rorschach Responses in Four Cultures," 4.

40. On creativity as a positive democratic value for social scientists in this period, see Cohen-Cole, *The Open Mind.* On twentieth-century history of creative thought in military-industrial contexts, see the work of Bregje van Eekelen, *Brainstorms: A Cultural History of Undisciplined Thought* (EUR- and NSF-funded multi-year investigation).

41. See, among others, Gregory Meyer and Joshua Eblin, "An Overview of the Rorschach Performance Assessment System (R-PAS)," *Psychological Injury and Law* 5 (2012): 107–121. See also the burgeoning field of the neuroscience of cultural differences.

42. Kaplan, "A Study of Rorschach Responses in Four Cultures," 32.

43. Bert Kaplan, typed unpublished memoir, 13, Bert Kaplan Papers, Santa Cruz, CA.

44. Kaplan memoir, 14.

45. Ibid.

46. Quotation from Robert LeVine interview with author, December 16, 2013, Berlin, Germany; others have said this too.

47. Kaplan, "A Study of Rorschach Responses in Four Cultures," 4.

48. G. William Domhoff interview with author, July 27, 2011, Santa Cruz, CA.

Chapter 6. The Double Experiment

1. As of 2012, announced the Library of Congress, the global jukebox was finally built or almost completed, digitized from some seventeen

thousand hard-won music tracks Lomax had gathered over three decades, between the 1930s and 1960s. See Larry Rohter, "Folklorist's
Global Jukebox Goes Digital," *New York Times,* January 30, 2012.

2. On current data-sharing practices, see Gary King, "Restructuring the
Social Sciences: Reflections from Harvard's Institute for Quantitative
Social Science," *PS: Political Science and Politics* 47 (2014): 165–172,
and R. Ankeny and Sabina Leonelli, "Valuing Data in Postgenomic
Biology: How Data Donation and Curation Practices Challenge the
Scientific Publication System," in *Post-Genomics,* ed. Hallam Stevens
and Sarah Richardson (Durham, NC: Duke University Press, 2015).

3. "Perhaps 20,000 8 × 10 sheets of paper": letter from Bert Kaplan to
Mr. Henry C. Wilde, March 7, 1955, Kaplan Papers, Santa Cruz, CA.

4. James Waldram, *Revenge of the Windigo: The Construction of the Mind
and Mental Health of North American Aboriginal Peoples* (Toronto:
University of Toronto Press, 2004), 44.

5. Letter from Melford Spiro to A. I. Hallowell, May 9, 1948, Series I,
Hallowell Papers, American Philosophical Society, Philadelphia.

6. Hallowell, Transcript of Anne Roe interview, January 2, 1963, Series
II (Subject Files), Hallowell Papers.

7. Letter from Bert Kaplan to A. I. Hallowell, January 19, 1955, Kaplan
Papers.

8. Ibid. Kaplan invited Clyde Kluckhohn to take part, but Kluckhohn
declined saying, "I feel badly about this but I really must say no because I have promised myself (and others) that I shall get out of
things—rather than take on new ones. I think your project has great
potential significance but it is not in the central line of my own interests for the coming years" (ibid., quoting Kluckhohn's letter to
Kaplan, February 17, 1955). In fact, Kluckhohn was in poor health
and would pass away not long after.

9. On HRAF, see Rebecca Lemov, "Filing the Total Human:
Anthropological Archives from 1928 to 1963," in *Social Knowledge in
the Making,* ed. Charles Camic, Neil Gross, and Michèle Lamont
(Chicago: University of Chicago Press, 2011), 119–150, and Rebecca
Lemov, "The Laboratory Imagination: Experiments in Human and
Social Engineering, 1929–1956" (PhD dissertation, University of
California, Berkeley, 2000), ch. 4.

10. Bert Kaplan, "A Newsletter, March 29, 1956, "Microcard Publications
of Primary Records in Culture and Personality," Committee on
Primary Records, A&P, NAS Archives, 1.

11. Letter from Bert Kaplan to Clyde Kluckhohn, November 15, 1954,
HUG 4490.5, folder K 1954, Clyde Kluckhohn Papers, Pusey

Library, Harvard University. That Kaplan still held out hopes for his "file" of data to be attached to the HRAF ongoing concern is clear from his strategizing, which appears somewhat later in the same letter: "Since the HRAF will probably be able to process materials from European, Asian and North African groups under its present contracts, I am concentrating on locating these first."

12. Letter from Bert Kaplan to Mr. Henry C. Wilde, March 7, 1955, Kaplan Papers.

13. Chicago and the U.S. Midwest more generally were hotbeds of microfilming and the seat of much of the American Documentation Institute's activities.

14. Bert Kaplan's report for NAS, April 1, 1957, p. 10; Committee on Primary Records, A&P, NAS Archives.

15. Bert Kaplan, editor's introduction, *Microcard Publications of Primary Records in Culture and Personality* 1, no. 1 (La Crosse, WI: Microcard, 1956): 9.

16. Letter from Bert Kaplan to John W. M. Whiting, May 30, 1955, Kaplan Papers.

17. Bert Kaplan, "Dissemination of Primary Research Data in Psychology," *American Psychologist* 13, no. 2 (1958): 53.

18. Letter from Bert Kaplan to John Whiting of the Laboratory on Human Development, May 30, 1955, Kaplan Papers.

19. Kaplan, "Dissemination of Primary Research Data in Psychology," 53.

20. Minutes of Committee on Primary Records, January 19, 1956; Committee on Primary Records, Meetings 1956, A&P, NAS Archives.

21. Kaplan, editor's introduction, *Microcard Publications of Primary Records in Culture and Personality*, 8–9.

22. See Whitney Laemmli, "The Choreography of Everyday Life" (PhD dissertation, University of Pennsylvania, forthcoming); it mentions Birdwhistell, among other things, encouraging Alan Lomax to collect the cross-cultural varieties of dance steps using the Laban notation system.

23. Letter from Ray Birdwhistell to A. I. Hallowell, May 2, 1956, Kaplan Papers.

24. George H. Davison, "Review of Progress," in *Microtext in the Form of Microcards and Other Micro-Opaques, Transparent Microfiches and Unitised Microfilm or Aperture Cards as Well as Reading and Enlarging Apparatus* (Moorgate, Rotherham, England: United Steel Companies, Research and Development Department, Swinden Laboratories, 1960), 67.

25. Minutes of Committee on Primary Records, November 9, 1956; Committee on Primary Records, Meetings 1956, A&P, NAS Archives.

26. Finch and Brogden in ibid.

27. John Whiting, *Becoming a Kwoma* (New Haven: Yale Institute of Human Relations, 1941), applied the Yale *Outline of Cultural Materials* code to psychological data of the New Guinea tribal group, and his second book, with I. L. Child, *Child Training and Personality* (New Haven: Yale University Press, 1953), calculated the frequency of different child-rearing strategies and techniques in cultures around the world.

28. Ariel Sabar, *Outsider: The Life and Times of Roger Barker* (Amazon Kindle, 2014), loc. 39–51 of 1109.

29. Bert Kaplan, "An Exploratory Study of the Need for the Preservation of Primary Scientific Records," "Exhibit A," attached to letter proposal to NSF, April 11, 1956, 3. Request for Funds 1956–1958, Committee on Primary Records, A&P, NAS Archives.

30. Cf. Vernon D. Tate, "Defrosting a Frozen Asset: Publication of Doctoral Dissertations," *College and Research Libraries* 14 (1953): 35–38, and Thompson Webb, Jr., "Microcopy, Near-Print and the New Film Composing Machines," *Library Quarterly* 25, no. 1 (1955): 114.

31. Quoted in Andreas Killen and Stephan Andriopoulos, "On Brainwashing," editors' introduction to "Brainwashing," special issue of *Grey Room* 45 (2011): 9. (This is a point originally made by, among others, Robert Tucker in *Stalin and the Uses of Psychology* [Santa Monica, CA: RAND Corporation, 1956].)

32. Minutes of Committee on Primary Records, January 19, 1956; Committee on Primary Records, Meetings 1956, A&P, NAS Archives.

33. Ibid.

34. Cf. Fernando Vidal and Nélia Dias, introduction, and Rebecca Lemov, "On the Second-Order Endangerment of Anthropological Objects," both in *Endangerment, Biodiversity and Culture*, ed. Fernando Vidal and Nélia Dias (London: Routledge, 2015). Claude Lévi-Strauss's *Tristes tropiques* (Paris: Librairie Plon, 1955) was translated as "world on the wane" in Lévi-Strauss, *World on the Wane*, trans. John Russell (New York: Criterion Books, 1961).

35. Summary of May 3, 1956, meeting of Committee on Primary Records at the La Salle Hotel, Chicago, p. 2; Committee on Primary Records, Meetings 1956, A&P, NAS Archives.

36. Quoted in minutes of Committee on Primary Records, May 3, 1956; Committee on Primary Records, Meetings 1956, A&P, NAS Archives.

37. Ibid.

38. Ibid.
39. Undated "Notice" from Committee on Primary Records; Committee on Primary Records, General, A&P, NAS Archives.
40. Ibid.
41. U.S. Bombing Survey, Pacific, http://www.anesi.com/ussbs01.htm# teotab.
42. Kaplan, editor's introduction, *Microcard Publications of Primary Records in Culture and Personality*, 9.
43. Letter from Finch to Dr. Detley Bronk, president of NAS, February 1, 1956; Committee on Primary Records, General, A&P, NAS Archives.
44. On disaster studies and their archives, see Cécile Stehrenberger, "Floods, Sociology and Cold War: On the History of Social Science Disaster Research, 1949–1979," paper given at First Annual Conference on the History of Recent Social Sciences, École Normale Supérieure de Cachan, Paris.
45. Minutes of Committee on Primary Records, November 9, 1956; Committee on Primary Records, A&P, NAS Archives.
46. Bert Kaplan, "Report of the Executive Secretary of the Committee on Primary Records" (Not for Distribution or Reproduction), April 1, 1957; Committee on Primary Records, General, A&P, NAS Archives. Quotations in this paragraph and the next two are from this report. Verner Clapp quotation: emphasis added.
47. Letter from Bert Kaplan to A. I. Hallowell, March 11, 1957, Kaplan Papers.
48. Bert Kaplan, "Report of the Executive Secretary," 12.
49. Gregory Crane, "Aristotle's Library: Memex as Vision and Hypertext as Reality," in *From Memex To Hypertext: Vannevar Bush and the Mind's Machine*, ed. James M. Nyce and Paul Kahn (San Diego, CA: Academic, 1992), 339–340. Other data-obsessives such as Philip E. Converse, Charles Y. Glock, and Myron J. Lefcowitz (who dealt with polling, survey, and public opinion data largely) were beginning by 1965 to speak of national data repositories, centralized or not. See, for example, "Data Archives: Problems and Promise," at the Proceedings of the Eighteenth Conference on Public Opinion Research, *Public Opinion Quarterly* 27, no. 1 (1963): 641–643. Specific proposals included Philip E. Converse, "A Network of Data Archives for the Behavioral Sciences," *Public Opinion Quarterly* 28, no. 2 (1964): 273–286, and Myron Lefcowitz and Robert O'Shea, "A Proposal to Establish: A National Archives for Social Science Survey Data," *American Behavioral Scientist* 6 (1963): 27–31.

50. Nick Thieberger, comment on thread in ASAONET discussion group, December 21, 2013, 11:08 a.m., "Fieldnotes and Notebooks (cassette version)."

51. Of subsequent installments: the second was significantly bigger than the first, but the third was smaller, while the fourth was very long and full of dreams.

52. "Microcards: A Practical Tool for Industry, Government, and Education," brochure, Microcard Corporation, West Salem, Wisconsin, n.d., Kaplan Papers.

53. Kaplan, "Dissemination of Primary Research Data in Psychology," 53.

54. Unless otherwise noted, all quotations in this paragraph are from Kaplan's public announcement, "Primary Records in Culture and Personality," NRC/NAS Archives.

Chapter 7. "I Do Not Want Secrets. . . . I Only Want Your Dreams"

1. "Television qualities": Dorothy Eggan, "The Manifest Content of Dreams: A Challenge to Social Science," *American Anthropologist* 54 (1952): 484. "I do not want [your] secrets": Eggan to informant [Joab] in Dorothy Eggan, "Hopi Dreams and a Life History Sketch," Series I, *Microcard Publications of Primary Records in Culture and Personality* 2, no. 16 (La Crosse, WI: Microcard, 1956): 24. Eggan continued explaining to Joab: "and to know how you grew up, how you learned to be a Hopi."

2. Self-description in letter from Dorothy Eggan to "Everybody," October 22, 1945, Folder 1, Box 3, Dorothy Eggan Papers, University of Chicago. [Hereafter: D. Eggan Papers.]

3. "I am *not* Dr.": letter from Dorothy Eggan to Bert Kaplan, March 4, 1956, Folder 25, Box 3, D. Eggan Papers; emphasis in original. "Combines accuracy": letter from David Riesman to Dorothy Eggan, May 15, 1956, Folder 18, Box 4, D. Eggan Papers. "You must realize": letter from Clyde Kluckhohn to Dorothy Eggan, March 4 (no year), Folder 3, Box 4, D. Eggan Papers. Robert LeVine confirmed this view: "Her articles stood out like jewels in the literature." Robert LeVine interview with author, December 23, 2013, Berlin, Germany.

4. Dorothy Eggan, "Cultural Factors in Dreams," paper presented at APA Meeting, San Francisco, 1958; Folder 10, Box 1, D. Eggan Papers, 5.

5. By "virtuosic," I refer to the sense developed in Paul Rabinow, *The Accompaniment: Assembling the Contemporary* (Chicago: University of

Chicago Press, 2011), especially "Humanism as Nihilism: The Bracketing of Truth and Seriousness in American Cultural Anthropology," and others of Rabinow's works that critique the single-author, struggling virtuoso model and explore the limits of "individualism as the subject position that has been at the core of anthropology's approach to research, pedagogy, publication, and, above all, thinking" (ibid., 202).

6. Letter from Dorothy Eggan to Bert Kaplan, March 4, 1956, Folder 25, Box 3, D. Eggan Papers; emphasis in original. Eggan's primary objection to Freudian dream analysis was the "seemingly arbitrary use of symbols in dream analysis" (which, she admitted, Freud had taken pains to counterbalance by warning that the importance of symbols can be overestimated). See Dorothy Eggan, "The Significance of Dreams for Anthropological Research," *American Anthropologist* 51, no. 2 (1949): 177.

7. This book is testimony to the fact that Kaplan's vision was data-centric and that he was in his modest way a pioneer of "big data," although not in the sense of blind aggregation. It can be argued that Lévi-Strauss viewed all of American anthropology as data. The Bibliothèque Lévi-Strauss in Paris has three components, one of which is the Yale Human Relations Area Files, which, according to the library's website, "represents the most important collection of ethnographic facts in the world"; http://las.ehess.fr/index.php?64. In 1966, Lévi-Strauss predicted that within a century, or sooner, "the last native culture will have disappeared from the Earth and our only interlocutor will be the electronic computer," with the remnants of those vanished cultures extensively documented. Claude Lévi-Strauss, "Anthropology: Its Achievements and Future," *Current Anthropology* 7 (1966): 124. See also John Whiting, *Becoming a Kwoma* (New Haven: Yale Institute of Human Relations, 1941), for a sense of how data collecting infused textual production.

8. "Massing of items" and "massing evidence": D. Eggan, "Cultural Factors in Dreams," Folder 10, Box 1, D. Eggan Papers, 9.

9. "But in dreams": Dorothy Eggan, "Confidential Notes and Comments on Don," Folder 7, Box 2, D. Eggan Papers, 41. "Social scientific documents": from Eggan's lecture to M. Singer's class on culture and personality, n.d., Folder 4, Box 1, D. Eggan Papers, 3.

10. Payment was in the form of billing, rather than a more casual arrangement, as one of Eggan's exchanges with Bert Kaplan made clear. At Eggan's request, Kaplan had administered a Rorschach test to her prime dream informant, and Eggan wanted to make sure the billing

was correctly done: "Don was very pleased with you and to do the test.... I will add an hour or so to his bill. I want him to welcome you back if you do more follow-up work with him on this or TATs." She mentioned in addition that she would like to reimburse Bert himself for more of his time: "Even though you only actually 'spent' two hours of time working with him, it is only fair for us to pay the expenses." She therefore expected a bill from Kaplan. Undated letter from Dorothy Eggan to Bert Kaplan (in response to letter of September 24, 1959), Folder 25, Box 3, D. Eggan Papers.

11. Talayesva's data, in addition to the 341 dreams he gave or sold to Eggan, included 350 hours of interviews with anthropologist Leo Simmons; 8,000 diary pages; a full Rorschach and other projective tests conducted by Bert Kaplan; a life-history collaboration with Simmons; and projects with many other anthropologists, including Leslie White, Fred Eggan, Edward Kennard, Volney Jones, and Harold Courlander. According to Dorothy Eggan, Talayesva's comprised "certainly the most complete record of any preliterate available." Letter from Dorothy Eggan to Bert Kaplan, July 15, 1959, Folder 25, Box 3, D. Eggan Papers.

12. "Talking about oneself": D. Eggan, "Collection of the Data," in "Hopi Dreams and a Life History Sketch," Series I, 9. "After ... *Sun Chief*": ibid., 11.

13. D. Eggan, "Joab's Comments on His Life," in Series I, "Hopi Dreams and a Life History Sketch," *Microcard Publications of Primary Records in Culture and Personality* 2, no. 16 (La Crosse, WI: Microcard, 1957): 2–3.

14. Joab comment at end of dream #43, in ibid., 56.

15. D. Eggan, "Background of the Study," in "Hopi Dreams and a Life History Sketch," 1.

16. Joab dreams and associations: #43 (butterflies), #45 (bed, gas in car), #46 (evil spirit); in D. Eggan, "Hopi Dreams and a Life History Sketch," Series I.

17. Dream #8, in D. Eggan, "Hopi Dreams: Second Series," 10.

18. Debbie's transcribed dream and Dorothy Eggan's description of Debbie are included in dream #2 in D. Eggan, "Hopi Dreams: Second Series," Series I, 20.

19. Joab, dream #34, in D. Eggan, "Hopi Dreams and a Life History Sketch," Series I.

20. Solicitation inaugurating their friendship: letter from Bert Kaplan to Dorothy Eggan, July 8, 1955, Folder 25, Box 3, D. Eggan Papers. "The 'dream' of my life": letter from Dorothy Eggan to Bert Kaplan,

undated, regarding the March 1956 visit to Lawrence; Folder 25, Box 3, D. Eggan Papers. "Your visit here": letter from Dorothy Eggan to Bert Kaplan, March 4, 1956, Folder 25, Box 3, D. Eggan Papers. In describing Eggan's speaking voice, I am drawing on Robert LeVine interview with author, December 23, 2013, Berlin, Germany. "I enjoy your chatty letters": letter from Bert Kaplan to Dorothy Eggan, November 29, 1956, Folder 25, Box 3, D. Eggan Papers.

21. "Looked very favorably": letter from Bert Kaplan to Dorothy Eggan, November 29, 1956, Folder 25, Box 3, D. Eggan Papers. On Hall's "comprehensive scoring system," developed along with Robert Van de Castle (1966) and Vernon Nordby (1972), see Barbara Tedlock, "Dreaming and Dream Research," in *Dreaming: Anthropological and Psychological Interpretations*, ed. Barbara Tedlock (Cambridge: Cambridge University Press, 1987), 21. Together with sociologists and anthropologists, they added a non-Western dimension to their database—and have "apparently felt satisfied that they have demonstrated meaningful cross-cultural similarities and differences in dream content relating to gender, developmental, and personality difference" (p. 21).

22. "I know you said": letter from Dorothy Eggan to Bert Kaplan, undated but ca. September 1957, Folder 25, Box 3, D. Eggan Papers. Barbara Tedlock makes the point about upending the rationalist hierarchy of evidence, citing Wendy Doniger, in Tedlock, *Dreaming*, 2.

23. Description of dream-collecting activities from Dorothy Eggan, "The General Problem of Hopi Adjustment," *American Anthropologist* 43, no. 3 (1943): 357–373. The two Microcard series include dreams from a variety of Hopi, but her papers include the full series of the Talayesva dreams, unpublished and largely unanalyzed.

24. She continued in intensive psychoanalysis for much of her adult life, first with French and later with Gerhard Piers (e-mail communication to author from Robert LeVine, December 18, 2013).

25. "Interactions between the entity": D. Eggan, "Background of the Study," 2. "A new psychic material": Sigmund Freud, "The Dream-Work," in *Interpretation of Dreams*, trans. James Strachey (New York: Avon, 1983), 295. "So we ignore it": letter from Dorothy Eggan to Bert Kaplan, March 4, 1956, Folder 25, Box 3, D. Eggan Papers. On dreams and their elaboration as part of a "projective process," see D. Eggan, "The Manifest Content of Dreams," 480. An earlier article, "The Significance of Dreams for Anthropological Research," referred to a dream as a "projective phenomenon" and likened dreams to Rorschach and TAT tests (p. 197 and n. 34). The discussion of the

neuroscience of dreaming is from J. Allan Hobson, "Film and the Physiology of Dreaming Sleep: The Brain as Camera-Projector," *Dreamworks* 1–2 (1980–1982): 14.

26. Yaw (pseudonym), dream #18, D. Eggan, "Hopi Dreams and a Life History Sketch," Series I, 82.

27. "Dreams from old, conservative Hopi": D. Eggan, "Cultural Factors in Dreams," 11. Yava quotation is from Harold Courlander, introduction to Albert Yava, *Big Falling Snow* (New York: Crown, 1978), ix–x. On middling modernism, see Paul Rabinow, *French Modern: Norms and Forms of the Social Environment* (Princeton, NJ: Princeton University Press, 1992).

28. Melford Spiro, "Comments on Symposium on Culture and Personality," American Anthropological Association 1958 meetings, Washington, DC; Folder 2, Box 3, D. Eggan Papers.

29. On Eggan's place in the 1940s, see Tedlock, *Dreaming*, 21. The Eggan paper laying out this approach is "The Manifest Content of Dreams." Tedlock (*Dreaming*, p. 21) claims that beginning in the 1940s, much anthropological work on dreaming had the goal of collecting manifest content. As support, the author claims an undergraduate thesis by Walter Sears on the "primitive dreams" of the Yir-Yoront and a 1957 article by Dittmann and Moore—which, as it happens, explicitly eschews the manifest-content approach after citing Eggan's views. See Walter E. Sears, "The Navaho and Yir-Yoront, Their Primitive Dreams" (BA thesis, Harvard University, 1948); Allan Dittmann and Harvey Moore, "Disturbance in Dreams as Related to Peyotism among the Navaho," *American Anthropologist* 59, no. 4 (1957): 642–649; Richard Griffith, Otoya Miyagi, and Akira Tago, "The Universality of Typical Dreams: Japanese vs. Americans," *American Anthropologist* 60, no. 6 (1958): 1173–1179; cf. Carl W. O'Nell, "A Cross-Cultural Study of Hunger and Thirst Motivation Manifested in Dreams," *Human Development* 8 (1965): 181–193.

30. "Discuss impossibility as well as *inadequacy*": from Eggan's pen-written notes on dream project, undated, accompanied by a typed note from "B" (possibly Bob LeVine), Folder 6, Box 1, D. Eggan Papers; emphasis in original. Quotations about excellence of Eggan's work are from Ed Bruner letter(s) to Dorothy Eggan, in which he asks, "In what way do you look at the data which makes your papers somewhat unique and so excellent?" and "My commenting upon this paper is rather like my going down to the Art Institute to add a few strokes to a Rembrandt." Bruner letter to Dorothy Eggan, August 1, 1954, Folder 6, Box 3, D. Eggan Papers. "Trouble is, there have": letter (undated

but circa June 1958) from Dorothy Eggan to Bert Kaplan, Folder 25, Box 3, D. Eggan Papers. "There is *Freud*": letter from Dorothy Eggan to David Schneider, April 17 [1958], Folder 6, Box 11, David Schneider Papers, University of Chicago Special Collections. [Hereafter: Schneider Papers.] "Each time I read it": letter from Bert Kaplan to Dorothy Eggan, December 10, 1960, Folder 25, Box 3, D. Eggan Papers.

31. Elliot Wolfson, "On His Book, *A Dream Interpreted within a Dream*," *Rorotoko*, December 19, 2011; http://rorotoko.com/interview/20111219_wolfson_elliot_on_dream_interpreted_within_a_dream/?page=4. Cf. Elliot Wolfson, *A Dream Interpreted within a Dream: Oneiropoesis and the Prism of Imagination* (New York: Zone Books, 2011).

32. Letters from Bert Kaplan to David Schneider, September 19, 1955, and May 4, 1955, Kaplan Papers, Santa Cruz, CA.

33. Nor did Mass Observation pull together materials from around the world. On Mass Observation, see Tyrus Miller, "In the Blitz of Dreams," in *Time-Images: Alternative Temporalities in Twentieth-Century Theory, Literature, and Art* (Cambridge: Cambridge Scholars, 2009), 60–81.

34. Charlotte Beradt, *The Third Reich of Dreams*, collected in the mid-to-late 1930s in Hitler's Germany, a mass of dreams foretelling the coming convulsions of violence (originally *Das Dritte Reich des Traums* [Munich: Nymphenburger Verlagshandlung, 1966]). As one reviewer observed in an otherwise positive account of Beradt's book, by strictly limiting access to the dreams and interpreting them only in relation to totalitarianism, "Beradt needlessly limits their usefulness." Betty Glad, review of Beradt, *American Political Science Review* 63 (1969): 546. Hall's subjects filled out standard forms that, when processed, revealed that of 2,668 actions people recorded in their dreams, the largest proportion (34 percent) involved bodily movement such as walking, running, or riding—but, contrary to popular belief, falling and floating were not at all common. Twenty-nine percent dreamed in "technicolor" versus black and white. Calvin Hall, "What People Dream About: In Which 10,000 Dreams are Statistically Investigated with Respect to Setting, Cast of Characters, Plot, Emotions and Coloring," *Scientific American* 184, no. 5 (1951): 60.

In the 1960s, contemporary with the last entries in Kaplan's Microcard archive, several other dream collections arose. In 1957 John Barker set up Bureaux of Premonitions in London and New York, collecting over a thousand premonitions (some dreams) in the hopes of avoiding disasters. Other dream collections were by H. F. Saltmarsh, Edith Lyttelton, and J. B. Priestley (precognitive dreams; see below).

35. Cf. Beradt, *The Third Reich of Dreams.*

36. On Hopi-white contact for four hundred years, see Frederick Dockstadter, *The Kachina and the White Man* (Albuquerque: University of New Mexico Press, 1990), and Don Fowler, *A Laboratory for Anthropology: Science and Romanticism in the American Southwest 1846–1930* (Albuquerque: University of New Mexico Press, 2000). "Friendlies" and "Hostiles": June 22, 1891, telegram from Supt. Collins to Hon. Commr. of Ind. Affrs., re conflict at "Arabi," in which "the rebellious faction," or "hostile" villagers, were described threatening the "friendly Arabi's" and their children, asking for U.S. troops to intervene. Reprinted in Peter Whiteley, *The Orayvi Split* (New York: AMNH, 2008), 2:851. See also Peter Whiteley, *Deliberate Acts: Changing Hopi Culture through the Oraibi Split* (Tucson: University of Arizona Press, 1988). On the Orayvi Split, see Matthew Saskiestewa Gilbert, *Education beyond the Mesas: Hopi Students at Sherman Institute, 1902–1929* (Lincoln: University of Nebraska Press, 2010), 51–70.

37. Joab's dream #19, in D. Eggan, "Hopi Dreams and a Life History Sketch," Series I, 44; Debbie's dream #5, in D. Eggan, "Hopi Dreams: Second Series," 22.

38. Don Talayesva, *Sun Chief* (New Haven: Yale University Press, 2013), 380.

39. Report of the Secretary of War, Fort Wingate, New Mexico; 52nd Cong., 1st sess., House Executive Documents, 1891, vol. 1, 258–260; reprinted in Whiteley, *The Orayvi Split*, 854–855.

40. Dream of bombed-out city: Lars's dream #1, in D. Eggan, "Hopi Dreams: Second Series," 66. Dreams of German planes: Asa's dream #3, in ibid., 53–54; "All of a sudden we heard the great roar of a plane; it was a German plane. . . . I have been listening to the war news every evening": Tammy, dream #45 in ibid., 46.

41. Letter from Dorothy Eggan to Hermia and Bert Kaplan, undated (replying to letter of July 27, 1957), Folder 25, Box 3, D. Eggan Papers.

42. Chad's dream #6, in D. Eggan, "Hopi Dreams: Second Series," 8. Chad's dream #7, in ibid., 8–9.

43. Eggan, "Cultural Factors in Dreams," 12.

44. "Potential information" and "explore more actively": from ibid., 4, 12, 1–2. "I'm so sure that": handwritten letter from Dorothy Eggan

to David Schneider, undated (but ca. mid-1958), Folder 6, Box 11, Schneider Papers. "For the study of dreams": D. Eggan, "Hopi Dreams and a Life History Sketch," Series I, 19; within this, Eggan quotes from Redfield, and the citation she gives is to Redfield lecture, University of Chicago, January 15, 1957.

45. Daf's husband's dream #3, in D. Eggan, "Hopi Dreams: Second Series," 58–59.
46. Tammy's dream #49, in ibid., 47. Chad's dream #1, in ibid., 4.
47. E. Aserinsky and N. Kleitman, "Regularly Occurring Periods of Eye Motility and Concurrent Phenomena during Sleep," *Science* 118 (1953): 273–274. W. Dement and N. Kleitman, "The Relation of Eye Movements during Sleep to Dream Activity: An Objective Method for the Study of Dreaming," *Journal of Experimental Psychology* 53 (1957): 89–97. More recently, James Foulkes and John Antrobus have (separately) shown that long narrative dreams with bizarre elements can happen in non-REM as well as in REM states (see Andrea Rock, *The Mind at Night: The New Science of How and Why We Dream* (New York: Basic Books, 2004), 28–31.
48. Modern television machines: Orrin E. Dunlap Jr., "Television Flashes Pictures through New York's Air," *New York Times*, November 15, 1936. "Man's strangest dream" refers to a statement made by Eduard Rhein in a 1929 essay: "Everything that interests us will be made visible on a little screen in our own home.... Our wildest dreams are now becoming marvelous reality." Rhein quoted in Stefan Andriopoulos, "Psychic Television," *Critical Inquiry* 31, no. 3 (2005): 620.
49. David Foster Wallace, "E Unibus Pluram: Television and U.S. Fiction," *Review of Contemporary Fiction* 13, no. 2 (1993): 159.
50. See also Anthony Wallace, "Dreams and the Wishes of the Soul: A Type of Psychoanalytic Theory among the Seventeenth Century Iroquois," *American Anthropologist* 60 (1958): 234–248.
51. Katy Price, "Testimonies of Precognition and Psychiatry in Letters to J. B. Priestley," *Studies in the History of Biological and Biomedical Sciences* (forthcoming).
52. This is not to say that Hopi life became unimportant for Eggan; it remained extremely important to contextualize the dreams, for they were (for her) deeply cultural artifacts. It is more that this cultural faculty became inherent to the dream function rather than something she attempted to view as if through a window the dream opened up.
53. "Dreams and Related Problems of Consciousness" (moderated by Montague Ullman), 1958 APA conference proceedings transcript, 3.

54. Ibid.

55. Sensory deprivation hallucinations "can be interpreted much as one would interpret a dream." They were ways of reflecting on the patient's real life. "In case 2, it is as though she said, 'if only this were a helicopter and if I could just dance again and have my husband and son with me.' . . . The little boy seemed to be saying, 'if only I could go home with my parents, I could then watch T.V. and go to sleep in my own bed.'" Ibid., 13.

56. "Dreams and Related Problems of Consciousness" (moderated by Montague Ullman), 1958 APA conference proceedings transcript.

57. Eggan answering question in discussion, ibid., 71.

58. Michael Chabon, "Head or Tale," *New York Review of Books,* September 27, 2012, 54.

59. "Practically all dream experience is forgotten." Much is not representable or even expressible in spoken or written language about dream states: "The spatiotemporal aspects of the dream experience are not those that could be inferred from recollection in the waking state." Hobson, "Film and the Physiology of Dreaming Sleep," 15.

Chapter 8. Not Fade Away (A History of the Life History)

1. Case 9's story is in George Spindler, "Autobiographic Interviews of Eight Menomini Indian Males," *Microcard Publications of Primary Records in Culture and Personality* 2, no. 12 (La Crosse, WI: Microcard, 1957): 17–31. Quotes in this and the following two paragraphs are from this work.

2. Ibid., 31.

3. The four sets are ibid., 1–105; George Spindler: "Rorschachs of Sixty-eight Menomini Men at Five Levels of Acculturation," *Microcard Publications of Primary Records in Culture and Personality* 2, no. 11 (La Crosse, WI: Microcard, 1957): 1–297, and "Personal Documents in Menomini Peyotism," *Microcard Publications of Primary Records in Culture and Personality* 2, no. 13 (La Crosse, WI: Microcard, 1957): 1–37; and Louise Spindler, "Sixty-one Rorschachs and Fifteen Expressive Autobiographic Interviews of Menomini Women," *Microcard Publications of Primary Records in Culture and Personality* 2, no. 10 (La Crosse, WI: Microcard, 1957): 1–302, plus a Rorschach addendum to J. S. Slotkin, "A Case Study and Autobiography of a Menomini Indian Paranoid Schizophrenic Man, with a Rorschach Analysis by George Spindler," *Microcard Publications of Primary Records in Culture and Personality* 3, no. 6 (La Crosse, WI: Microcard, 1961): 1–94.

4. "What mattered was getting the data": George Spindler, *Two Anthropologists Loose in North America* (DMF Publications, 2012), Kindle loc. 605. "We all have a stake": letter from George Spindler to Bert Kaplan, February 7, 1957, Kaplan Papers, Santa Cruz, CA.

5. George Spindler and Louise Spindler, *Dreamers without Power: The Menomini Indians* (New York: Holt, Rinehart and Winston, 1971), 14–15. Note that their Menominee friends and acquaintances protested this characterization vigorously by unfurling a banner across the main thoroughfare in town during one of the Spindlers' visits; the banner read, "Dreamers *with* Power," which in turn became the title of the book's second edition: George Spindler and Louise Spindler, *Dreamers with Power: The Menominee* (Long Grove, IL: Waveland Press, 1984).

6. As the Spindlers' own data amply testified; historical synopsis from L. Spindler, "Sixty-one Rorschachs and Fifteen Expressive Autobiographic Interviews," 8–9.

7. In *Dreamers without Power*, the Spindlers wrote of Menominee people "still struggling to make a living, to get an education, to gain self-respect" (p. 16). The Spindlers felt the "native oriented group," though only 3 percent of the population, best represented "traditional Menomini culture as it exists today in its attenuated but most visible form" (p. 25). On the "Woodlands type" of native personality, see A. I. Hallowell, "Some Psychological Characteristics of the Northeastern Indians," in *Men in Northeastern North America*, ed. F. Johnson (Andover, MA: Phillips Academy, 1946), vol. 3; Papers of the Peabody Foundation for Archaeology. Hallowell's argument about the basic personality type is complex; it is found to endure through time (among related tribes) and yet can be seen to deteriorate across certain event thresholds; note that Hallowell used historical documents, most often from Jesuits or missionaries, to reconstruct a personality configuration to which he could compare the results of the range of Rorschach protocols he secured in the late 1930s and early 1940s.

8. Case 1, L. Spindler, "Sixty-one Rorschachs and Fifteen Expressive Autobiographic Interviews," 11–28.

9. Spindler described Case 35 as a forty-year-old "alert, attractive" Menominee woman with five children, married to a poor white man and someone who drank too much. L. Spindler, Case 35, "Sixty-one Rorschachs and Fifteen Expressive Autobiographic Interviews," 186.

10. Case 32, "Lower Status Acculturated," fifty-eight-year-old woman, a "true marginal," known as a witch (bear cub story); L. Spindler,

Case 32, "Sixty-one Rorschachs and Fifteen Expressive Autobiographic Interviews," 166–172.

11. By "structure of feeling" Williams referred to "a particular sense of life, a particular community of experience hardly needing expression." Raymond Williams, *The Long Revolution* (New York: Broadview, 2001), 64. Card VI response from Rorschach records of a "transitional" man with a drinking problem who dreamed of reviving the Menominee ways, a "casualty of the acculturation process"; quoted in George Spindler and Louise Spindler, "Fieldwork among the Menomini," in *Being an Anthropologist: Fieldwork in Eleven Cultures*, ed. George Spindler (Prospect Heights, IL: Waveland Press, 1970), 288. (Also in G. Spindler Rorschach data set.)

12. Freud quotes Schelling: the *Unheimlich* is "the name for everything that ought to have remained . . . hidden and secret and has become visible" (Sigmund Freud, "The Uncanny," first published in 1919, trans. Alex Strachey, in *On Creativity and the Unconscious* [New York: Harper Colophon Books, 1958], 129).

13. Description of Shumaysen by hitchhiker and "Of course, . . ." are from Spindler and Spindler, "Fieldwork among the Menomini," 275. H. David Brumble III, "Social Scientists and American Indian Autobiographers: Sun Chief and Gregorio's 'Life Story,'" *Journal of American Studies* 20 (1986): 280.

14. G. Spindler, *Two Anthropologists Loose in North America*, loc. 576 ebook.

15. "Being the first attempt" is from Henry Mayhew, preface, *London Labour and the London Poor* (London: Griffin, Bohn and Company, 1851), 1:xv. "In the dark confusion" is from Peter Quennell, introduction to *Mayhew's London: Being Selections from London Labour and the London Poor* (London: Spring Books, 1969 [orig. 1857]), 18. Description of rat killer ("The Destroyer[] of Vermin") is from Mayhew, *London Labour and the London Poor* (London: Griffin, Bohn and Company, 1861), 3:2–3. Available at https://archive.org/stream/londonlabourando1mayhgoog#page/n4/mode/2up.

16. Of course, this was not an entirely new empirical spirit applied to the intimate corners of personal life. Such things had been collected in Western Europe since the Enlightenment at least, when one could find between 1784 and 1790 Jacob Friedrich Abel's *Collection and Explanation of Curious Phenomena Drawn from Human Life* in three volumes, the second devoted to seemingly randomly arrived at life-history subjects—the case of a male and a female thief who worked together. These were scientific but in a non-modern sense: they were

to testify to man's place in the next world, via the working out of his fate in this world. Life histories were called *Lebensbeschreibungen*, said by Abel to show "the growth and development of the faculties of the human soul and, through these, of man's happiness, if not in this world, then at least in the next." Quoted in Fernando Vidal, *The Sciences of the Soul: Early Modern Origins of Psychology* (Chicago: University of Chicago Press, 2011), 327.

17. Being caught between two cultures was and remains a common way of describing the condition encountered by twentieth-century immigrants in transition. However, this description relies on a particular historical understanding of what "culture" is, how cultures are bounded, and how one might move between them. Park's view was sharply dualistic: people moved, necessarily but painfully, from traditional to modern. In contrast, see Aihwa Ong on "flexible citizens of the late twentieth century," a term that describes a more fluid identity. Aihwa Ong, *Flexible Citizenship* (Durham, NC: Duke University Press, 1999). On "suspended identity," see Harry Liebersohn and Dorothee Schneider, introduction, " *'My Life in Germany before and after January 30, 1933': A Guide to a Manuscript Collection at Houghton Library, Harvard University* (Philadelphia: Transactions of the American Philosophical Society, 2001). On the afterlife of Park's concept and its prefiguring of the dynamics of globalization, see Chad Alan Goldberg, "Robert Park's Marginal Man: The Career of a Concept in American Sociology," *Laboratorium* 2 (2012); http://soclabo.org/index.php/laboratorium/article/view/4/119.

18. Robert E. Park, "Human Migration and the Marginal Man," *American Journal of Sociology* 33 (1928): 893. Note that Park referred to autobiographies and had not yet used the language of the "life history" that would soon be commonly adopted. An early use of "life-history documents" was Ernst T. Krueger, "The Value of Life History Documents for Social Research," *Journal of Applied Sociology* 9 (1925): 196–201; https://www.brocku.ca/MeadProject/Krueger/Krueger_1925a.html. See also Robert Park and Herbert Miller, *Old World Traits Transplanted* (New York: Holt, 1921), which relied on personal narrations to illustrate immigrant attitudes.

19. William I. Thomas and Florian Znaniecki, *The Polish Peasant in Europe and America*, 5 vols. (New York: Dover Publications, 1958 [orig. 1918–1920]). Alternatively, the title "father of life history" has at least once been given to Clifford Shaw, the Chicago criminologist who began collecting life histories of delinquents around 1921 and was influential in the Chicago School movement discussed below (see

Judy Long, "The Sociological Life History," in Long, *Telling Women's Lives: Subject, Narrator, Reader, Text* [New York: New York University Press, 1999]). However, James Bennett argues that Shaw, an activist to the core, was a master of the method but was constitutionally unlikely to have been its inventor (Bennett, *Oral History and Delinquency: The Rhetoric of Criminality* [Chicago: University of Chicago Press, 1988], 167).

20. "The Polish peasant . . .": Thomas and Znaniecki, *The Polish Peasant in Europe and America*, 1: 303. "*Perfect* type of sociological material": ibid., 3: 6. In addition to the Thomas and Znaniecki volumes, Herbert Blumer's Social Science Research Council (SSRC)–sponsored evaluation of that work was also prominent: Herbert Blumer, "An Appraisal of Thomas and Znaniecki's *The Polish Peasant in Europe and America*," in *Critiques of Research in the Social Sciences*, no. 1, ed. Herbert Blumer (New York: SSRC, 1939); and W. I. Thomas and Florian Znaniecki, "Comment on Blumer's Analysis," in Blumer, *Critiques of Research in the Social Sciences.*"

21. Long, "The Sociological Life History," 74.

22. "Totality of their objective complexity": Thomas and Znaniecki, preface, *The Polish Peasant in Europe and America*, 1:vii; "the Polish peasant was selected": ibid., viii.

23. Gordon Allport, *The Use of Personal Documents in Psychological Science* (New York: SSRC, 1942), 6; Social Science Research Council Bulletin 49. Freud drew on Judge Schweber's paranoiac account, completed the case study "Analysis of a Phobia in a Five-Year-Old Boy," and utilized the controversial girlhood diary *Tagebuch eines halbwüchsiges Mädchens*; Hall compiled adolescent records constituting an "ephebic literature" and supplemented them by nine hundred brief questionnaire-generated protocols, eventually grouping these into twenty-five types of fears.

24. Martin Bulmer, *The Chicago School of Sociology: Institutionalization, Diversity, and the Rise of Sociological Research* (Chicago: University of Chicago Press, 1984), 54. Herbert Blumer, *An Appraisal of Thomas and Znaniecki's The Polish Peasant in Europe and America* (New York: SSRC, 1939), 1–98; Social Science Research Council Bulletin 44; Allport, *The Use of Personal Documents in Psychological Science*; Louis Gottschalk, Clyde Kluckhohn, and Robert Angell, *The Use of Personal Documents in History, Anthropology and Sociology* (New York: SSRC, 1945); Social Science Research Council Bulletin 53. Liz Stanley, "To the Letter: Thomas and Znaniecki's *The Polish Peasant* and Writing a Life, Sociologically," *Life Writing* 7, no. 2 (2010): 139. See also

Bulmer, *The Chicago School of Sociology*, 45–63, on *The Polish Peasant* as a landmark if neglected classic, an important origin point of Chicago School sociology.

25. Thomas left Chicago owing to the scandal generated by an extramarital affair he was discovered conducting in a Chicago hotel with a prominent and married antiwar activist. Thomas lost his job at the University of Chicago, and his professional career never quite recovered.

26. "A bit of information about the saloons": quoted in Bulmer, *The Chicago School of Sociology*, 45. Wright studied at Tuskegee Institute with Park, who inspired him to collect for his documentary book, *12,000,000 Black Voices* (New York: Basic Books, 2002 [orig. 1941]). Wright quoted in Bennett, *Oral History and Delinquency*, 155.

27. Howard Becker interview with author, July 2011, San Francisco. Other "own stories" were incorporated into Chicago School works such as Frederic Thrasher's *The Gang: A Study of 1,313 Gangs in Chicago* (Chicago: University of Chicago Press, 2013). A re-study of the jack-roller has been followed by a re-restudy: Shadd Maruna and Amanda McTravers, "N=1: Criminology and the Person," special issue of *Theoretical Criminality* 11, no. 4 (2007): 427–442 and 1362–4806.

28. "Method of data collecting": Daniel Bertaux and Martin Kohli, "The Life Story Approach: A Continental View," *Annual Review of Sociology* 10 (1984): 217. "Original research": Mark McGurl, *The Program Era: Postwar Fiction and the Rise of Creative Writing* (Cambridge, MA: Harvard University Press, 2011), 85. Writing programs were a way of accommodating immigrant influx and rendering citizen material. In turn the roots of progressive education and its experiential imperative extend back to Europe and the United States in the nineteenth century. Parallels between the nascent institution of the writing program and the life history "own story" as social-scientific tool are rich but have not much been explored.

29. Brumble, "Social Scientists and American Indian Autobiographers," 273, 279.

30. Viktor Barnouw, "The Phantasy World of a Chippewa Woman," *Psychiatry: Journal for the Study of Interpersonal Relations* 12 (1949): 67–76; Edward Bruner, "The Life History of a Fort Berthold Indian Psychotic," *Microcard Publications of Primary Records in Culture and Personality* 2, no. 9 (La Crosse, WI: Microcard, 1957). Barnouw was also a contributor to the "database of dreams," to which he added thirteen Nepalese Rorschach tests (ibid. 1, no. 2 [1956]). See also Slotkin,

"A Case Study and Autobiography of a Menomini Indian Paranoid Schizophrenic Man, with a Rorschach Analysis by George D. Spindler."

31. "Egodocument" was coined in 1955 by historian Jacques Presser to describe a document in which the "I" is constantly present, ranging from an autobiography to a letter. A few additional examples of such centers include the Center for the Study of Egodocuments and History at Erasmus University, Rotterdam, The Netherlands, and the University of Sussex's Centre for Life History and Life Writing Research. (Note that the widely known British anthropological aversion for in-depth psychological research no longer applies.) An overview of European centers along these lines is at http://www.iaba-europe.eu/organisations. Thanks to Marijke Huisman for the references.

32. Spiro quoted in Louis Langness, *The Life History in Anthropological Science* (New York: Holt, Rinehart and Winston, 1965), 51.

33. Jennifer Platt, *A History of Sociological Research Methods in America: 1920–1960* (Cambridge: Cambridge University Press, 1999), 20; see Langness, *The Life History in Anthropological Science*, 7. Kluckhohn's contribution, discussing the sources and ways of collecting personal documents, is found in one of the SSRC volumes commissioned to respond to the methodological challenge posed by *The Polish Peasant: The Use of Personal Documents in Anthropology* (New York: SSRC, 1955). It would arguably be revived once more with the great popularity of oral history in the 1960s and 1970s. Today there are "definite signs that the life history method is making a comeback"; see Maruna and McTravers, "N=1," 430.

34. John Dollard, *Criteria for the Life History: With Analysis of Six Notable Documents* (New Haven: Yale University Press, 1936), 1. Quotations that follow in this paragraph are from pp. 6, 2, 4, 10, and 34.

35. Thomas quoted in Bulmer, *The Chicago School of Sociology*, 54.

36. On the history of interviewing techniques, see Raymond Lee: "David Riesman and the Sociology of the Interview," *Sociological Quarterly* 49 (2008): 285–307, and "The Most Important Technique: Carl Rogers, Hawthorne and the Rise and Fall of Nondirective Interviewing in Sociology," *Journal of the History of the Behavioral Sciences* 47 (2011): 124. See also Anke Te Heesen, "Hermann Bahr and the Interview: Between Journalism and the Social Sciences," colloquium, February 25, 2014, Max Planck Institute for the History of Science, Berlin. On the focused interview and focus groups, see Rebecca Lemov, "Everywhere and Nowhere: Focus Groups as All-Purpose Devices," *Limn* 2 (2012): 32–35, http://limn.it/issue/02/.

37. Sandra Epperson Wolf, "Finding Her Power through Collaboration: A Biography of Louise Spindler" (PhD dissertation, University of Texas, Austin, 2002), 102.

38. Louise Spindler, *Menomini Women and Culture Change* (Menasha, WI: American Anthropological Association, 1962), and Spindler and Spindler, "Fieldwork among the Menomini." The EAI continues to be widely used, especially in conditions of time constraint or limited resources—for example, as discussed in David Fetterman, *Ethnography Step-by-Step* (London: SAGE, 2010), 54–55.

39. George Spindler, personal communication to Sandra Wolf, 2001; quoted in Wolf, "Finding Her Power through Collaboration," 106.

40. William Henderson and B. W. Aginsky, "A Social Science Field Laboratory," *American Sociological Review* 6 (1941): 42.

41. Elizabeth Colson, "Collecting the Life Histories," in "Autobiographies of Three Pomo Women," *Microcard Publications of Primary Records in Culture and Personality* 1, no. 16 (La Crosse, WI: Microcard, 1956): 1. Quotations that follow in this paragraph are from pp. 2–9.

42. The "verbatim interview" (Eugene Burgess, "What Case Records Should Contain to be Useful for Sociological Investigation," *Social Forces* 6 [1928]: 526–528, and Ruth Schonle Cavan, "Topical Summaries of Current Literature: Interviewing for Life-History Material," *American Journal of Sociology* 35 [1929]: 100–115) was a recollected "report of the interview, in anecdotal form, including gestures, facial expressions, questions, and remarks of the interviewer" (Cavan, p. 107). See Raymond M. Lee, "Recording Technologies and the Interview in Sociology, 1920–2000," *Sociology* 38, no. 5 (2004): 869–889.

43. On non-directive methods, see Carl Rogers, "The Non-Directive Method as a Technique for Social Research," *American Journal of Sociology* 50, no. 4, (1945): 279–283. Note that the 1941 Allport study of anti-Nazis also employed interviews to experimentally test the consistency of personalities as registered in the life history. Gordon Allport, Jerome Bruner, and Ernest Jandorf, "Personality under Social Catastrophe: Ninety Life-Histories of the Nazi Revolution," in Clyde Kluckhohn and Henry A. Murray, *Personality in Nature, Society and Culture* (New York: Alfred Knopf, 1941).

44. Some even considered oral history as a form of interview itself, specifically a "very unique kind of interview situation" utilizing storytelling, "empowerment," and self-awareness. Sharlene Nagy Hesse-Biber and Patricia Leavy, "Oral History," in Hesse-Biber and Leavy, *The Practice of Qualitative Research, Part II: Methods of Data Collection*

(London: Sage, 2005); on using "live history interviews" and "giving a voice to those who had been silenced," see interview with Phillip Bonner, http://www.goethe.de/ins/za/prj/wom/orh/enindex.htm. For a nuanced discussion of the role of the recorder in "told-to" oral-history narratives, see Sophie McCall, "Collaboration and Authorship in Told-to Narratives," in *First Person Plural: Aboriginal Storytelling and the Ethics of Collaborative Authorship* (Toronto: University of British Columbia Press, 2011), esp. 6–8.

45. Platt quoted in Lee, "Recording Technologies and the Interview in Sociology," 869.

46. Quoted in H. Russell Bernard, *Research Methods in Anthropology* (Lanham, MD: Altamira, 2006), 414.

47. Jennifer Platt observes that after 1945 serious methodological treatment of the life history, case materials, and personal document declined in favor of a focus on participant observation and fieldwork. See Platt, *A History of Sociological Research Methods in America*, 37. Allport's postwar contributions (including the personal-document-centered *Letters from Jenny* [New York: Harcourt, 1965]), following his 1942 SSRC compendium, may have marked a way forward in this new climate. During the war, the life history, with its demand for patient listening and relatively long-term commitment, flagged, although many collected first-person accounts. See Peter Bearman and Katherine Stovel, "Becoming a Nazi: A Model for Narrative Networks," *Poetics* 27 (2000): 69–90, esp. on wartime sources. For an evaluation of Allport's contribution, see Vincent Hevern, "Allport's (1942) *Use of Personal Documents:* A Contemporary Reappraisal," paper presented at the annual meeting of the American Psychological Association, Boston, August 1999.

48. Harry Moore, "Review of Blumer's Appraisal," *Social Forces* 18 (1940): 580–583. Cf. Joel Isaac, "Epistemic Design: Theory and Data in Harvard's Department of Social Relations," in *Cold War Social Science*, ed. Mark Solovey and Hamilton Cravens (New York: Palgrave Macmillan, 2012), 79–98.

49. George Spindler and Louise Spindler, foreword to Langness, *The Life History in Anthropological Science*, v. As the Spindlers point out, this work was the only integrative and systematic publication on the life history in anthropology since Kluckhohn's *The Personal Document in Anthropological Science* in 1945. On the "different style" of American sociology and its positivist, structuralist slant, see Stanley, "To the Letter," 142.

50. Recollections by Robert Edgerton, "Prophets *with* Honor: The Early Rorschach Research of George and Louise Spindler," in *The Psychoanalytic Study of Society: Essays in Honor of George and Louise Spindler,* ed. Bryce and Ruth Boyer (Hillsdale, NJ: Analytical Press, 1992), 17:61.

51. The series was a set of formal methodological field guides addressing such topics as the life-history collection method (Langness), kinship analysis (Ernest Schusky), collecting and interpreting data on values (Ethel Albert), data on social structure (A. Kimball Romney), and the participant-observation method (Arthur Rubel).

52. G. Spindler, *Two Anthropologists Loose in North America,* v.

53. "The people thought it was fun": Spindler and Spindler, "Fieldwork among the Menomini," 281. On the number who turned them down, see Spindler and Spindler, "Fieldwork among the Menomini," 281: one turned them down. G. Spindler, "Rorschachs of Sixty-eight Menomini Men at Five Levels of Acculturation," 2: four turned them down. "A landmark": Edgerton, "Prophets *with* Honor," 60. "There was a time when it was considered quite unusual to use psychological tests and categories in anthropological research. When we started out [in 1948], the use of Freudian terms, or the use of Rorschachs, or even the use of terms like 'the self' which I used in my dissertation in 1956, was heavily criticized." (The Spindlers are quoted in Wolf, "Finding Her Power through Collaboration," 119).

54. "All I know is Jesus save you" or "Hey, no, no, no, no": L. Spindler, introduction, "Sixty-one Rorschachs and Fifteen Expressive Autobiographic Interviews"; G. Spindler, "Personal Documents in Menomini Peyotism," 11. Also see Omer Stewart, *Peyote Religion: A History* (Norman: University of Oklahoma Press, 1993). Note that Skinner and Hoffman, around the turn of the century and early twentieth century respectively found the Metäwin, a secret society to prolong life that was not mentioned by Jesuits from an earlier period in the *Jesuit Relations* account. (Also known as *Relations des Jésuites de la Nouvelle-France, The Jesuit Relations,* the earliest published in 1632, contained priests' accounts of how their attempts to convert American Indians to Christianity were going and served as a source of ethnographic information about native life in New France; these became standard bases of comparison for twentieth-century ethnographers of the area. Available in facsimile at http://puffin.creighton.edu/jesuit/relations/.) Metäwin was probably a syncretic result of white and Indian cultures.

55. "Bulk of the old values": L. Spindler, introduction to "Sixty-one Rorschachs and Fifteen Expressive Autobiographic Interviews." Or, as the Spindlers put it elsewhere, these groups were "the people who tried to keep their culture" (*Dreamers without Power,* 18).

56. "Menominee Princess Is Tribal Ruler at 19," *Milwaukee Journal,* March 17, 1929, http://news.google.com/newspapers?nid=1499&dat=19290317&id=tPUjAAAAIBAJ&sjid=uiEEAAAAIBAJ&pg=7016,4594263.

57. Except as otherwise noted, the account and quotations in this paragraph and the next come from Case 20, G. Spindler, "Autobiographic Interview of Eight Menomini Indian Males," 4–20.

58. Theodor Adorno, afterword to Walter Benjamin, *Berliner Kindheit um Neunzehnhundert* (Berlin Childhood around 1900) (Frankfurt A.M.: Suhrkamp, 1950). Cf. "These fragments I have shored against my ruins"; T. S. Eliot, "The Waste Land."

59. Jonathan Lear, *Radical Hope: Ethics in the Face of Cultural Devastation* (Cambridge, MA: Harvard University Press, 2008).

60. "An endeavor to discern": *Through a Glass Darkly;* quoted in Jill Lepore, "Historians Who Love Too Much: Reflections on Microhistory and Biography," *Journal of American History* 88, no. 1 (2001): 132. Howard Becker, introduction to Clifford Shaw, *The Jack-Roller: A Delinquent Boy's Own Story* (Chicago: University of Chicago Press, 1930, 1966), vi–vii.

Chapter 9. New Encyclopedias Will Arise

1. The library card catalog was first invented in revolutionary France to track displaced books but emerged in force at the Harvard library in 1840 with the first "slip catalog." Each slip was 9″ × 2″. In 1877 at the inaugural American Library Association meeting (under the motto "The best reading for the greatest number, at the least cost") two standard sizes (2″ × 5″ and 3″ × 5″) emerged. The latter size (the Standard Catalogue Card) won out, and the fact that the Microcard's inventor was a librarian indicates that it was no coincidence he chose this size. On the standardization of library cards, see Markus Krajewski, *Paper Machines: About Cards and Catalogs, 1548–1929* (Cambridge, MA: MIT Press, 2011), 91ff.

2. Fremont Rider, *The Scholar and the Future of the Research Library: A Problem and Its Solution* (New York: Hadham Press, 1944), 107. On Rider's early career, see Fremont Rider, *And Master of None: An Autobiography in the Third Person* (Middletown, CT: Godfrey

Memorial Library, 1955), 146–147. Rider experimented at Wesleyan's library with the "Wesleyan Order System," a method maintaining records of all aspects of library bookkeeping on standard slips the same size as Library of Congress cards, and filing order slips along with bibliographic cards; he also developed a service of pre-printing standard cards (*And Master of None*, 161, 173–174).

3. Kaplan brochure for *Microcard Publications of Primary Records in Culture and Personality*, vol. 1. Kaplan Papers, Santa Cruz, CA.

4. Council on Library Resources, for release December 12, 1958, Recent Developments, 10, District 7–8877, "First Scientific Journal in Microform Established"; Kaplan Papers. Foster E. Mohrhardt, librarian of the USDA, was enlisted to observe and report on the experiment's impact.

5. Thomas Biolsi, "The Birth of the Reservation: Making the Modern Individual among the Lakota," *American Ethnologist* 22, no. 1 (1995): 30. Royal Hassrick, "152 Rorschach Tests of Sioux Indian Boys and Girls," *Microcard Publications of Primary Records in Culture and Personality* 3, no. 4 (La Crosse, WI: Microcard, 1961). Laura Thompson, head of this project and the Indian Personality Project of which it was a part, was a good friend of Kaplan's.

6. Kaplan brochure, n.d., Kaplan Papers.

7. Murdock's group produced multiple iterations of its *Outline of World Cultures* and *Outline of Cultural Materials;* the description in the text is from George P. Murdock, "The Processing of Anthropological Materials," in *Anthropology Today: An Encyclopedia Inventory*, ed. Alfred Kroeber (Chicago: University of Chicago Press, 1953), 477.

8. Jorge Luis Borges, "Nightmares," in *Seven Nights*, trans. Eliot Weinberger (New York: New Directions, 2006), 26–27. On the use of dreams in anthropological studies, see Roy d'Andrade, "Anthropological Studies of Dreams," in *Psychological Anthropology: Approaches to Culture and Personality*, ed. Francis L. K. Hsu (Homewood, IL: Dorsey, 1961), 296–332; Erica Bourgignon, "Dreams and Dream Interpretation in Haiti," *American Anthropologist* 184, no. 5 (1954): 266.

9. Calvin Hall, "What People Dream About: In Which 10,000 Dreams are Statistically Investigated with Respect to Setting, Cast of Characters, Plot, Emotions and Coloring," *Scientific American* 184, no. 5 (1951): 60.

10. D'Andrade, "Anthropological Studies of Dreams," 299.

11. Proceedings of the convocation (held in 1961) are found in Gustav Von Grunebaum and Roger Caillois, eds., *The Dream and Human*

Societies (Berkeley: University of California Press, 1966), including work by Dorothy Eggan. Global understanding is invoked in Ivan Mensh and Jules Henry, "Direct Observation and Psychological Tests in Anthropological Fieldwork," *American Anthropologist* 55, no. 4 (1953): 461.

12. Nils Gilman, *Mandarins of the Future: Modernization Theory in Cold War America* (Baltimore: Johns Hopkins University Press, 2003), 24–72. On the valuing of openness, see Jamie Cohen-Cole, *The Open Mind: Cold War Politics and the Sciences of Human Nature* (Chicago: University of Chicago Press, 2014).

13. "Show Man to Man": from Dorothea Lange's 1953 letter of invitation to participants, quoted in Fred Turner, "*The Family of Man* and the Politics of Attention in Cold War America," *Public Culture* 24, no. 1 (2012): 60. My analysis in indebted to Turner's against-the-grain take on this exhibit.

14. Kaplan referred to the collection as "raw data," although it is today a contested term. Bert Kaplan, "Dissemination of Primary Research Data in Psychology," *American Psychologist* 13, no. 2 (1958): 54. Cf. Lisa Gitelman, ed., *Raw Data Is An Oxymoron* (Cambridge, MA: MIT Press, 2013), which argues that "raw data" is "anything but raw."

15. Reported in George H. Davison, "Review of Progress in 1960," in *Microtext in the Form of Microcards and Other Micro-Opaques, Transparent Microfiches and Unitised Microfilm or Aperture Cards as Well as Reading and Enlarging Apparatus* (Moorgate, Rotherham, England: United Steel Companies Research and Development Department, Swinden Laboratories, 1960).

16. Microcard Corporation brochure, n.d., Kaplan Papers.

17. "Elbow": Fremont Rider, "Microcards," *Bulletin of the Medical Librarians Association* 37, no. 1 (1947): 12. Murdock, "The Processing of Anthropological Materials," 476, 485.

18. "Pooling of data": letter from Bert Kaplan to A. F. C. Wallace, November 24, 1954, Kaplan Papers. "Astonished and delighted": letter from Bert Kaplan to A. I. Hallowell, March 11, 1957, Kaplan Papers. "Certainly be interested": letter from Raymond Firth to Bert Kaplan, August 2, 1956, Kaplan Papers. Letter from Rupert C. Woodward to Bert Kaplan, November 4, 1957, Kaplan Papers. "Although like many of my colleagues . . .": letter from Omer Stewart to Hallowell, May 16, 1956. See also letter from Robert A. McKennan to A. I. Hallowell, June 18, 1956; but McKennan is leery of microfilm and suspects Microcards would come under the same heading. "Revolutionary" micropublication: Kaplan, "Dissemination of Primary Research Data in

Psychology," 53. Fifty-eight universities: handwritten list, n.d., Kaplan Papers. A count of more than seventy-five is mentioned in Bert Kaplan, unpublished draft version of "Dissemination of Primary Research Data in Psychology," 4, Kaplan Papers.

19. "Wonderful progress": letter from Sol Tax to Kaplan, April 4, 1956, Kaplan Papers. "Search and retrieval": Kaplan, "Dissemination of Primary Research Data in Psychology," 53. Letter from Bert Kaplan to Evan L. Wolfe, U.S. Naval Hospital, Oakland, May 14, 1956 (in response to Wolfe's letter of May 6, 1956). From Wolfe's letter, it does not look like he has received the Microcards yet, however.

20. Megan Prelinger, *Another Science Fiction: Advertising the Space Race 1957–1962* (New York: Blast Books, 2010).

21. Vannevar Bush, "As We May Think," *Atlantic Monthly* 176 (July 1, 1945). According to some recent Memex commenters such as Microsoft senior researcher Gordon Bell, Bush's invention was in essence a critique of indexing as an inadequate way of finding pertinent material in a large set of self-generated data. Thus it's likely Bush intended "Memex" to refer to "Memory Extender." On the other hand, Bush was offering a vision for a new kind of indexing, and in this sense he may have intended Memex as "Memory Index." See Gordon Bell and Jim Gemmell, *Your Life, Uploaded* (New York: Penguin, 2010), 42–43. According to Bell, Bush's dismissal of indexing was a rare failure of imagination. With modern computers, every single word and phrase can be indexed and searched in an instant. Thus the index is the whole document (just as the map is the territory in Borges). It is this expanded sense of indexing Kaplan may have purveyed within his own lineage.

22. Vannevar Bush, "MEMEX II," 1958 essay included in *From Memex to Hypertext: Vannevar Bush and the Mind's Machine*, ed. James Nyce and Paul Kahn (San Diego: Academic Press, 1992), 167.

23. Bush, "MEMEX II," 166.

24. Bush quoted in G. Pascal Zachary, *Endless Frontier: Vannevar Bush, Engineer of the American Century* (Cambridge, MA: MIT Press, 1999), 71. On Bush's engineering vision, see Mark Bowles, "U.S. Technological Enthusiasm and British Technological Skepticism in the Age of the Analog Brain," *IEEE Annals of the History of Computing* 18, no. 4 (1996): 5–15.

25. Bush, "As We May Think."

26. Note that Bush pushed for an atomic bomb research program and oversaw it, while Leslie Groves ran the program directly; Bush was in charge of the National Defense Research Committee (NDRC). Colin

Burke, "The Other Memex: The Tangled Career of Vannevar Bush's Information Machine, the Rapid Selector," *Journal of the American Society for Information Science* 43, no. 10 (1992): 648.

27. "Internet's first visualization": Michael Buckland, "Emanuel Goldberg, Electronic Document Retrieval, and Vannevar Bush's MEMEX," *Journal of the American Society for Information Science* 43, no. 4 (1992): 284–294 (accessed online: http://people.ischool.berkeley.edu/~buckland/goldbush.html on 7/3/11). "If there was a more eerily prescient": William Gibson, "Googling the Cyborg," in Gibson, *Distrust That Particular Flavor* (New York: G. P. Putnam's Sons, 2012), 252.

28. Burke, "The Other Memex," 648.

29. Verner Clapp, foreword to J. C. R. Licklider, *Libraries of the Future* (Cambridge, MA: MIT Press, 1969). Memex was "the private memory device in which all a man's records may be stored, linked by associative indexing and instantly ready for his use" (p. vii). On "pooling of data," see letter from Bert Kaplan to A. F. C. Wallace, November 24, 1954, Kaplan Papers.

30. The *OED* record for "index" includes the following: "Some men of very great Commerce and trading keep a Kalender, Register, or an Alphabeticall Index, of the names of Men, Wares, Ships." Macaulay, *Johnson* in Misc. (1860) II. 273.

31. Jane Bennett, "The Agency of Assemblages and the North American Blackout," *Public Culture* 17, no. 3 (2005): 446. Relevant here are several bodies of theoretical writings, including the new materialism, especially the contributions of Karen Barad (which locate more agency in things and less in people than usually assumed and find a kind of distributed or distributional arrangement), and the new technological determinism. On "intra-activity" of matter see Barad, "Posthumanist Performativity: Toward an Understanding of How Matter Comes to Matter," *Signs: Journal of Women in Culture and Society* 28, no. 3 (2008): 801–831. Anthropologist Jason Pine likewise argues that technological determinism has new life if one accepts a broadened sense of agency within technological systems. Jason Pine, "Meth Labs, Alchemical Ontology, and Homespun Worlds," Harvard University Anthropology Department Colloquium, March 2013. Also see "thing theory," for which a good starting place is Lorraine Daston, ed., *Things That Talk: Object Lessons from Art and Science* (New York: Zone Books, 2004).

32. On unbuilt versus built machines, compare Tracy Kidder's description of the successful struggle to build 1980s computing machines

versus Bruno Latour on the never-built Paris rail system, Aramis. This comparison is suggested by Howard S. Becker, "How Much Is Enough?" *Public Culture* 25, no. 3 (2013): 378. On pooling of data, see ibid., and letter from Bert Kaplan to A. F. C. Wallace, November 24, 1954, Kaplan Papers.

33. Gibson, "Googling the Cyborg," 252.

34. Sherry Turkle and Seymour Papert, "Epistemological Pluralism," http://www.papert.org/articles/EpistemologicalPluralism.html. A version of this article appeared in the *Journal of Mathematical Behavior* 11, no. 1 (1992): 3–33.

35. Claude Lévi-Strauss, *The Savage Mind* (Chicago: University of Chicago Press, 1968); Jacques Derrida, "Structure, Sign and Play in the Human Sciences" (1968), in *The Structuralist Controversy: The Languages of Criticism and the Sciences of Man*, ed. Richard Macksey and Eugenio Donato (Baltimore: Johns Hopkins University Press, 1970). Sherry Turkle, *Life on the Screen: Identity in the Age of the Internet* (New York: Simon and Schuster, 1997). Recent Science and Technology Studies (STS) literature discusses the improvisational quality of data-based scientific exchange. See, especially, Paul N. Edwards, Matthew S. Mayernik, Archer L. Batcheller, Geoffrey C. Bowker, and Christine L. Borgman, "Science Friction: Data, Metadata, and Collaboration," *Social Studies of Science* 41, no. 5 (2011): 667–690.

36. "Elements of risk": Krajewski, *Paper Machines*, 5, quoting Friedrich Kittler. Innovation "always includes recombination": ibid., 65.

37. On the place of the human within developing technological systems, see Bernard Dionysius Geoghegan, "After Kittler: On the Cultural Techniques of Recent German Media Theory," *Theory Culture Society* 30, no. 6, (2013): 66–82.

38. "Mental revolution" and "this time thoughts": Bush, "MEMEX II," in Nyce and Kahn, 7.

39. This is not to say that dreams and subjective materials had never been collected en masse before or that micro-photographic techniques lacked a rich history. It is to assert, however, that these two desires— in the form of a self-conscious two-pronged "experiment"—had not been linked before.

40. "A number of workers": Bert Kaplan, "A Newsletter," *Microcard Publications of Primary Records in Culture and Personality*, dated March 29, 1956, NAS/NRC Archives. Discussion of "delicate issues" and instructions about how to use the data respectfully are found in Bert Kaplan, editor's introduction, *Microcard Publications of Primary Records in Culture and Personality* (La Crosse, WI: Microcard Foundation, 1956), 13.

41. Judith Williams and Herbert Williams, "Sixteen Autobiographical Dream Series of Moslem and Maronite Men and Women," *Microcard Publications of Primary Records in Culture and Personality* 4, no. 2 (La Crosse, WI: Microcard, 1962): 9.

42. Catherine Crump and Matthew Harwood, "The Net Closes around Us," *Outlook India*, March 25, 2014, http://www.outlookindia.com/article/The-Net-Closes-Around-Us-/289923.

43. See Gitelman, *Raw Data Is an Oxymoron.*

44. Alan Rosen, *The Wonder of Their Voices: The 1946 Holocaust Interviews of David Boder* (London: Oxford University Press, 2010), 193. See also Adina Hoffman and Peter Cole, *Sacred Trash: The Lost and Found World of the Cairo Geniza* (New York: Schocken, 2011).

Chapter 10. Brief Golden Age

1. Letter from Bert Kaplan to A. I. Hallowell, March 11, 1957, Kaplan Papers, Santa Cruz, CA. Kaplan mentions he has not heard from John Whiting about whether he could make the upcoming meeting. "I assume he is putting off answering because of some uncertainty in the situation. I am inclined to agree that we should have the meeting in any case, with the possibility of having another later on."

2. Interview with Hans Wallach in Bert Kaplan, "Report of the Executive Secretary of the Committee on Primary Records" (Not for Distribution or Reproduction), April 1, 1957; Committee on Primary Records, General, A&P, NAS Archives, 3.

3. Asch interview quoted in Kaplan, "Report of the Executive Secretary," 3. Asch also pointed out that when key studies such as the Smith-Bruner-White report (Brewster Smith, Jerome Bruner, and Robert White, *Opinions and Personality* [New York: John Wiley and Sons, 1956]) or Theodor Adorno, Else Frenkel-Brunswik, Daniel Levinson, and Nevitt Sanford's *The Authoritarian Personality: Studies in Prejudice* (New York: Harper and Row, 1950) appeared, he and likely many other researchers yearned to set eyes on the original data, as "personality data is susceptible to various interpretations" (ibid.).

4. Michael Cox and David Ellsworth laid out the challenges of visualizing large data sets and put the two words together as a phrase in 1997. Theirs is the first use found in the Association for Computing Machinery (ACM) database (aside from an article earlier the same year by the same authors that uses the term in a less deliberate manner). Michael Cox and David Ellsworth, "Application-Controlled Demand Paging for Out-of-Core Visualization," *Proceedings of the*

8th IEEE Visualization '97 Conference, 1997, 235; emphasis in original. This is not to say that the words "big data" had never been put together before; an earlier usage dates to 1970 (Gali Halevi and Henk Moed, "The Evolution of Big Data as a Research and Scientific Topic," *Research Trends* 30 [2012]).

5. Senior psychologist (Asch) quoted in Kaplan, "Report of the Executive Secretary," 4.

6. *Science* 331 (February 11, 2011); online survey reported on p. 693.

7. Russel and Boring discussed a "central archive" at the American Psychological Association; reported in Kaplan, "Report of the Executive Secretary," 1. Quotation on state of Pacific data is from Harold Coolidge, executive director of the Pacific Science Board, ibid. Robert Holt of New York University admitted he had long fantasized about a "central registry with a lot of T.A.T.s" (ibid., 5). Another area where Kaplan's project was ahead of its time was in giving access to the "data themselves" that lay behind the interpretations given in the studies where they ended up.

8. Quotations in these two sentences are from Maria Rickers-Ovsiankina in Kaplan, "Report of the Executive Secretary," 7.

9. Letter from George Spindler to Bert Kaplan, February 7, 1957, Kaplan Papers.

10. G. William Domhoff interview with author, July 27, 2011, Santa Cruz, CA.

11. Letter from Bert Kaplan to A. I. Hallowell, March 11, 1957, Kaplan Papers.

12. Robert K. Merton et al., *The Focused Interview: A Manual of Problems and Procedures* (Glencoe, IL: Free Press, 1956), 22–23.

13. Ibid. See also Mark Levy, "The Lazarsfeld-Stanton Program Analyzer: An Historical Note," *Journal of Communication* 32, no. 4 (1982): 30–38.

14. See Rebecca Lemov, "'Hypothetical Machines: The Science-Fiction Dreams of Cold War Social Science," *Isis* 101 (2010): 401–411.

15. Regarding appointing Herbert Phillips as successor: letter from Bert Kaplan to Melford Spiro, December 10, 1962, Kaplan Papers.

16. Hermia Kaplan interview with author, July 25, 2011, Santa Cruz, CA. Kaplan was sanguine about the move to Houston and pleased to be asked to preside over an ambitious new project, as he wrote proudly to Hallowell (see letter from Bert Kaplan to A. I. Hallowell, February 16, 1963, Hallowell Papers, Series I: Correspondence).

17. Bert Kaplan proposal to Ford Foundation, "Proposed Center for Research in Ethnopsychology"; document attached to a letter dated

April 24, 1963, to A. I. Hallowell; Hallowell Papers, American Philosophical Society, Philadelphia.

18. Quotations in this paragraph and the following one are from Saul Friedman, "The Sick Navahos," *Newsweek*, July 27, 1964, 72.

19. Richard Randolph, e-mail correspondence with author, August 10, 2011.

20. Bert Kaplan letter to editor, *Newsweek*, August 17, 1964, 2. Cover: "Vietnam: Widening War?"

21. Richard Randolph, e-mail correspondence with author, August 10, 2011. Randolph, who was on the Navajo reservation with Kaplan in 1964 when the Friedman article came out, believes Kaplan stopped publishing after the *Newsweek* incident. He did continue to lecture widely and to publish versions of some of these lectures; for example, a talk given at the inaugural election in 1966–1967 at the University of California, Santa Cruz, titled "The Method of the Study of Persons," was published in Edward Norbeck, Douglass Richard Price-Williams, and William Maxwell McCord, *The Study of Personality: An Interdisciplinary Appraisal* (New York: Holt, Rinehart and Winston, 1968). An (untitled) list of Bert Kaplan's papers, annotated by Kaplan himself, is in the Kaplan Papers. In the 1980s, Kaplan attempted to revive the project, seeking funding from the NSF. Renewal grant application, U.S. Public Health Service (USPHS), Kaplan Papers.

22. Bert Kaplan and Dale Johnson, "The Social Meaning of Navaho Psychopathology and Psychotherapy," in *Magic, Faith and Healing: Studies in Primitive Psychiatry Today*, ed. Ari Kiev (New York: Free Press, 1964), 203. Note that Kaplan's mental health approach began with a comparative 1956 study of Wellesley and Roxbury, two economically contrasting towns in Massachusetts: Bert Kaplan, Robert Reed, and Wyman Richardson, "A Comparison of Hospitalized and Non-Hospitalized Cases of Psychosis in Two Communities," *American Sociological Review* 21, no. 4 (1956): 472–479.

23. Gordon MacGregor, with Royal Hassrick and William Henry, *Warriors without Weapons: A Study of the Society and Personality Development of the Pine Ridge Sioux* (Chicago: University of Chicago Press, 1946), and George Spindler and Louise Spindler, *Dreamers without Power: The Menomini Indians* (New York: Holt, Rinehart, Winston, 1971).

24. Bert Kaplan, "A Cross-Cultural Study of Psychopathology," Application for Research Grant (privileged communication), U.S. Department of Health, Education, and Welfare, Public Health Service, February 7, 1963, Kaplan Papers, 11.

25. Harry Harootunian, "'Modernity' and the Claims of Untimeliness," *Postcolonial Studies* 13, no. 4 (2010): 367. Claims of the "untimeliness" of the third world, the unmodern and the underdeveloped, tend, according to Harootunian, to dematerialize time, support global capitalism, and reinforce "the domination of the ever changing new" (p. 367). See also Reinhart Kossellek, *Futures Past: On the Semantics of Historical Time* (New York: Columbia University Press, 2004).

26. Bert Kaplan, ed., *The Inner World of Mental Illness: A Series of First Person Accounts of What It Was Like* (New York: Harper and Row, 1964).

27. Letter from anthropologist Evon Vogt, one of the Five Cultures "Values Project" directors (and Kaplan's colleague and collaborator from graduate school), to John Otis Brew, head of the Harvard Peabody Museum, January 20, 1952, UAV 801.2010, Harvard Archives. Kaplan gathered much of the data that Vogt used in his monograph, *Navaho Veterans: A Study of Changing Values* (Cambridge, MA: Harvard University Peabody Museum of Archeology and Ethnology, 1951). In fact, Vogt wrote to Kaplan letting him know he was using his (Kaplan's) interpretations of projective test results in the case of each Navajo case history. Letter from Evon Vogt to Bert Kaplan, March 29 [no year], Kaplan Papers.

28. Letter from Evon Vogt to Allan Harper of U.S. Indian Service, Window Rock, Arizona, February 10, 1952, UAV 801.2010, Harvard Archives.

29. Irving Telling (office coordinator) memorandum to Values Project, February 27, 1952, UAV 801.2010, Harvard Archives.

30. Raymond Williams, *The Long Revolution* (New York: Broadview, 2001), 64.

31. The quotations in the remainder of this paragraph are from Louise Spindler, "Sixty-one Rorschachs and Fifteen Expressive Autobiographic Interviews of Menomini Women," *Microcard Publications of Primary Records in Culture and Personality* 2, no. 10 (La Crosse, WI: Microcard, 1957): 1–302.

32. A. I. Hallowell, Transcript of Anne Roe interview, Hallowell Papers.

33. Stephanie Gänger, *Relics of the Past: Creole Antiquarianism and the pre-Columbian Past in Chile and Peru, c. 1830–1910* (Oxford: Oxford University Press, 2014).

34. Louise Spindler, "Sixty-one Rorschachs and Fifteen Expressive Autobiographic Interviews of Menomini Women," *Microcard Publications of Primary Records in Culture and Personality* 2, no. 10 (La Crosse, WI: Microcard, 1957).

35. Viktoria Tkaczyk describes twentieth-century acoustic recordings resulting in large data sets of transcriptions from conferences (colloquium comments, June 25, 2014, Max Planck Institute for the History of Science, Berlin).

36. "I don't know what David Riesman saw in Bert but he did see something": Hermia Kaplan interview with author, July 25, 2011, Santa Cruz, CA. "Hard-nosed empiricists": Page Smith, "Founding Cowell College and UCSC, 1964–1973," Oral History Interview, conducted 1996, 15. UCSC online at http://escholarship.org/uc/item/9hr6t6b3. Permission for publication granted from University of California, Santa Cruz. McHenry Library, Regional History Project.

37. Page Smith, Oral History Interview, 15–18.

38. This and the preceding quote from Dean E. McHenry, "The University of California, Santa Cruz: Its Origins, Architecture, Academic Planning, and Early Faculty Appointments, 1958–1968," vol. 2, Oral History Interview, conducted 1974; quotations in this paragraph are from pp. 6–8; online at http://www.escholarship.org/uc/item/3ws3p49b?query=dean%20mchenry#.

39. McHenry quotation is in ibid. Discussion of collecting rat and primate data is in Kaplan, "Report of the Executive Secretary," 9, 11, 15.

40. Ibid.

41. Letter from Clyde Kluckhohn to Dorothy Eggan, undated (ca. 1956), Folder 3, Box 4, Dorothy Eggan Papers, University of Chicago.

42. Alan Liu, *Laws of Cool: Knowledge Work and the Culture of Information* (Chicago: University of Chicago Press, 2004).

43. Jon Agar, "What Difference Did Computers Make?" *Social Studies of Science* 36, no. 6 (2006): 870.

44. Bernal quoted in ibid., 872. Quotations in this paragraph are from ibid., 876.

45. Ibid., 871.

46. The history of analog technologies also is revealing of presentist flaws in the standard narrative, even within the history of the computing field. On "data journeys" as "anything but smooth," see Sabina Leonelli, "What Difference Does Quantity Make? On the Epistemology of Big Data in Biology," Proceedings of Collecting, Organizing and Storing Big Data Conference, Swiss STS-UNIL, February 2014, Lausanne; podcast available at http://wp.unil.ch/stsbigdata/podcasts/. On the neglected history of analog/hybrid computing, see James S. Small, *The Analogue Alternative: The Electronic Analogue Computer in Britain and the USA, 1930–1975* (London: Routledge, 2001).

47. Nicholas Christakis, "Experiments with Online Networks," http://nicholaschristakis.net/research/. See also Nicholas Christakis and James Fowler, "The Spread of Obesity in a Large Social Network over 32 Years," *New England Journal of Medicine* 357 (2007): 370–379. Anne-Sophie Godfroy, "Challenging the Benchmarking Culture: New Epistemologies for Large Scale International Cross Comparisons in Human and Social Sciences," Proceedings of Collecting, Organizing and Storing Big Data Conference, Swiss STS-UNIL, February 2014, Lausanne. Used in conjunction with Computer-Aided Qualitative Data Analysis Software (CAQDAS).

48. Jo Guldi and David Armitage, *The History Manifesto* (Cambridge: Cambridge University Press, 2014), esp. ch. 4, "Big Questions, Big Data," 89.

49. Eytan Bakshy, Jake Hofman, Winter A. Mason, and Duncan J. Watts, "Everyone's an Influencer: Quantifying Influence on Twitter," Proceedings of the Fourth International Conference on Web Search and Data Mining, 2011, Hong Kong. See Duncan Watts website for the paper.

50. Matt Raymond, "Library of Congress Acquires Entire Twitter Archive," Library of Congress blog entry for April 14, 2010, http://blogs.loc.gov/loc/2010/04/how-tweet-it-is-library-acquires-entire-twitter-archive/ (accessed June 30, 2011). Online comments included the following skeptical response: "Archiving the ephemeral, the meaningless and the lulzy."

51. Steven King, *On Writing: A Memoir of the Craft* (New York: Scribner, 2010), 44.

52. On the facsimile machine as an understudied "old" technology that has the potential to help understand new media despite its failure to live up to expectations, see Jennifer Light, "Facsimile: A Forgotten 'New Medium' from the 20th Century," *New Media and Society* 8, no. 3 (2006): 355–378.

53. Preface to Elizabeth Colson, republication of *Autobiographies of Three Pomo Women* (Berkeley: Archeological Research Facility, 1974).

54. David Brumble, e-mail to author, April 7, 2014. H. David Brumble III, *American Indian Autobiography* (Lincoln: University of Nebraska Press, 2008), references Kaplan's Microcard archive. The following quotation is also from Brumble e-mail.

55. Otlet quoted in W. Boyd Rayward, "The Legacy of Paul Otlet, Pioneer of Information Science," *Australian Library Journal* 41, no. 2 (1992): 91.

56. The Microcard's eclipse came as the direct result of the fact that the U.S. military and government—in particular NASA, the Atomic Energy Commission, and the Department of Defense—began to favor the easier-to-use microfiche over the Microcard. Military requirements drove the outcome, emphasizing the fact that military sources had different needs from libraries. More sophisticated information retrieval, for example, was demanded, as well as copiers for high-volume printing of micropublished documents. See Susan Cady, "Machine Tool of Management: A History of Microfilm Technology" (PhD dissertation, Lehigh University), 1994, 14.
57. Michael Gabriel and Dorothy Ladd quoted in Martin Jamison, "The Microcard: Fremont Rider's Precomputer Revolution," *Libraries and Culture* 23, no. 1 (1988): 14.
58. Cady, "Machine Tool of Management," 16.
59. Some of these capacities preexisted digital databases. As noted above, Markus Krajewski describes Konrad Gessner's sixteenth-century card catalog, in which carefully stored slips of paper allowed long-lasting use and respective materials were kept "in a flexible form so as to enable the creation of ever new and different arrangements" (*Paper Machines: About Cards and Catalogs, 1548–1929* [Cambridge, MA: MIT Press, 2011], 10). Here is a difference from Kaplan's Microcard archive, which did not have much flexibility at all. The Primary Records were qualitative data, text-based, and therefore not resolvable into textual units of type, a problem Optical Character Recognition would resolve.
60. NRC Committee on Disaster Studies, *Field Studies of Disaster Behavior: An Inventory* (Washington, DC: NRC/NAS, 1957), Disaster Study No. 14, vii.
61. On the tendency of new innovations in data to exist in "suites of technology," see Martin Hand, *Ubiquitous Photography* (Cambridge: Polity Press, 2012), 98, and on the inseparability of epistemic objects from technical objects in data-driven science, see Sarah de Rijcke and Anne Beaulieu, "Networked Neuroscience: Brain Scans and Visual Knowing at the Intersection of Atlases and Databases," in *Representations in Scientific Practice Revisited*, ed. Catelijne Coopmans et al. (Cambridge, MA: MIT Press, 2014).

Conclusion

1. Joseph Mitchell, from his memoir, recently published as "A Place of Pasts: Finding Worlds in the City," *New Yorker*, February 15, 2015.
2. On "total archive," see above, ch. 3, n. 1.

3. Ryan Maguire, "The Ghost in the MP3," Proceedings of the ICMC/ SMC, September 14–20, 2014, Athens. To listen to "Tom's Diner" as ghost sounds: http://theghostinthemp3.com.

4. Claude Lévi-Strauss, *The Savage Mind* (Chicago: University of Chicago Press, 1966 [1962]), 17.

5. Hallam Stevens, *Life Out of Sequence: A Data-Driven History of Bioinformatics* (Chicago: University of Chicago Press, 2013), 59. "Message in Bottle Arrives after 101 Years," *The Guardian*, April 7, 2014.

6. Website of S-T Technologies, http://www.microrecord.com/solutions_ st_imaging.html, accessed April 17, 2015. "All-new, all-digital solution" is from the promotional video embedded on the website page.

7. On Google books as total library, see Erez Aiden and Jean-Baptiste Michel, *Uncharted: Big Data as a Lens on Human Culture* (New York: Riverhead, 2013), 15. Google books' digitized texts only became "dat-ified" (around 2006) using optical character recognition (pp. 83–84), and then it became possible for computers to process the data and algorithms to analyze them.

8. Kaplan, "Report of the Executive Secretary of the Committee on Primary Records" (Not for Distribution or Reproduction), April 1, 1957, 2; Committee on Primary Records, A&P, NAS Archives.

9. Lisa Gitelman is one of the few scholars (to my knowledge) to ask, "What difference does it make [to read] digital scans of microfilmed images?," and her recent work offers many new angles of approach to this and like questions: *Paper Knowledge: Toward a Media History of Documents* (Durham, NC: Duke University Press, 2014), 80.

10. Thomas Pynchon, *Against the Day* (New York: Penguin, 2006), 250. On Rorschach's special dream, see Henri Ellenberger, "Hermann Rorschach, M.D., 1884–1922," *Bulletin of the Menninger Clinic* 18, no. 5 (1954): 199.

11. On this theme see, among others, Rob Horning, "Permanent Recorder," *New Inquiry*, March 5, 2015, http://thenewinquiry.com/ blogs/marginal-utility/permanent-recorder, and Natasha Schüll, *Addiction by Design* (Princeton, NJ: Princeton University Press, 2013).

Index

AAA. *See* American Anthropological Association

Abel, Jacob Friedrich, 314n16

Aboriginal blood samples, 68

Acculturation, destructiveness of, 164–165

ACM (Association for Computing Machinery) database, 328n4

Adair, John, 99, 103, 108, 111–113, 257

Adolescent Fantasy Test, 39

Adorno, Theodor, 56, 118, 205, 295n7

Advanced Research Projects Agency (ARPA), 273n9

African Americans, 39, 112

Agar, Jon, 84, 246, 273n10

Airmail service, 78

Akavia, Naamah, 42

Algonkian people, 50, 58, 65, 203, 240

Allport, Gordon, 206, 296n10, 319n43

American Anthropological Association (AAA), 51, 132, 136, 141, 275n20

American Anthropologist, 164

American Documentation Institute, 301n13

American Indian tribes: studies of, 13, 289n6; Assimilation Era of 1880s and, 102, 104; choice as research subject, 239–240; feedback and pushback to Hallowell, 64–65;

Heizer's research, 249; inner lives and, 186–187, 209–210; loss of essence, self, or soul in, 237; mental health and, 234–237; violence attributed to Indians, 104, 234–235, 237, 291n25. *See also specific tribes*

American Psychological Association conference (1958), 177–178

Analog computing and techniques, 70, 91, 222, 224, 246–248, 262n9, 285n36, 332n46

Anderson, Nels, 192

Anonymity, 13

Anthropology: anti-freedom-of-information drive, 238–239; British school of social anthropology, 195, 215, 318n31; dream files in, 156; insights of, 125, 128–129; psychological anthropology, 125, 321n53; purpose of, 254; Rorschach and TAT testing and, 44–45, 54, 55, 272n6; southwest United States and, 13, 103–104, 238; Spindlers' changes to, 201–202; uniform system for classification of societies and cultures in, 210; Zuni and, 5, 103–108, 114, 292n35. *See also specific anthropologists and techniques*